Studies in Classification, Data Analysis, and Knowledge Organization

Springer
Berlin
Heidelberg
New York
Barcelona
Hong Kong
London
Milan
Paris
Singapore
Tokyo

Titles in the Series

H.-H. Bock and P. Ihm (Eds.)
Classification, Data Analysis, and Knowledge Organization. 1991
(out of print)

M. Schader (Ed.)
Analyzing and Modeling Data and Knowledge. 1992

O. Opitz, B. Lausen, and R. Klar (Eds.)
Information and Classification. 1993
(out of print)

H.-H. Bock, W. Lenski, and M.M. Richter (Eds.)
Information Systems and Data Analysis. 1994
(out of print)

E. Diday, Y. Lechevallier, M. Schader, P. Bertrand, and B. Burtschy (Eds.)
New Approaches in Classification and Data Analysis. 1994
(out of print)

W. Gaul and D. Pfeifer (Eds.)
From Data to Knowledge. 1995

H.-H. Bock and W. Polasek (Eds.)
Data Analysis and Information Systems. 1996

E. Diday, Y. Lechevallier and O. Opitz (Eds.)
Ordinal and Symbolic Data Analysis. 1996

R. Klar and O. Opitz (Eds.)
Classification and Knowledge Organization. 1997

C. Hayashi, N. Ohsumi, K. Yajima, Y. Tanaka, H.-H. Bock, and Y. Baba (Eds.)
Data Science, Classification, and Related Methods. 1998

I. Balderjahn, R. Mathar, and M. Schader (Eds.)
Classification, Data Analysis, and Data Highways. 1998

A. Rizzi, M. Vichi, and H.-H. Bock (Eds.)
Advances in Data Science and Classification. 1998

M. Vichi and O. Opitz (Eds.)
Classification and Data Analysis. 1999

W. Gaul and H. Locarek-Junge (Eds.)
Classification in the Information Age. 1999

H.-H. Bock and E. Diday (Eds.)
Analysis of Symbolic Data. 2000

H.A.L. Kiers, J.-P. Rasson, P.J.F. Groenen, and M. Schader (Eds.)
Data Analysis, Classification, and Related Methods. 2000

W. Gaul, O. Opitz, and M. Schader (Eds.)
Data Analysis. 2000

R. Decker and W. Gaul
Classification and Information Processing at the Turn of the Millenium. 2000

Simone Borra · Roberto Rocci
Maurizio Vichi · Martin Schader (Eds.)

Advances in Classification and Data Analysis

With 66 Figures and 94 Tables

Springer

Dr. Simone Borra
University of Rome "Tor Vergata"
Department SEFEMEQ
Via di Tor Vergata
00133 Rome
Italy
E-mail: borra@economia.uniroma2.it

Professor Maurizio Vichi
University of Rome "La Sapienza"
Department Statistica,
Probabilità e Statistiche Applicate
Piazza le Aldo Moro
00185 Rome
Italy
E-mail: vichi@dmqte.unich.it

Professor Roberto Rocci
University of Molise
Department SEGeS
Via de Sanctis
86100 Campobasso
Italy
E-mail: rocci@pow2.sta.uniroma1.it

Professor Dr. Martin Schader
University of Mannheim
Lehrstuhl für
Wirtschaftsinformatik III
Schloß
68131 Mannheim
Germany
E-mail:
mscha@wifo.uni-mannheim.de

ISBN-13: 978-3-540-41488-9 Springer-Verlag Berlin Heidelberg New York

Cataloging-in-Publication Data applied for
Die Deutsche Bibliothek – CIP-Einheitsaufnahme
Advances in classification and data analysis: with 94 tables / Simone Borra ... (ed.). – Berlin; Heidelberg; New York; Barcelona; Hong Kong; London; Milan; Paris; Singapore; Tokyo: Springer, 2001
(Studies in classification, data analysis, and knowledge organization)

ISBN-13: 978-3-540-41488-9 e-ISBN-13: 978-3-642-59471-7
DOI: 10.1007/978-3-642-59471-1

Springer-Verlag Berlin Heidelberg New York
a member of BertelsmannSpringer Science+Business Media GmbH

Softcover-Design: Erich Kirchner, Heidelberg

SPIN 10793752 43/2202-5 4 3 2 1 0 – Printed on acid-free paper

PREFACE

This volume contains a selection of papers presented at the biannual meeting of the Classification and Data Analysis Group of Società Italiana di Statistica, which was held in Rome, July 5-6, 1999. From the originally submitted papers, a careful review process led to the selection of 45 papers presented in four parts as follows:

CLASSIFICATION AND MULTIDIMENSIONAL SCALING
- Cluster analysis
- Discriminant analysis
- Proximity structures analysis and Multidimensional Scaling
- Genetic algorithms and neural networks

MULTIVARIATE DATA ANALYSIS
- Factorial methods
- Textual data analysis
- Regression Models for Data Analysis
- Nonparametric methods

SPATIAL AND TIME SERIES DATA ANALYSIS
- Time series analysis
- Spatial data analysis

CASE STUDIES

INTERNATIONAL FEDERATION OF CLASSIFICATION SOCIETIES

The International Federation of Classification Societies (IFCS) is an agency for the dissemination of technical and scientific information concerning classification and data analysis in the broad sense and in as wide a range of applications as possible; founded in 1985 in Cambridge (UK) from the following Scientific Societies and Groups: British Classification Society - BCS; Classification Society of North America - CSNA; Gesellschaft für Klassifikation - GfKl; Japanese Classification Society - JCS; Classification Group of Italian Statistical Society - CGSIS; Société Francophone de Classification - SFC. Now the IFCS includes also the following Societies: Dutch-Belgian Classification Society - VOC; Polish Classification Society - SKAD; Associação Portuguesa de Classificação e Análise de Dados - CLAD; Korean Classification Society - KCS; Group-at-Large.

BIANNUAL MEETING OF THE CLASSIFICATION AND DATA ANALYSIS GROUP OF SIS

The biannual meeting of the Classification and Data Analysis Group of Società Italiana di Statistica was held in Rome, July 5-6, 1999.
Seventy-five papers have been presented during this meeting, divided in nineteen sessions. Each session was organised by a chairperson and generally with two invited speakers and two contributed papers from a call for papers. All the works were referred. Furthermore, during the meeting a discussant was provided for each session. A short version of the papers (4 pages) were published before the meeting.

The scientific program covered the following topics:

CLASSIFICATION THEORY

Fuzzy Methods - Hierarchical Classification - Non Hierarchical Classification - Pattern Recognition - Bayesian Classification - Classification of Multiway Data - Probabilistic Methods for Clustering - Consensus and Comparison Theories in Classification - Spatial Clustering - Validity of Clustering - Neural Networks and Classification - Genetic Algorithms - Classification with Constraints.

MULTIVARIATE DATA ANALYSIS

Categorical Data Analysis - Correspondence Analysis - Biplots - Factor Analysis - Discrimination and Classification - Multiway methods - Symbolic Data Analysis - Non Linear Data Analysis - Bayesian Multivariate Analysis - Multilevel Analysis - Covariance Structure Analysis.

PROXIMITY STRUCTURE ANALYSIS

Multidimensional Scaling - Similarities and Dissimilarities - Unfolding and Other Special Scaling Methods - Multiway Scaling.

SOFTWARE DEVELOPMENTS

Algorithms for Classification - Data Visualisation - Algorithms for Multivariate Data Analysis.

APPLIED CLASSIFICATION AND DATA ANALYSIS

Classification of Textual Data - Data Diagnostics Using Classification Tree - Data Analysis in Economics Data Analysis in Quality Control - Data Analysis in Environmental Science - Classification Approaches in Medical Science - Cognitive Sciences and Classification - Classification in Biology and Ecology - Data Analysis in Demography.

AKNOWLEDGEMENTS

The Editors would like to extend their sincere thanks to the authors whose enthusiastic participation made the meeting possible. We are very grateful to the reviewers for the time spent in their professional reviewing work. We are also grateful to the chairpersons and discussants of the sessions that also provided very useful suggestions to the authors.

We thanks the National Research Council (CNR) that partially supported the publication of this volume.

Special thanks are due to the local organizing committee: S. Borra, G. Bove, R. Rocci, D. Vicari and M. Vichi.

We would finally like to thank Dr. M. Bihn and her colleagues from Springer-Verlag, Heidelberg, for the excellent cooperation in publishing this volume.

Roma, Campobasso, Mannheim
November 2000

Simone Borra
Roberto Rocci
Maurizio Vichi
Martin Schader

CONTENTS

PART I

CLASSIFICATION AND MULTIDIMENSIONAL SCALING

Genetic Algorithms and Neural Networks

PART II MULTIVARIATE DATA ANALYSIS

Factorial Methods

Textual Data Analysis

Regression Models for Data Analysis

PART IV
APPLICATIONS AND CASE STUDIES

PART I

CLASSIFICATION AND MULTIDIMENSIONAL SCALING

Galois Lattices of Modal Symbolic Objects

D. Bruzzese, A. Irpino
Dipartimento di Matematica e Statistica - Facoltà di Economia
Università "Federico II" Napoli
bruzzese@dms.unina.it irpino@dms.unina.it

Abstract: In this paper we propose a classification structure of modal symbolic objects based on Galois lattice, and on the concept of credibility and capacity. This classification structure, which is hierarchical and non-partitive, has been developed in the framework of formal concept analysis in order to modelize the knowledge in form of context. The graph representation of a Galois lattices allows us to associate homogeneity measures in the nodes and transition measures on the edges.

Keywords: Modal Symbolic Object, Galois lattices, credibility, capacity, T-norm.

1. Introduction

1.1 Modal symbolic objects

Let E be a set of 'objects' u described by a set of symbolic variables $Y_1,\dots,Y_p$ of different type: quantitative real or at interval, discrete multivalued or modal variables.
According to (Boch & Diday, 2000) we have the following definitions of symbolic variable, modal variable, boolean and modal symbolic object and modal assertion.
A generical symbolic variable $Y(u)$ is termed set-valued with domain $\mathcal{Y}$, as it takes its values in the system $\mathcal{B}=\mathcal{P}(\mathcal{Y})=\{U|U\subseteq\mathcal{Y}\}$ of all non-empty subsets U of the set $\mathcal{Y}$:$Y(u)\subseteq\mathcal{Y}$ for all $u\in E$.
In the present paper the objects are mainly described by modal variables.
A modal variable Y on a set E of objects whit domain $\mathcal{Y}$ is a mapping
$Y(u)=(U(u),\pi_u)$ for $u\in E$
where

- π_u is a measure or a (frequency, probability or weight) distribution on the domain $\mathcal{Y}$ of possible observation values (completed by a suitable σ-field) and
- $U(u)\subseteq\mathcal{Y}$ is the support of π_u in the domain $\mathcal{Y}$.

Then a modal variable with domain $\mathcal{Y}$ is nothing more than a mapping $Y:E\rightarrow\mathcal{B}=\mathcal{M}(\mathcal{Y})$ from E into the family $\mathcal{M}(\mathcal{Y})$ of all non-negative measures π on $\mathcal{Y}$, with values $Y(u)=\pi_u$.

A *Boolean symbolic object* is characterized by a data vector Y, a description $d \in D$, *where* D belongs to the cartesian product of all the domains $\mathcal{Y}_i$ of the components of Y, i.e. Y_i, and a relation R that compares Y and d. The result of this comparison is described by a binary (Boolean) function a:E→{0,1} with $a(u)=1$ if $Y(u)$ fulfils the stated requirements and $a(u)=0$ otherwise. Formally a Boolean symbolic object is defined as a triple $s=(a,R,d)$. The function a is called the *extension mapping* of the symbolic object s. It defines the extension of s in E given by
$Ext_E(s):=Ext(s|E):=\{u \in E | a(u)=1\}=a^{-1}$,
i.e., the set of elements of E which fulfill the properties stated in s.
If R is the fuzzy relation $\Phi(\cdot,\cdot)$ indicating the degree of fitting between two description vectors, defined on $\mathcal{X}x\mathcal{X}$, where $D \subseteq \mathcal{X}$, and values in [0,1], the most general definition of a 'symbolic object' is the following:
A symbolic object is a triple $s=(a, \Phi, d)$ where d is a given description from a description space D, Φ a fuzzy relation between descriptions, and a is the mapping from E to interval [0,1] defined by $a(u):= \Phi(Y(u),d)$ for $u \in E$.
A symbolic object is termed modal symbolic object if Φ is a fuzzy relation, i.e., if extension function a takes values $a(u)$ strictly between 0 and 1.
For each $u \in E$, the value $a(u)$ measures the extent to which the object $u \in E$ complies with the description d. The extension of a modal object s is computed fixing threshold $\alpha \in [0,1]$, so that the extension of level α in E is given by:
$Ext_\alpha(s):= Ext_\alpha(s|E):=\{u \in E|\ a(u) \geq \alpha\}=a^{-1}([\alpha,1])=\{u \in E|\ a_s(u)= \Phi(Y(u),\pi) \geq \alpha\}$.
A particular kind of object is the *assertion object* q is the conjunction of events of the type: $q=\wedge_{v=1,..,r}[Y_v \mathcal{R}_v D_v]$ where r is the number of descriptor of q.
Moreover, several order symbolic object are obtained by means of generalization and specialization. Given two objects s_r and s_t: if $Ext(s_r|E) \supseteq Ext(s_t|E)$ then s_r is a generalization of s_t, otherwise, if $Ext(s_t|E) \supseteq Ext(s_r|E)$, then s_r is a specialization of s_t.

1.2 Galois Lattice of symbolic objects

Galois Lattices have been introduced by Birkhoff (Birkhoff, 1967) and defined by Barbut and Monjardet (Barbut-Monjardet, 1970), in the framework of a Galois correspondence, i.e. a couple of decreasing mappings between two ordered sets f:E→F (intension) and g:F→E (extension), with h=gof and k=fog extensive (i.e. $A \leq_E h(A)$, $B \leq_F k(B)$), E and F being two lattices with a partial order denoted by $\leq_E$, $\leq_F$. According to this correspondence, the sets $I_h=\{A \in E|\ gof(A)=A\}$ and $I_k=\{B \in F|\ fog(B)=B\}$ are defined. The elements of I_h (resp. of I_k}) are called closure of E (resp. of F).
The set of couples $(A,B) \in E \times F$, where $A \in I_h$ and $B \in I_k$, B=f(A) and A=g(B), defines a lattice termed "of Galois", associated to this correspondence. The lattice elements are called Galois closures.

In the framework of Symbolic Data Analysis (SDA), the set of the parties of symbolic objects is isomorphic to a Galois lattice.
There are two kinds of Galois lattices, an intersection lattice and a union lattice.
Let $s_1=(a_1,R,d_1)$ and $s_2=(a_2,R,d_2)$ be two symbolic objects. The symbolic union between s_1 and s_2, denoted by $s_1 \cup s_2$, is defined as the conjunction of all the symbolic objects whose extension on E contains Ext(s_1|E)$\cup$Ext(s_2|E). The symbolic intersection between $s_1$and s_2, denoted by $s_1 \cap s_2$, is defined as the conjunction of all the symbolic objects whose extension on E contains Ext(s_1|E)$\cap$Ext(s_2|E) (Diday, 1995).
The intersection lattice is an algebraic structure given by the intersections of the subsets of the power set of E. The union lattice, is an algebraic structure given by the unions the subsets of the power set of E.

2. Construction of a Galois lattice of modal symbolic object

Let Π be the matrix of the symbolic descriptions of n objects $u_t \in E$ $(t=1,..,n)$, along the rows, with respect to k *nominal* modal variables y_i $(i=1,...,k)$, along the columns. Each variable Y_i has support in $D_i=\{d_{i,1}, d_{i,2}, ..., d_{i,j}, ..., d_{i,p_i}\}$ where $d_{i,j}$ is the j-th category of the set of descriptions of Y_i and p_i the number of categories of Y_i, Each cell describes the distribution of the generic variable Y_i for the generic object u_t, by means of a system of weights (frequencies, probabilities, ...) denoted as $\pi_{t,i,j}$, that is the generic weight of the category $d_{i,j}$ for the symbolic object s_t. In particular if $d_{i,j}$ is not in the description of the symbolic object s_t the associated weight $\pi_{t,i,j}$ is equal to zero.
Let $\pi_{t,i,j}$ and $\pi_{r,i,j}$ be the weights of the same category $d_{i,j}$ of the generic variable y_i for two different objects u_t e u_r. Defined a *T-Norm* as a mapping from $[0,1]^2$ to $[0,1]$, $(\pi_{t,i,j},\pi_{r,i,j}) \rightarrow \pi_{t,i,j} T \pi_{r,i,j}$, the dual of T, T^* is a *T-Conorm*, such that $\pi_{t,i,j} T^* \pi_{r,i,j} = 1-[(1-\pi_{t,i,j}) T (1-\pi_{r,i,j})]$. We use $\pi_{t,i,j} T \pi_{r,i,j} = \min(\pi_{t,i,j},\pi_{r,i,j})$ and $\pi_{t,i,j} T^* \pi_{r,i,j} = \max(\pi_{t,i,j},\pi_{r,i,j})$ as the corresponding *T-Conorm*. According to Diday (Diday, 1997) these measures can be interpreted, respectively as the degree of credibility and the degree of capacity of two independent discrete random variables. The credibility can be interpreted as the probability that all the elements of a specialized object present a certain category, while the capacity as the probability that at least one element of a generalized object presents that category.
Let us define $\overline{\pi_{t,i,j}} = \max_{s_t \in E} \pi_{t,i,j}$ and $\underline{\pi_{t,i,j}} = \min_{s_t \in E} \pi_{t,i,j}$. The criterion we use in building the Galois Lattice is based upon the degree of credibility of subsets of objects. It is an extension of the Godin algorithm for the generation of an intersection lattice. According to this criterion, we define maximal and minimal object as follows:

- The maximal assertion object q_M is defined in intension by all the categories of the symbolic descriptors with the lowest weights associated. The extension of q_M is given by all the objects of the set E. The weights associated to the categories of each descriptor of q_M can be considered as measures of credibility of the object generalizing all the objects:

$$q_M = \wedge_{i=1,..,k}\left[y_i \in \left\{d_{i,1}\left(\underline{\pi_{t,i,1}}\right),\ldots,d_{i,j}\left(\underline{\pi_{t,i,j}}\right),\ldots,d_{i,p_i}\left(\underline{\pi_{t,i,p_i}}\right)\right\}\right], \quad (1)$$

$$Ext\left(q_M \mid E\right) = \left\{u_t \in E \middle| \forall d_{i,j}; i=1,\ldots,k; j=1,\ldots,p_i; \pi_{t,i,j} \geq \underline{\pi_{t,i,j}}\right\} = E \quad (2)$$

- The minimal assertion object q_m is defined in intension by all the categories of the symbolic descriptors with the highest weights associated. The extension of q_m is an empty set. The weights associated to the categories of each descriptor of q_m can be considered as measures of credibility of the object specializing all the objects of *E*:

$$q_m = \wedge_{i=1,..,k}\left[y_i = \left\{d_{i,1}\left(\overline{\pi_{t,i,1}}\right),\ldots,d_{i,j}\left(\overline{\pi_{t,i,j}}\right),\ldots,d_{i,p_i}\left(\overline{\pi_{t,i,p_i}}\right)\right\}\right], \quad (3)$$

$$Ext\left(q_m \mid E\right) = \left\{u_t \in E \middle| \forall d_{i,j}; i=1,\ldots,k; j=1,\ldots,p_i; \pi_{t,i,j} \geq \overline{\pi_{t,i,j}}\right\} = \varnothing \quad (4)$$

The procedure to build the lattice consists of the following steps:
1. Generation of the maximal and of the minimal object.
2. Selecting for each category $d_{i,j}$, the second minimum value $\pi_{t,i,j}$, for $u_t \in E$, among the weights associated to the categories of each descriptor of the objects in E. A set of new objects is generated.
3. A *m*eet between the objects generated in the previous step is performed until the whole construction of a complete lattice

Figure 1. *Line diagram of a Galois lattice*

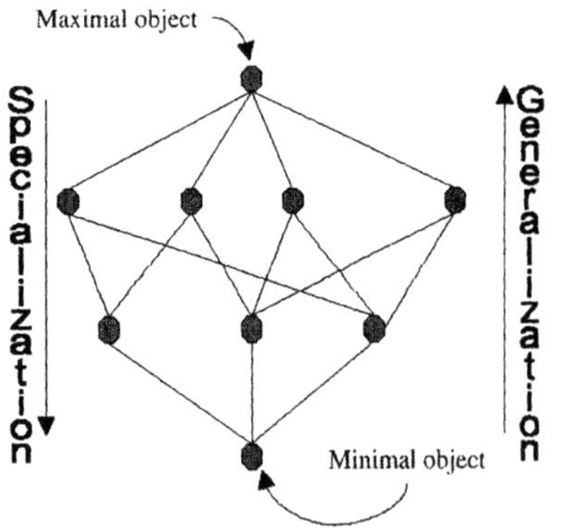

2.1 Measures associated to the nodes and to the edges

In the lattice building we have only used the concept of credibility, we can enrich the description of a node of the lattice introducing information in terms of credibility and the capacity. In other words, being C a class of objects, $\overline{\pi_{t,i,j}}^{C} = \max_{s_t \in C} \pi_{t,i,j}$ and $\underline{\pi_{t,i,j}}^{C} = \min_{s_t \in C} \pi_{t,i,j}$, the assertion corresponding to the intention of the node generated by C is described as follows:

$$q_C = \wedge_{i=1,..,k}\left[y_i = \left\{d_{i,1}\left[\underline{\pi_{t,i,1}}^{C};\overline{\pi_{t,i,1}}^{C}\right]...,d_{i,j}\left[\underline{\pi_{t,i,j}}^{C};\overline{\pi_{t,i,j}}^{C}\right]...,d_{i,p_i}\left[\underline{\pi_{t,i,p_i}}^{C};\overline{\pi_{t,i,p_i}}^{C}\right]\right.\right] \quad (5)$$

Each class C of objects can be described by the set categories $d_{i,j}$ of the variable Y_i with associated the minimum and the maximum weight $\pi_{t,i,j}$ of the objects $u_t \in C$.
The minimum weight for a category represents the common information to all the objects of the class related to that category. In fact, if we choose an object in the class C, its weight for that category will be equal or greater then the minimum. As in (Diday, 1995), it represents the degree of credibility for that category in C, while the maximum, , measures of degree the capacity of such class. In other words, the degree of credibility is the information obtained by the intersection of modal objects; while the degree of capacity is the information obtained by the union of modal symbolic objects.
If the categories of a variable are weighted by relative frequency or probability distributions, it is easy to show that the sum of the degree of credibility for the categories of a modal variable increases from 0 to 1, while the degree of capacity decreases from a number greater than 1 to 1, when the objects are more and more specialized.
Being q_1 and q_2 two modal objects, y_l a modal variable described by p_l categories, $d_{1,1}$, $d_{l,2}, \ldots$, d_{1,p_1}, and $\pi_{t,1,1}$, $\pi_{t,1,2}$, ..., $\pi_{t,1,p_1}$, $t=1,2$, the associated weights (frequencies or probabilities) the follow constraint are verified:

$$\pi_{1,1,1}+\pi_{1,1,2}+\ldots+\pi_{1,1,p_l}=1; \quad \pi_{2,1,1}+\pi_{2,1,2}+\ldots+\pi_{2,1,p_l}=1$$

for the internality property of the mean follows that:

$$\Sigma_{j=1,..,p_i}\min(\pi_{1,1,j};\pi_{2,1,j}) \leq \Sigma_{j=1,..,p_i}(\pi_{1,1,j}+\pi_{2,1,j})/2;$$

$$\Sigma_{j=1,..,p_i}\max(\pi_{1,1,j};\pi_{2,1,j}) \geq \Sigma_{j=1,..,p_i}(\pi_{1,1,j}+\pi_{2,1,j})/2$$

Where $\Sigma_{j=1,..,p_i}(\pi_{1,1,j}+\pi_{2,1,j})/2=1$

Furthermore:

$$lim_{\pi_{1,1,j}\to\pi_{2,1,j}} \min(\pi_{1,1,j};\pi_{2,1,j}) = \pi_{2,1,j} \quad lim_{\pi_{1,1,j}\to\pi_{2,1,j}} \max(\pi_{1,1,j};\pi_{2,1,j}) = \pi_{2,1,j}$$

this means that the more the weights of two or more objects are similar for the distribution of Y_l; the more they have similar distribution for Y_l .
It is worth noting, as is shown in (Shafer, 1975), credibility and capacity spaces have not the same properties of the probability space, one for all, the degrees of credibility or capacity for the categories of a variable does not (necessarily) add to 1.
In each node, and for each variable, the interval between the credibility and the capacity can be interpreted as the convex hull where the theoretical probability distribution of the class represented by the node can be found.

Each edge of lattice represents an association rule between the descriptors of the descriptors of the son one (more specialized) and those of the father node (more general), defined $r: d \to d'$ (*if d then d'*), where *d* is the description of the object of the son node, described by the assertion q_{son}, the cause, and *d'* the description of father node, the consequence, described by the assertion q_{fat}. The evaluation of the rule is usually done by means of two measure called support and confidence. The first one, the support, denoted by $\sigma(r)$, defined (Agrawal et al.(1993)) as the ratio $Card(\text{Ext}(q_{son}|E))/Card(E)$. The second measure is the confidence, denoted by $\varphi(r)$, defined as the ratio $Card(\text{Ext}(q_{son}|E))/Card(\text{Ext}(q_{fat}|E))$.
If E is the set of all the units, commonly known as Ω, these measures can evaluate the portion of applicability of the rules in the total population of units (the support) or the portion of those units that are described by q_{son} whit respect to those described by q_{fat} (the confidence). If E is not Ω, but, for example, collection of classes, other measures have to be used in order to evaluate rules between descriptors of objects. We can observe that moving from an upper node to a lower one, according to the meet operator, the minima increase and the maxima decrease. This means that more the class is specialized, more the objects belonging to the class present similar variable distributions. In the nodes, which describe the single objects, the maximum and the minimum converge to the same values, in fact each object is described by single values for each category (its weight). According to such information a measure can be associated on each edge. This measure evaluates, for the variable Y_i, the reduction of the sum of the degrees of capacity and the increase of sum of the degrees the credibility when we move from a level of the lattice to another one. A simple proposal is the *increase of Total credibility ratio* for the variable Y_i when we move from an upper node described by the class C_l to a lower linked node described by the class C_k in the following way:

$$\textit{Increase of Total credibility ratio}(y_i) = \frac{Cred(Y_i(C_k)) - Cred(Y_i(C_l))}{Cred(Y_i(C_l))} \tag{6}$$

Where $Cred(Y_i(C_k)) = \Sigma_{j=1,\dots,p_i} \min_{u_t \in C_k}(\pi_{t,i,j})$. Alternatively we can focus the attention on the *decrease of Total capacity*

$$\textit{Decrease of Total capacity ratio}(y_i) = \frac{Cap(Y_i(C_l)) - Cap(Y_i(C_k))}{Cap(Y_i(C_k))} \tag{7}$$

Where $Cap(Y_i, C_k) = \Sigma_{j=1,\dots,p_i} \max_{u_t \in C_k}(\pi_{t,i,j})$.

3. Example

Let Π be a symbolic table where on the rows we have four objects descriptions (Professionals, Managers, Students, Housewives) deriving from a survey on the purchase behaviors of classes of consumers in a general store.
On the columns we have two nominal variables describing the purchase items

(Purchases), with by three categories (Products for house, Products for free time, Books and reviews), and the value of purchase (Purchase value), with two categories (Low, High) on the base of a threshold of 100$. In the cells there are the relative frequencies.

Table 1. *Table Π*

Variables	Purchases			Purchase value	
	Prod. for house	*Free time*	*Books & reviews*	*Low*	*High*
Professionals	**0,1**	**0,3**	**0,6**	**0,2**	**0,8**
Managers	**0,5**	**0,3**	**0,2**	**0,1**	**0,9**
Students	**0,2**	**0,1**	**0,7**	**0,5**	**0,5**
Housewives	**0,8**	**0**	**0,2**	**0,6**	**0,4**

Figure 2. *Line diagram of the intersection Galois lattice of the table Π.*

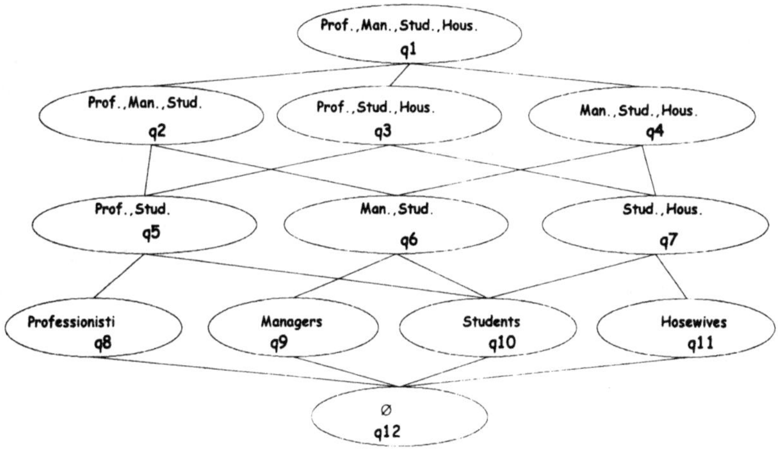

The nodes have two labels: the bottom one represents the intensions described by the assertions *q*, the top one, the extension of the assertion *q*. The edges represent the rules between the descriptors of father node and the descriptors of the son node.

Among the several rules that can be extracted by the Galois lattice, let's take, for example, the rule that links the node described by the assertion *q2* and the *q5* node, i.e. of the two objects:

q2= [Purchases⊇{(0,1) Prod.for house,(0,1) Free time,(0,2) Books & Review}]
∧ [Purchase value⊇{(0,1) Low,(0,5) High}]
Ext(q2|E)={ Professionals,Managers,Students}

q5= [Purchases⊇{(0,1) Prod.for house, (0,1) Free time, (0,6) Books & Review}]
∧ [Purchase value⊇{(0,2) Low,(0,5) High}]
Ext(q5|E)={ Professionals, Students}

The rule is

*R*1: *if* Purchase(Free time)≥**0,1** *and* PurchaseValue(High)≥**0,5** *then* Purchase(Books & R.)≥ **0,6**

the measures associated to this rule, considering the variation of the degree of

credibility, for the two variables, are:
Increase of Total Credibility ratio(Purchase)=0,4/0,4=1,0
Increase of Total Credibility ratio(Purchase *V*alue)=0,1/0,6=0,19
These measures express the importance of the rule In particular R1 is very important from the point of view of the purchases, in fact, it allows to double the degree of information linked to the credibility, while such rule is less important for the Purchase value, in fact the information given by the degree of credibility is increased only of 19%.

4. Conclusions

Galois lattice is a useful tool to discover association rules between objects, but they present some computational problems. In fact, the number of nodes and edges of the lattice exponentially increases, more and more objects and variables are considered.
Solutions to the computational problem can be based either on the experience or on the properties of Galois Lattice. From the first point of view, we cannot generate the nodes or edges which are not coherent with some a priori information, such as taxonomies between variables, or rules, given for instance by experts. Therefore, such logical constraints can be obtained by fixing a threshold of minimal increase of credibility. From the second point of view, based on the properties of Galois Lattice (Wille, 1982), we can decompose the initial context in simpler sub-context, i.e. described by few objects or variables, in order to find local rules, in the sub-lattices, and global rules between sub-lattices, without loss of information, because is always possible to move from the sub-lattices to the complete lattice and vice-versa.

References

Agrawal, R., Imielinski, T. & Swami, A.(1993). Mining association rules between sets of items in large databases. *Proc. of ACM-SIGMOD*, Washington D.C.

Barbut M., Monjardet B. (1970). *Ordre et Classification*, Hachette, Paris.

Birkhoff, G., (1967) Lattice Theory , *AMS Colloq. Public.* Vol. XXV.

Boch H.H., Diday E., (2000), *Analysis of Symbolic Data,* Springer-Verlag, Berlin.

Diday E.(1995) Probabilist, possibilist and belief objects for knowledge analysis. *Annals of Operations Research*. 55, 227-276.

Diday E., Emilion R., Hillali Y.(1996) Symbolic Data Analysis of Probabilistic objects by capacities and credibilities. *Atti della XXXVIII Riun. SIS.* Rimini.

Shafer G.(1976) *Mathematical Theory of Evidence*. Princeton University Press.

Wille, R., (1982) Restructuring lattice theory: an approach based on hierarchies of concepts. *Ordered Sets*. Rival, I., D. Reidel Publishing Company, Dordrecht-Boston. 445-470.

Exploratory Methods for Detecting High Density Regions in Cluster Analysis

Andrea Cerioli and Sergio Zani
Dipartimento di Economia, Sezione di Statistica, Università di Parma
Via Kennedy 6, 43100 Parma, Italy; E-mail: Statec1@unipr.it

Abstract: In this paper we propose some simple diagnostics which can prove useful for detecting high density regions in $\Re^p$, for $p \geq 2$. Our approach does not require full estimation of the multivariate density and exploits the spatial contiguity information which can be attached to objects in $\Re^p$. The suggested method could be routinely applied as a preliminary step in nonhierarchical cluster analysis, where it provides useful guidance both in choosing the appropriate number of clusters and in selecting the values of initial cluster seeds.

Keywords: Local statistics, Multivariate histograms, Nonhierarchical classification, Spatial clustering.

1. Introduction

It is well-known that applying a clustering algorithm to a data set of n objects produces a classification irrespective of the existence of a genuine structure in the data. This has been long recognized as a major issue of potential concern in cluster analysis and has led to the development of several confirmatory techniques based on suitable probability models for clustering in multidimensional space, with the aim of validating the computed classification (see Bock, 1996, for a review). Significance testing plays a role also prior to the computation of the selected classification rule, because acknowledgment of the absence of any group structure would make cluster analysis unnecessary.

In cluster analysis population homogeneity is usually described in terms of either the Poisson model, where the observation vectors $\mathbf{x}_1, \ldots, \mathbf{x}_n$ are conceived as a random sample from the uniform distribution on a finite domain $S \subset \Re^p$, or the unimodality model, where $\mathbf{x}_1, \ldots, \mathbf{x}_n$ have the same unimodal density $f(\mathbf{x})$ on $\Re^p$. Tests of the Poisson model follow from the methodology established for the investigation of spatial point patterns (Bock, 1996, sect. 6.2; Gordon, 1996, sect. 7.1), while the adequacy of the unimodal model can be verified via a number of multimodality tests (Bock, 1996, sect. 6.3).

In the present paper we propose a diagnostic technique which can be useful for exploratory purposes at the preliminary stage of cluster analysis. Our approach does not rely on a specific model of homogeneity, but tries to take advantage of the ideas underlying both standard null models. Indeed we perform only the first step of the whole process of density estimation, i.e. computation of a multivariate

histogram, but we do not require the additional effort of choosing a kernel and estimating $f(\mathbf{x})$ for all $\mathbf{x} \in S$.
We enhance the histogram diagnostic power through the definition of a neighbourhood structure in $\Re^p$ and the computation of "local" statistics. Spatial information is shown to be important for the purpose of improving the identification of local maxima of the density and might also help to overcome some inherent difficulties in the interpretation of results from global validity tests (see Bock, 1996, p. 430). Furthermore, it can reduce the random noise that typically affects the construction of multivariate histograms as the number of bins increases, thus producing a more stable representation of the structure of the data.
In two and three dimensions, results are displayed through simple plots which can be very informative on the actual structure of the data and provide useful guidance at subsequent steps of the classification process. The definition of formal multimodality indices when $p > 3$ will be the subject of future research and will not be addressed in detail in this article. The potential of the methodology is shown both in a standard example (the well-known Fisher's iris data set) and in an application to financial data concerning the performance of investment funds.

2. High density diagnostics

Our technique for exploring high density regions is quite simple. Given p variables, say $X_1, \ldots, X_p$, we first define a regular grid of (multidimensional) quadrats which is overlaid to the p-variate scatterplot. Let Q denote the total number of quadrats, indexed in lexicographical order. For simplicity we restrict our attention to the situation where a fixed number of quadrats, say q, is chosen in each dimension, so that $Q = q^p$.
Let n_j be the number of points falling in quadrat j, $j = 1, \ldots, Q$, and let $N \equiv \sum_j n_j/Q = n/Q$. Of course, the value of q controls the amount of smoothing implied by the procedure. If the underlying density were known, optimal selection of quadrat edges would follow from the theory of multivariate histograms (Scott, 1992, sect. 3.4; Wand, 1997). In practice, however, $f(\mathbf{x})$ is unknown and strict optimality criteria are not applicable. In our examples we found that choosing q in such a way that $N \approx 1$ is usually satisfactory for exploratory purposes. In fact this choice implies that under the Poisson model $\mathrm{E}(n_j) = \mathrm{Var}(n_j) \approx 1$ for $j = 1, \ldots, Q$. Furthermore, it will be shown that general conclusions drawn from our density diagnostics are largely insensitive to the actual value of q.
Variable width quadrats (based e.g. on univariate deciles) can also be adopted, followed by appropriate standardization of quadrat counts. This option is not to be recommended in general, however, due to possible distortion in the perceived contiguity between clusters which are in fact well separated. The current version of our algorithm also supports the input of a user-defined origin for the grid, thus allowing repeated analyses from different starting points.
Histogram computation can be seen as a first step in the whole process of nonparametric density estimation. For many applications of cluster analysis,

however, full estimation of the multivariate density may not be required. In fact, at a preliminary stage, recognition of potential clusters and approximate identification of their shapes and modes usually suffice. To improve the exploratory power of our method, we suggest to take the geometrical structure of $\Re^p$ into account and supplement the knowledge given by quadrat counts with spatial information. We thus define a multidimensional neighbourhood structure in $\Re^p$ and compute *local clustering statistics*, i.e. diagnostic measures which can detect local concentrations of high counts n_j.

Let n_j^* be the average of quadrat counts in the p-dimensional neighbourhood of quadrat j, i.e. $n_j^* = \sum_{l=1}^{Q} w_{jl} n_l / \sum_{l=1}^{Q} w_{jl}$, where $w_{jj} \equiv 0$ and, for $j \neq l$, $w_{jl} = 1$ if quadrats j and l are neighbours in $\Re^p$ and $w_{jl} = 0$ otherwise (Cerioli and Riani, 1996). To set up the neighbourhood structure, in our algorithm we have implemented the multimensional versions of both the rook and the queen definitions of contiguity, which are the most widely adopted when dealing with a regular grid.

Our first diagnostic is defined as

$$m_j = n_j - n_j^*, j = 1, \ldots, Q. \tag{1}$$

High values of m_j correspond to quadrat counts which are disproportionally large if compared to the surrounding area, and hence reveal the presence of local peaks in the underlying density. The computation of (1) is thus directed towards the identification of outlying quadrat counts (in a local sense). Note that for the purpose of outlier detection the m-diagnostics are usually to be preferred to the raw counts n_j, because they do not provide information on a global scale but only refer to a restricted portion of the point cloud.

For depicting high density regions, however, it is also important to spot local concentrations of large values of n_j across contiguous quadrats. Local indicators of spatial association have been recently developed in the special case where $p = 2$ and the coordinate system has a geographical meaning (see Unwin, 1996, and Wilhelm and Steck, 1998, for a review). In what follows, we extend that line of research to the general case of a multidimensional space where axes represent substantive variables, and define a number of clustering statistics based on quadrat counts n_j.

Specifically, we focus on the following diagnostics:

$$I_j = |n_j - N| \sum_{l=1}^{Q} w_{jl} (n_l - N) / \sum_{l=1}^{Q} w_{jl}, j = 1, \ldots, Q, \tag{2}$$

and

$$Z_j = [\sum_{l=1}^{Q} w_{jl}^* n_l/(QN) - W_j^*/Q] \times [\kappa^2 W_j^* (Q - W_j^*)/(Q^3 - Q^2)]^{-1/2}, j = 1, \ldots, Q, \tag{3}$$

where $w_{jl}^* \equiv w_{jl}$ if $j \neq l$ and $w_{jj}^* \equiv 1$, $W_j^* \equiv \sum_l w_{jl}^*$, and κ is the variation coefficient of the quadrat counts.

The diagnostic I_j is a modified version of the local Moran autocorrelation coefficient originally suggested by Anselin (1995) and computed on quadrat counts. In the present context the local measure has been scaled and suitably modified in order to distinguish between spatial concentrations of low and high counts. In particular, we are interested in the peaks of the distribution of I_j: a large positive value of I_j occurs when a high count in quadrat j is associated to similar counts in neighbouring quadrats and hence indicates local clustering of points in the multidimensional cloud. In a similar vein, Z_j is a standardized statistic which originates from the attempt to quantify clustering through the sum of local counts, including the count n_j itself (see Getis and Ord, 1992, for a formal justification of expression (3)). Again, we focus on large positive values of this index.

Visual inspection of the distribution of I_j and Z_j across the quadrats can be helpful in the detection of local maxima of the underlying density. At an exploratory stage identification of these maxima is usually the most important information and full estimation of the multivariate density is not required. For instance, looking at the modes of the distribution of our diagnostics provides a plausible criterion for choosing the number of groups in nonhierarchical classification and also gives information about potential cluster shapes. Furthermore, a set of initial cluster seeds can be obtained by taking the variable centroids in those quadrats where the selected measure has its local maxima.

Density diagnostics I_j and Z_j are also potentially resistant to the inclusion of clusters of p-variate outliers in the data set. In fact, extreme observations falling in the same quadrat are typically downweighted in the computation of (2) and (3) by consideration of the surrounding values. This has the additional bonus of providing results that are less sensitive to the specific value of q. In the present setting, outlier detection might instead be performed through the plot of m_j.

The above properties of I_j and Z_j should be compared with those of the "data sharpening" technique of Choi and Hall (1999), which amounts to a perturbation of the original data in order to make them more clustered than before. Sharpening aims at reducing the bias of traditional density estimators, but this is at the expense of a constant-factor increase in variance. The article by Choi and Hall (1999) was published when the present paper was almost completed and thus a detailed comparison of the two approaches will be the subject of future research.

3. Applications

3.1 Iris data

We first apply the diagnostic tools suggested in the previous section to the familiar Fisher's iris data set, which contains information on four variables (sepal and petal length and width) for 150 flowers belonging to three different species of iris (Setosa, Versicolor and Virginica). It has been noted by many authors that in the 4-dimensional space the Setosa variety does not overlap with the other two species, which, on the contrary, are not as well separated. To obtain simple 3-dimensional representations of our diagnostics, we first project the 4-dimensional point cloud on the first two principal axes (which explain more than 95% of the total variation). We then compute the statistics I_j, Z_j and m_j on component scores.

Figure 1 shows the plot of I_j obtained with q = 12 and the rook definition of contiguity. From this diagram the existence of two main clusters of irises is clearly apparent. In addition, the local maximum of I_j in the upper-right corner of the grid suggests the possible splitting of the biggest cluster into two subgroups. In fact, this peak is attributable to flowers of the Virginica variety and shows, without the need of estimating the density in full, that the Versicolor and Virginica data can be distinguished in the density cloud (cfr. Scott, 1992, p. 25). Similar displays are also obtained through the calculation of (1) and (3) but are not reported here due to space constraints.
There is empirical evidence that local clustering statistics provide a picture of the underlying density that is more robust to the choice of the number of quadrats than the standard histogram. For example, Figure 2 gives the plots of I_j and Z_j for the first two component scores of the iris data for q = 20 and the rook definition of contiguity. The basic structure of the density is still apparent even in this situation where the number of quadrats might be deemed excessive for the size of the original data set (Silverman, 1986, p. 80), with a higher degree of smoothing implied by the standardized statistic Z_j. These histograms are to be compared with that based on raw quadrat counts, also shown in Figure 2, where random noise effects clearly produce a number of spurious modes within each group of irises.
With three or more principal axes slicing is possible, and a series of bivariate plots similar to those reported in Figures 1 and 2 can be drawn for each pair of component scores (possibly after conditioning on the remainder). As an alternative, when p = 3, a 3-dimensional contour can be produced adopting a discretized version of the α-level surfaces described by Scott (1992, sect. 1.4).

Figure 1: *Plot of I_j for the first two component scores of the iris data and q = 12.*

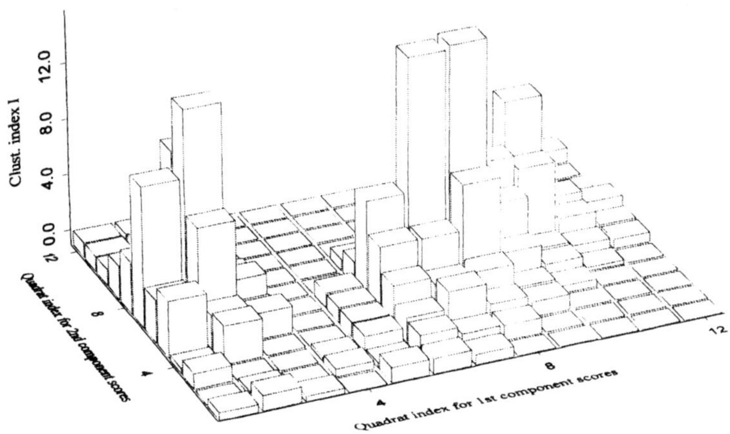

3.2 Investment funds

The methodology is applied to explore the performance of 103 investment funds operating in Italy since April 1996. For this purpose we consider the following variables: X_1 = short term (12 month) performance; X_2 = medium term (36 month) performance; X_3= medium term (36 month) volatility. The data are taken from the italian financial journal *Il Sole-24 Ore*, n. 123, May 7th, 1999.

Figure 2: *Plots of* I_j, Z_j *and* n_j *for the first two component scores of the iris data and* $q = 20$.

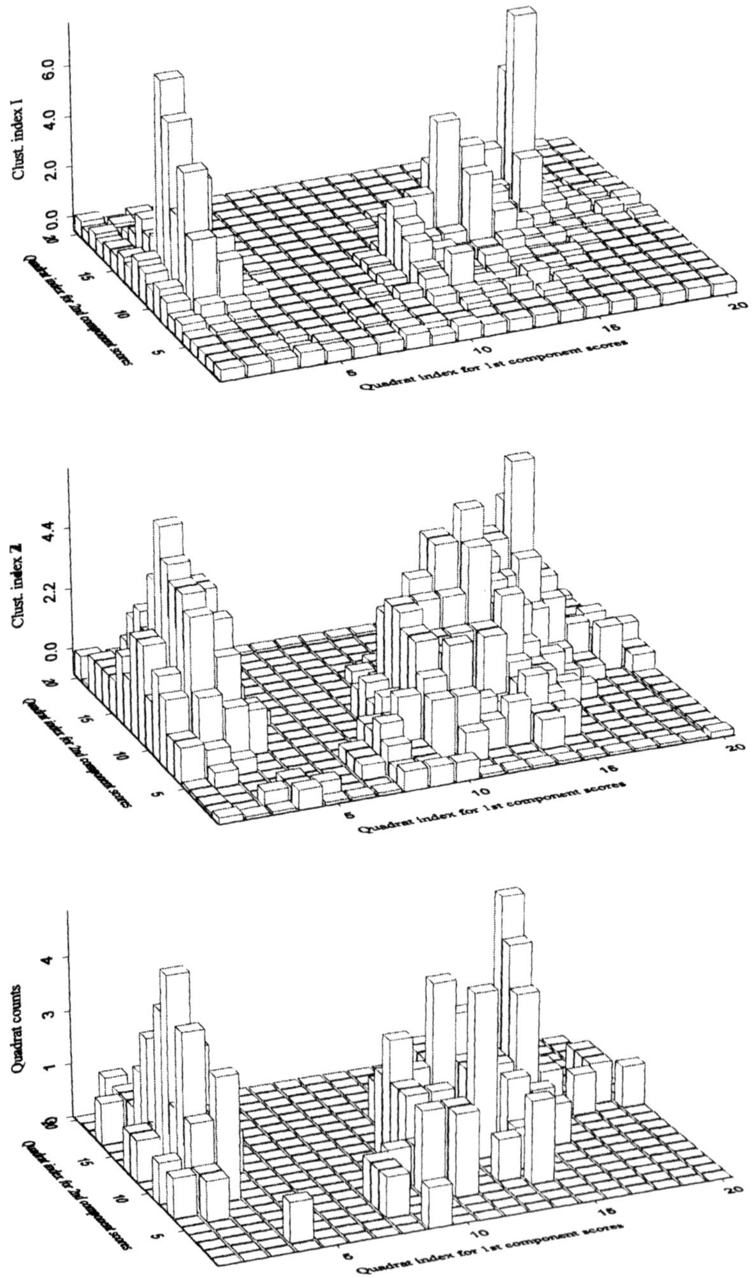

Figure 3: *Plots of I_j for all pairs of performance variables of the investment funds and $q = 10$. First plot: variables $X_1 - X_2$; second plot: variables $X_1 - X_3$; third plot: variables $X_2 - X_3$.*

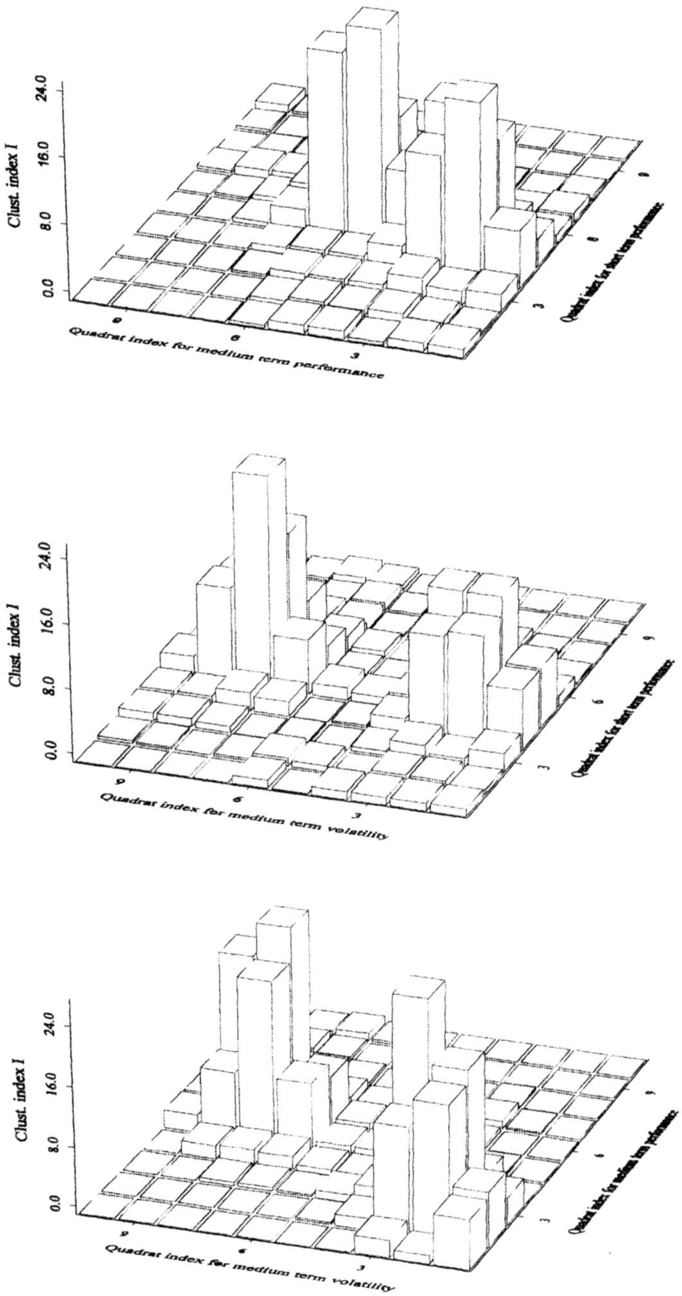

In this application we choose the simple option to display the 3-dimensional histograms corresponding to pairs of financial indicators, instead of performing the analysis on component scores. The plots obtained with respect to the local clustering measure I_j are given in Figure 3, for $q = 10$ and the rook definition of contiguity. For each pair of variables our index clearly shows two pronounced modes in the corresponding bivariate distribution, more striking in the second and third plot of Figure 3 but still apparent in the first display. This is not unexpected, however, because short and medium term performance provide highly correlated information and therefore define clusters which are partially overlapping.
The bimodality of the distribution of the performance indicators is due to the fact that the investment funds actually belong to two different groups (stock funds and balanced funds). Through our exploratory technique we have thus made obvious an important feature of the data, without the need of estimating the multivariate density in full. This information can prove useful in subsequent steps of the classification process, such as the identification of the appropriate number of groups and the computation of preliminary cluster centroids.

References

Anselin L. (1995) Local Indicators of Spatial Association — LISA, *Geographical Analysis*, 27, 93-115.

Bock H. H. (1996) Probability Models and Hypotheses Testing in Partitioning Cluster Analysis, in: *Clustering and Classification*, Arabie, P., Hubert, L. J. and De Soete, G. (Eds.), World Scientific, Singapore, 377-453.

Cerioli A. and Riani M. (1996) Metodi per l'individuazione di valori anomali nelle analisi spaziali, *Atti della XXXVIII Riunione Scientifica della Società Italiana di Statistica*, vol. 1, Maggioli, Rimini, 487-498.

Choi E. and Hall P. (1999) Data sharpening as a prelude to density estimation, *Biometrika*, 86, 941-947.

Getis A. and Ord J. K. (1992) The Analysis of Spatial Association by Use of Distance Statistics, *Geographical Analysis*, 24, 189-206 (Erratum: 25, 276).

Gordon A. D. (1996) Hierarchical Classification, in: *Clustering and Classification*, Arabie, P., Hubert, L. J. and De Soete, G. (Eds.), World Scientific, Singapore, 65-121.

Scott D. W. (1992) *Multivariate Density Estimation*, Wiley, New York.

Silverman B. W. (1986) *Density Estimation for Statistics and Data Analysis*, Chapman and Hall, London.

Unwin A. (1996) Exploratory Spatial Analysis and Local Statistics, *Computational Statistics*, 11, 387-400.

Wand M. P. (1997) Data-Based Choice of Histogram Bin Width, *The American Statistician*, 51, 59-64 (Correction: 53, 174).

Wilhelm A. and Steck R. (1998) Exploring Spatial Data by Using Interactive Graphics and Local Statistics, *The Statistician*, 47, 423-430.

A *k*-means Consensus Classification

Giuseppina Damiana Costanzo
Dipartimento di Economia Politica, Università degli Studi della Calabria,
Via P. Bucci, 87036 Rende (CS), Italy. e-mail: dm.costanzo@unical.it

Abstract: The aim of this paper is to detect social and territorial structures from the census data of the administrative units in Italy. Variables describing each administrative unit can be divided in three distinguishable groups. Three approaches to analyze these data are proposed and compared. The first two approaches are essentially two-step "tandem analysis" consisting of a preliminary factorial analysis and then a subsequent cluster analysis on the first few factors. The third approach suggests three different classifications, one for each group of variables, and then establishes a consensus classification *via* the *k*-means-clustering criterion.

Keywords: *k*-means, consensus classification, MFA.

1. Introduction

Many research projects generate data, which are multivariate. In some cases the variables are divided in two or more conceptually different sets. For example in environmental studies there might be: a set of variables **Y**, such as species abundance; a set of environmental **X** variables; a set of contaminant levels **Z**. Variables that describe statistical units (for example families) in a socioeconomic survey, can be classified into many groups for example: demographic (e.g., age of the head of the family, number of children in the family), economic (e.g., income), those that provide a general description of the social position of the family, and so on.

Statistical models and methods to analyze this kind of data strictly depend on the aim of the analysis and of the distinctive features of the data. Usually the objective of an exploratory analysis is to discover relationships between the sets of variables, and then to describe any significant relationship. When we have just two sets of variables, various multivariate statistical approaches are available. Here we will focus on a practical situation where there are more than two (distinguishable) groups of variables, none of them can be considered a set of response variables, and relationships between sets of variables are relevant in finding significant groups of statistical units.

2. Case study

The aim of this paper is to detect "social and territorial structures" using the census data of the administrative units in Italy (ISTAT, *Censimento della Popolazione e delle Abitazioni,* 1991). The term "social and territorial structures" is meant to refer to the relationships between social, economic and demographic characteristics of a population in a territory. These relationships give rise to specific social structures.
To find out these "structures" (groups of administrative units) the following variables have been considered:

a) **socio-economic variables** that provide an exhaustive description of the economic situation and the employment position of the resident population in each administrative unit.

b) **social** and **demographic variables** such as the kind of family: single-parent, nuclear and extended family, single and house-sharing family.

c) **urban** and **territorial variables** such as the time of building (six categories) and the type of ownership of the houses.

The data matrix looks like a contingence table; on the columns there are 85 features describing each row (administrative unit) of the table. They can be distinguished in three distinct groups, each group is referring to a different population: the population of individuals, the population of families and the population of houses in each administrative unit; in each cell of the table there is the number n_{ik} of objects (individuals, families or houses) belonging to the administrative unit i and "having" the variable k (k=1,...,85). In the following, the analysis is confined only to the administrative units of the district Calabria (altogether 409 administrative unit), and in the next session three different kinds of statistical approaches proposed to analyse the data are illustrated.

3. Methods

The first approach of analysis used is a Multiple Factor Analysis (see Escofier B. and Pagès J., 1984, 1994) and a subsequent Cluster Analysis on the most important (in terms of explained variability) factorial axis.
Multiple Factorial Analysis (MFA) was developed by the French School in the first years of 80's to study situations in which several groups of variables are defined on the same set of individuals. MFA is based on a Principal Component Analysis (PCA) applied to the whole set of variables in which each group of variables is weighted. Weighting of variables groups is necessary to make the influence of each group comparable in a global analysis. Let P be the number of groups, and let $\lambda(p,1)$ denote the first eigenvalue of the principal component analysis applied to the single group p. Weights of the groups are specified by $1/\lambda(p,1)$. MFA then consists of a principal component analysis of the whole set of variables, where each variable of the group p is weighted by the coefficient $1/\lambda(p,1)$.

In the second approach groups of variables are analyzed separately using PCA. The most important factors of each analysis are then retained and conjointly used in a subsequent Cluster Analysis.
These two step procedures, called "tandem analysis" or "tandem clustering", have been warned against by several authors (Arabie and Hubert, 1994, De Soete e Carrol, 1994), because the first few components or factors of **X** do not necessarily define a subspace that is the most informative about the cluster structure in the data but on the contrary it may obscure or mask the taxonomic information in **X**. These warnings suggest the third approach to directly classify units on the basis of each group of variables. Then the three classifications are "summarized" by establishing a consensus classification. Various authors have been engaged in the problem of the comparison and consensus of classifications; an overview of the main approaches is outlined by Barthélemy, Leclerc and Monjardet (1986), while a bibliographic review of this topic is presented by Leclerc and Cucumel (1987).
The next section describes how the *k*-means criterion (MacQueen, 1967) can be directly used to establish a consensus classification.

4. The *k*-means consensus classification

Let $\mathbf{X}=(x_{ik})$ contains measurements of *n* individuals or objects on *K* variables ($k=1, ..., K$). If the aim of the analysis of the data matrix **X** entails clustering the *n* objects into *L* ($L<n$) homogenous (non overlapping) clusters starting from **X**, let the $L \times K$ matrix **C** specify the centroids of the *L* clusters and let **U** be an $n \times L$ binary indicator matrix that specifies for each object the cluster membership:

$$u_{il} = \begin{cases} 1 & \text{iff object } i \text{ belongs to cluster } l \\ 0 & \text{otherwise} \end{cases}$$

Since in *k*-means cluster analysis each object is assigned to one and only one cluster $\sum_{l=1}^{L} u_{il} = 1$ holds for $i=1,...,n$.
In *k*-means clustering, the matrices **C** and **U** are determined such that the sum of the squared Euclidean distances between the object and the centroids of the clusters to which they belong, is minimal; that is **C** and **U** are determined such that the least-squares loss function $F(\mathbf{C},\mathbf{U}) = \| \mathbf{X}-\mathbf{UC} \|^2 = tr(\mathbf{X}-\mathbf{UC})'(\mathbf{X}-\mathbf{UC})$ is minimal. The loss function $F(\mathbf{C},\mathbf{U})$ is usually minimised by means of an Alternating Least-Squares (ALS) procedure that alternates between minimising $F(\mathbf{C},\mathbf{U})$ with respect to **U** given the current estimate of the cluster centroids **C**, and minimising $F(\mathbf{C},\mathbf{U})$ with respect to **C** given the current cluster assignments **U**, until no more changes occur in **U** and henceforth in **C**. To start the algorithm an initial estimate of **C** is required. A variety of suggestions have been offered as to how to form the *L* starting points used as initial estimates of cluster centres. An

implementation of ALS procedure which considers the first L administrative units that are mutually furthest apart is used in this paper.
The cluster membership variable resulting from a cluster analysis is actually a categorical variable that can be reformulated into a binary matrix. Consider, as an example, the three classifications, C_1, C_2, C_3 of three classes each and the corresponding binary matrices $\mathbf{Y}_1$, $\mathbf{Y}_2$, $\mathbf{Y}_3$ in Table 1:

Table 1: *Classifications and their representation by binary*

Objects	Classifications			Binary matrices		
	C_1	C_2	C_3	$\mathbf{Y}_1$	$\mathbf{Y}_2$	$\mathbf{Y}_3$
O_1	1	2	3	1 0 0	0 1 0	0 0 1
⋮	⋮	⋮	⋮	⋮	⋮	⋮
O_i	2	1	3	0 1 0	1 0 0	0 0 1
⋮	⋮	⋮	⋮	⋮	⋮	⋮
O_j	3	2	1	0 0 1	0 1 0	1 0 0
⋮	⋮	⋮	⋮	⋮	⋮	⋮
O_n	3	2	1	0 0 1	0 1 0	1 0 0

It is well known that for just one classification the Euclidean distance between objects i and j is $d^2(i,j)=2(1-c(i,j))$ where $c(i,j)$ is the Jaccard coefficient (Everitt, 1977) taking values 0 or 1 in this case. Considering the three partitions in Table 1 all together, the Euclidean distance between i and j is $d^2(i,j)=3\cdot 2(1-c(i,j))$; $c(i,j)$ takes the values of the set $\{0, 1/3, 2/3, 1\}$.
This "result" can be generalised to the case of M classifications and obtain $c(i,j)=\dfrac{M-m}{M}$, with $m = 0,1, \ldots, M$ denoting the number of partitions in which the pair (i,j) belongs to different clusters.
It follows that:

- $c(i,j)=0$ if (i,j) belongs to different clusters in all of the classifications (i.e. m=M);
- $c(i,j)=1$ if (i,j) belongs to the same cluster in all of the partitions (i.e. m=0);
- $c(i,j)$ decreases as m increases, that is as the number of partitions in which (i,j) belongs to different clusters increases.

The Euclidean distance on the other hand is $d^2(i,j)=2\cdot M(1-c(i,j))=2M$.
It follows that:

- $d^2(i,j)=0$ if (i,j) belongs to the same cluster in all of the classifications;
- $d^2(i,j)=2M$ if (i,j) belongs to different cluster in all of the partitions;
- $d^2(i,j)$ increases as the number of partitions in which (i,j) belongs to different clusters increases.

Note that the number of cluster of the partitions does not affect distance.
Given M partitions, let $\mathbf{Y}=[\mathbf{Y}_1 \mid \mathbf{Y}_2 \mid \ldots \mid \mathbf{Y}_M]$ be the matrix obtained concatenating horizontally the M binary matrices corresponding to the M partitions; let $\mathbf{C}=[\mathbf{C}_1 \mid \mathbf{C}_2 \mid \ldots \mid \mathbf{C}_M]$ be the corresponding matrix of centroids for the M partitions $\mathbf{Y}_l$ l=1, …, M; the least-square k-means loss function $F(\mathbf{C},\mathbf{U})$ written down for $\mathbf{Y}$ matrix is:

$$
\begin{aligned}
F(\mathbf{C},\mathbf{U}) &= \|\mathbf{Y}-\mathbf{U}\mathbf{C}\|^2 = \mathrm{tr}[(\mathbf{Y}-\mathbf{U}\mathbf{C})'(\mathbf{Y}-\mathbf{U}\mathbf{C})] \\
&= \mathrm{tr}\{[(\mathbf{Y}_1-\mathbf{U}\mathbf{C}_1) \mid \cdots \mid (\mathbf{Y}_M-\mathbf{U}\mathbf{C}_M)]'[(\mathbf{Y}_1-\mathbf{U}\mathbf{C}_1) \mid \cdots \mid (\mathbf{Y}_M-\mathbf{U}\mathbf{Y}_M)]\} \\
&= tr[\sum_{l=1}^{M}(\mathbf{Y}_l-\mathbf{U}\mathbf{C}_l)'(\mathbf{Y}_l-\mathbf{U}\mathbf{C}_l)] \\
&= \sum_{l=l}^{M} tr[(\mathbf{Y}_l-\mathbf{U}\mathbf{C}_l)'(\mathbf{Y}_l-\mathbf{U}\mathbf{C}_l)] \\
&= \sum_{l=l}^{M}\|\mathbf{Y}_l-\mathbf{U}\mathbf{C}_l\|^2 = F(\mathbf{C}_l,\mathbf{C}_2,\cdots,\mathbf{C}_M,\mathbf{U})
\end{aligned}
$$

where $\|\mathbf{Y}_l - \mathbf{UC}_l\|^2$ measures the distance between each partition $\mathbf{Y}_l$ and the consensus partition **U**. The consensus classification is by this way determined by minimising a (least-square) loss function between the set of observed classification and the consensus classification (*optimisation approach*, Barthelémy, Leclerc and Monjardet, 1986). Since **U** belongs to the set of all partitions of the set of *n* objects, the classification determined by means of the *k*-means loss function is a *central* (or *median*) *heuristic partition* (Régnier, 1965). Note that in the consensus partition the new cluster centroid co-ordinates are represented by the weight of the clusters of the *M* original partitions.
Since comparison of different methods used on the same data can sometimes be a useful strategy to deep insight some aspects of the methods themselves, in the application presented in the next section the *k*-means clustering method is used in all the approaches previously outlined.

5. Results and comparisons

In order to meaningfully interpret results of the different analyses, all the row profiles in each group of variables of the data illustrated in section 2 are divided by the resident population, the number of families and the stock of houses of the corresponding administrative unit respectively. The first numerical results that deserve comment are the eigenvalues and the percentage of the explained variance of MFA; the first three axis predominate, they account for about 70% of the total variance: the first and the second eigenvalues account for about 46% and 16% respectively, the third one account for about 6%. In order to perform the second analysis, when each group of variables is separately analyzed by the ordinary PCA it appears that in the group of variables pertaining to the population of individuals, the first three axis account for about 72% of the total variance; in the group of variables pertaining to the population of family, the first axis alone accounts for about 82%; in the group pertaining to the population of houses the first three axis account for about 85%.
Tables 2, 3 and 4 summarize the results of the three analyses, that is: results of the standard *k*-means cluster analysis of the object scores on the first three MFA factorial axis (Tab. 2); results from the standard *k*-means cluster analysis conjointly performed on the seven principal components from the separate

analysis previously mentioned (Tab. 3); results of the *k*-means consensus classification analysis consisting of three *k*-means cluster analysis of each group of variables and of a *k*-means cluster analysis of the resulting three classifications as outlined in the previous section.

Plots of the criterions (least-square loss functions) against different values of the number of the clusters, together knowledge from previous analyses of the likely range of the number of clusters suggest that six clusters is an acceptable value for each of the three analysis. For each cluster in the partitions along with the cluster frequencies, the root-mean square distance between observations in the cluster (column (2)), the maximum distance from the cluster centroids to any observation in the cluster (column (3)), the distance between the centroids of the current cluster and the nearest cluster (column (4)) are listed in the tables. Values of the loss function are enclosed in brackets at the top of the tables.

Projection techniques of objects that have been clustered into mutually exclusive sets have been proposed (see, e.g. Gnanadesikan et al., 1982) as informal data-oriented methods to assess and interpret the results of cluster analyses. One of such tecnique is borrowed from discriminant analysis. In fact, since a clustering algorithm yields groups (albeith data-based ones), one can study the configuration of these groups just as one would investigate the prespecified groups in a (multigroup) discriminant analysis.

Figures 1, 2 and 3 show the two dimensional discriminant spaces obtained by performing a (non linear) canonical discriminant analysis (see, e.g. Gifi, 1990) on each of the three classifications. The visual inspection of the three plots shows that the *k*-means consensus analysis results in a better solution both in terms of the separation and in terms of the tightness (homogeneity and cohesiveness of individual clusters) of recovered clusters. Further, albeith in all the three approaches the nonlinear discriminant analysis yields almost optimal solutions (the first two dimensions explain about 90% of the relationship), the two tandem analysis show a tendency towards misclassification of the administrative units.

	Table 2				**Table 3**				**Table 4**			
	k-means clustering of MFA factorial axis (***loss=0.74***)				*k*-means clustering of PC's (***loss=0.84***)				*k*-means consensus classification (***loss=0.29***)			
Cl	**Freq.**	**(1)**	**(2)**	**(3)**	**Freq.**	**(1)**	**(2)**	**(3)**	**Freq.**	**(1)**	**(2)**	**(3)**
1	37	0.869	4.247	2.594	36	0.939	6.492	3.246	43	0.304	1.793	1.412
2	46	0.712	3.560	2.878	39	0.677	3.680	3.224	49	0.273	1.804	1.391
3	55	0.795	3.713	2.411	70	0.756	5.053	2.875	50	0.293	1.456	1.391
4	75	0.750	3.293	2.169	79	0.873	6.518	2.611	57	0.285	1.875	1.298
5	84	0.720	3.420	2.131	91	0.853	5.769	2.506	89	0.296	1.708	1.298
6	112	0.707	3.503	2.131	94	0.886	5.165	2.506	121	0.297	1.423	1.393

Figure 1: *Discriminant plot: PCA classification*

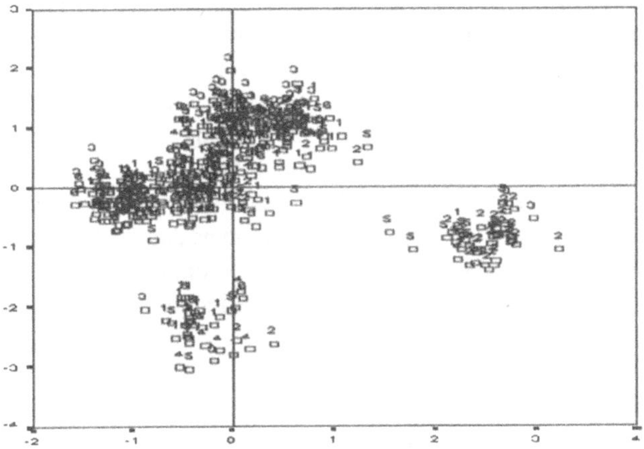

Figure 2: *Discriminant plot: MFA classification*

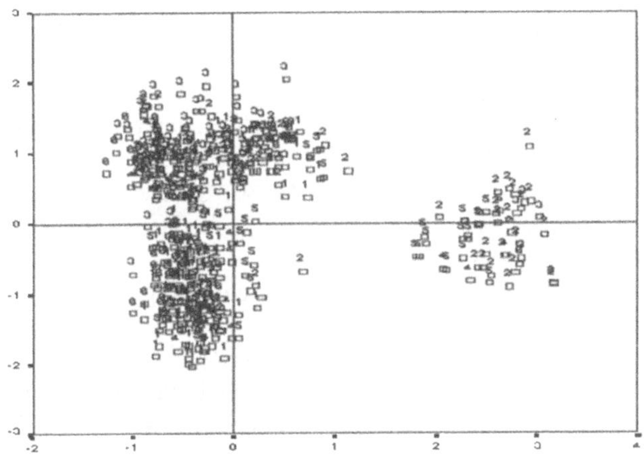

Figure 3: *Discriminant plot: k-means consensus classification*

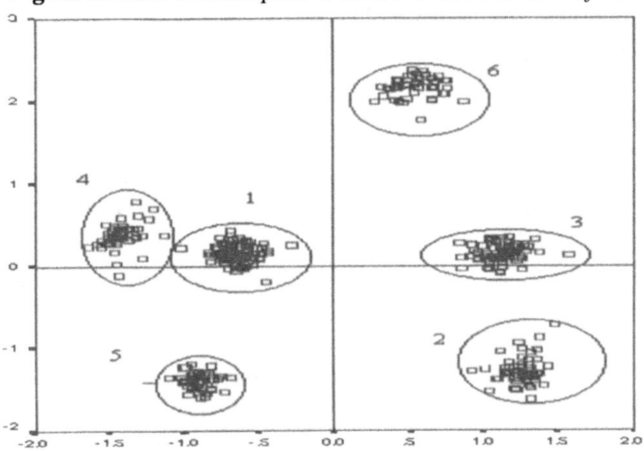

This proves on the one hand that performing a factorial analysis (MFA or PCA) as first step of a cluster procedure may really.obscure the original taxonomic information, and on the other hand, that the performance of the *k*-means method in discovering a classification of the set of the administrative units is affected by the starting point of the analysis, that is co-ordinate axis or classifications.
Relative importance of the variables in accounting for the separateness and tightness in discriminant analysis, and interpretation of cluster profiles aid to identify the "social and territorial structures" from the *k*-means consensus analysis. For example: cluster 3 contains urban towns and middle size towns with a strong presence of the tertiary sector; cluster 2 contains largely coastal towns and cluster 4 mostly upland towns; cluster 1, the biggest one, contains little size towns having a mean profile of the variables.

Acknowledgements

The author thanks M. Vichi and R. Rocci for helpful suggestions and comments on an earlier version of this paper.

References

Arabie P., Hubert L. (1994) Cluster analysis in marketing research. In: P. Bagozzi (ed.), *Handbook of marketing research,* Blackwell, Oxford.

Barthelémy J.P., Leclerc B., Monjardet B. (1986) On the use of ordered sets in problems of comparison and consensus of classifications, *Journal of Classifications,* 3, 187-244.

De Soete G., Carrol J. D., (1994) *K*-means clustering in a low-dimensional Euclidean space. In: Diday et al. (eds), *New approaches in classification and data data analysis,* Springer, Heidelberg, 212-219.

Escofier B., Pagès J. (1984) L'analyse factorielle multiple, *Cahiers du Buro*, 42.

Escofier B., Pagès J. (1994) Multiple factor analysis (AFMULT package), *Computational Statistics &Data Analysis*, 18, 121-144.

Everitt, B. (1977) Cluster Analysis. Heinemann Educational Books, London.

Gifi, A. (1990) Nonlinear multivariate analysis. New York: Wiley.

Gnanadesikan R., Kettenring J. R., Ldwehr, J. M. (1982) Projection plots for displaying clusters. In *Statistics ad Probability: Essays in Honor of C. R. Rao* (G. Kalliapur, P. R. Krishnaiah & J. K. Ghosh, eds), North-Holland, Amsterdam, 269-80.

Leclerc B., Cucumel G. (1987) Consensus en classification: une revue bibliographique. *Mathématiques et Sciences Humaines*, 100, 109-128.

MacQueen, J. (1967) Some methods for classification and analysis of multivariate observations. In: L.M. LeCam and J. Neyman (eds): *5th Berkeley Symposium on Mathematic, Statistics and Probability.* University of California Press, Berkeley, vol. 1, 281-298.

Régnier, S. (1965) Sur quelques aspects mathématiques des problems de classification automatique, *International Computation Centre Bulletin,* 4, 175-191. Reprinted in *Mathématiques et Sciences Humaines*, 82, 13-29 (1983).

MIXISO: a Non-Hierarchical Clustering Method for Mixed-Mode Data

Agostino Di Ciaccio
Faculty of Economics, University of Urbino, diciaccio@uniurb.it

Abstract: In this paper we consider the approach proposed in the hierarchical clustering method MIXCLAS to obtain a new non-hierarchical method which we named MIXISO. This method can analyze mixed mode data with a large number of units including in the presence of missing data. We demonstrate the application of this method on artificial and real data-set with missing values.

Keywords: Mixed mode data, non-hierarchical clustering, MIXCLAS, MIXISO.

1. Introduction

When we have to analyze mixed-mode data looking for a partition of the statistical units, several approaches may be adopted:
transform quantitative variables into qualitative variables or vice-versa. This approach introduces arbitrary choices in the analysis.
Use a proximity index for mixed-mode data: the most famous is Gower's general coefficient of similarity. In practice this coefficient has seldom been used and seems not to be so effective. Another possibility is the combined resemblance matrix approach (Bloxom 1984) or a generalized measure of variability (for a non-hierarchical approach see Chiodi 1990).
Apply the Optimal Scaling approach: it allows to jointly analyze variables with all measurement levels (Van Buuren & Heiser 1989, Di Ciaccio 1992). A drawback of this approach is the tendency to merge the categories of the qualitative variables into 2-3 classes.
Use the finite mixture model (*Everitt-Merette 1990*): the heavy assumptions made in this approach and the prohibitive computational problem render this method difficult to apply. The proposal of Everitt-Merette is limited to the analysis of only 2 variables.
Perform separate analyses: it implies the arbitrary choice of a method capable of summarizing the separate analyses.
The MIXCLAS/MIXISO approach (*Di Ciaccio 1992,1999*): this approach is based on the introduction of an isolation measure of a set of units which is a function of conditional proximities between units. It gives good results if the clusters are well-separated; it can also manage missing data in a very convenient way. MIXCLAS is a hierarchical clustering method which requires heavy computational resources. MIXISO is a non-hierarchical clustering method which is able to analyze several thousand cases. These methods are included in the

statistical package StatEasy (free download of beta 2 release at http://space.tin.it/scuola/adiciacc).
In this paper we consider the latter approach in detail, showing its theoretical properties. In the last paragraph we consider the performance of these clustering methods on artificial and real data sets including in the presence of missing data.

2. Well-separated partitions

Consider a data-set with n statistical units and J observed variables. Our aim is to assign the units to L clusters in order to obtain a *well-separated partition* **P**, as defined by Jardine & Sibson (1971) (we will extend this definition later).
Indicate with a,b,h generic units, Q a cluster and $d(a,b)$ a measure of distance between the units a and b; in a well-separated partition we have, for each $a,b \in Q$ and $h \notin Q$:

$$min_{a,h}\, d(a,h) > max_{a,b}\, d(a,b) \quad \text{for each } Q \in \mathbf{P} \tag{1}$$

Then for each cluster Q and for each couple of units $a,b \in Q$ and $h \notin Q$, in a well-separated partition the distance $d(a,b)$ is less than $d(a,h)$ or $d(h,b)$ and each Q is defined a *ball-cluster*. A *well-structured-partition* is also an L^**-partition* and a *well-separated partition* but not vice-versa (for these definitions see Di Ciaccio 1992). It could be proven that several well-separated partitions, with the same number of groups, could exist for a given set of units. Jardine & Sibson showed that the Single Link method always detects ball-clusters if they exist, but it can miss detection of a well-separated partition (Di Ciaccio 1992).
Expression (1) means that, for each couple belonging to Q and for each unit h not belonging to Q, the distance 'inside' the couple is lower than the distance 'outside' the couple, with respect to the unit h.
This interpretation suggests the need to introduce conditioned measures to individuate ball-clusters. For example, if we define $d(a,b/h)$ as the ratio between $d(a,b)$ and the minimum of $d(a,h)$ and $d(b,h)$, we can rewrite (1) as

$$max_{a,b,h}\, d(a,b/h) < 1 \tag{2}$$

Since we are interested in analyzing mixed-mode data, we will extend (2) using, in place of the distances $d(a,b/h)$ (defined only for quantitative variables), the proximities $p(a,b/h)$. We have now to define such proximities for each measurement level.

Quantitative variable
Given a variable X which assumes the values $\{x_1, x_2, \ldots, x_n\}$, let $d_{a,b}=d(x_a,x_b)$, $d_h=min(d(x_a,x_h);d(x_b,x_h))$, then $p(a,b/h)$ is defined as

$$p(a,b/h)=\begin{cases} 0 \quad if \quad d_h = d_{a,b} \\ 1-\dfrac{d_{a,b}}{d_h} \quad if \quad d_h > d_{a,b} \\ \dfrac{d_h}{d_{a,b}}-1 \;\; if \;\; d_h < d_{a,b} \;\; with \;\; x_h \notin [x_a, x_b] \\ -1 \;\; if \;\; d_{a,b} > 0 \;\; with \;\; x_h \in [x_a, x_b] \end{cases} \tag{3}$$

Then *p(a,b/h)* assumes a positive value if $d_h > d_{ab}$ and this value is near to the maximum value 1 when d_h is much larger then d_{ab}. The minimum value –1 is obtained when x_h is a value between x_a and x_b.

Qualitative ordinal variable

For an ordinal variable X, we have only to consider the order of the modalities, so *p(a,b/h)* can be defined as

$$p(a,b/h)=\begin{cases} 1 \;\; if \;\; x_h \notin [x_a, x_b] \\ -1 \;\; if \;\; x_h \in [x_a, x_b] \\ 0 \;\; if \;\; x_a = x_b = x_h \end{cases} \tag{4}$$

Qualitative nominal variable

For a nominal variable X we can only observe whether two units have the same or a different category therefore, indicating with x_a the category assumed on the unit *a*, *p(a,b/h)* can be defined as

$$p(a,b/h)=\begin{cases} 1 \;\; if \;\; x_a \neq x_h \;\text{ and }\; x_b \neq x_h \\ -1 \;\; if \;\; x_a \neq x_b = x_h \;\text{ or } \\ \qquad\quad x_b \neq x_a = x_h \\ 0 \;\; if \;\; x_a = x_b = x_h \end{cases} \tag{5}$$

We can now rewrite the expression (2) using the conditional proximities *p(a,b/h)*:

$$min_{a,b,h}\, p(a,b/h) > 0 \tag{6}$$

For one quantitative variable we can note that, applying definition (3), this property implies that we have a well-separated partition following the Jardine & Sibson definition. This is true as all clusters have to be ball-clusters. Considering qualitative variables we obtain a natural extension of the definition of well-separated partitions for mixed-mode data by applying definitions (4) (5),.

Consider one ordinal variable: applying (4), a partition is well-separated if for each cluster Q, given x_a and x_b the extreme modalities assumed in Q, no other cluster includes the modalities in the interval $[x_a, x_b]$.

If we consider only one ordinal variable X and the modalities are more than the number of clusters, then we can easily individuate several partitions satisfying (6).

For one nominal variable, applying (5), a partition is well-separated if the categories assumed in a cluster Q are not present in any other cluster. If we consider only one nominal variable X and the categories are more than the number of clusters, then several partitions satisfy (6).

Definition 1.
Given a data-set with mixed-mode data and a partition **P** of the units, we say **P** constitutes a well-separated partition if, for each variable, the following property is true

$$min_{a,b,h}\, p(a,b/h) > 0$$

for each cluster $Q \in \mathbf{P}$ and for each couple of units $a,b \in Q$ and $h \notin Q$, with $p(a,b/h)$ defined in (3)(4)(5) depending on the measurement level of the variable.

We must note a main difference from the original definition of Jardine & Sibson: in that definition, which is valid only for quantitative variables, with 2 or more quantitative variables we do not consider the variables separately. Instead, we compute a unique distance (usually a Euclidean distance) between two units which we can use to verify the property (2). In definition 1 the property (6) is verified separately for each variable. When we only have quantitative variables, this makes definition 1 stronger than Jardine & Sibson's proposal.

3. Isolation of a cluster

In a well-separated partition all clusters can be defined 'isolated', while in other cases we should introduce a measure of lack of isolation.
Using conditional proximities we can define an index which measures the 'isolation' of a cluster.
Consider one variable X, then the index

$$mean_h\, p(a,b/h) \qquad (7)$$

is a mean, with respect to all units except a and b, of the conditional proximities of the couple (a,b). This index can be considered an isolation measure of the couple (a,b). For one quantitative variable the index (7) assumes a value near to 1.0 when the value $|x_a-x_b|$ is much lower than $|x_a-x_h|$ and $|x_b-x_h|$. For an ordinal variable it assumes the value 1.0 when no unit, except a and b, assumes a modality between x_a and x_b. For a nominal variable it assumes the value 1.0 when no unit, except a and b, assumes the categories x_a or x_b.
From (7) we can obtain an isolation measure for a cluster g_l, with respect to one variable, in two different ways:

1) $$\delta(g_l) = \underset{\substack{a \neq b \\ h \notin g_l}}{mean}\; p(a,b/h) \qquad (8)$$

which defines as isolation of cluster g_l the mean isolation of the couples belonging to g_l.

2) $$\delta(g_l) = \min_{\substack{a \neq b \\ h \notin g_l}} p(a,b/h) \tag{9}$$

which defines as isolation of cluster g_l the worst isolation of the couples belonging to g_l.
Considering J variables and indicating with X_j the generic variable, we can construct an overall isolation measure of a cluster as a mean of $\delta_j(g_l)$ with respect to the J variables.

Definition 2.
Given J variables $X_1, X_2, .., X_j, .. , X_J$, and a partition $G=\{g_1, g_2, ..., g_l, ... g_L\}$, we define isolation of a cluster g_l as the value of the index

$$\varphi(g_l) = \frac{1}{\sum_j w_j} \sum_{j=1}^{J} w_j \, \delta_j(g_l) \tag{10}$$

Where w_j is the weight of j-th variable.
When $\delta_j(g_l)$ is defined as in (8) we name $\varphi(g_l)$ *mean isolation*, when $\delta_j(g_l)$ is defined as in (9) it is named as the *minimal isolation*.
If we do not use a shortcut, it becomes evident that the great number of calculations needed to obtain the isolation of a cluster makes the index inapplicable to data-sets having a large number of units. The main advantage of this approach is that it is easy to define simple and coherent conditional proximities for each measurement level. This simplifies the construction of an index for mixed mode data. It is simple to prove that making a linear transformation of quantitative variables or making a monotone transformation of ordinal variables or changing the codes for nominal variables does not alter the value of the index (10). It is also possible to calculate the index (10) when some variable has missing values (Di Ciaccio 1992). In fact in this case it is sufficient to exclude from our calculation only the values *p(a,b/h)* which can not be calculated.

3. MIXCLAS and MIXISO

MIXCLAS (Di Ciaccio 1992) is a hierarchical method which, at each step, tries to maximize the *minimal isolation* of the clusters. The method can not analyze more than 200-300 units. MIXISO is a recent proposal (Di Ciaccio 1999) of a non-hierarchical method which tries to maximize the *mean isolation* of the clusters.
The objective function of MIXISO is:

$$max \;\; Iso(\mathrm{G}) = \max \frac{1}{\sum \sqrt{n_{.l}}} \sum_{l=1}^{L} \sqrt{n_{.l}}\, \varphi(g_l) \leq 1 \tag{11}$$

Weighting the isolation of a cluster with its cardinality avoids assigning too much importance to small clusters, while the square root avoids the attribution of excessive weight to great clusters.
MIXISO is able to analyze data with several thousands of units, using a shortcut in the computational procedure.
Given a partition G={g_1, g_2, ..., g_l, ... g_L} of a set of n statistical units, for each qualitative variable X we can construct the following classification table:

Table 1: *Classification table for one variable*

	G_1	g_2	....	g_l	....	g_L	tot.
x_1	N_{11}	n_{12}	....	n_{1l}	....	n_{1L}	$n_{1.}$
....	....	....	....	....	....	....	....
x_i	N_{i1}	n_{i2}	....	n_{il}	....	n_{iL}	$n_{i.}$
....	....	....	....	....	....	....	....
x_m	N_{m1}	n_{m2}	....	n_{ml}	....	n_{mL}	$n_{m.}$
tot.	$n_{.1}$	$n_{.2}$	....	$n_{.l}$	....	$n_{.L}$	n

where X has m categories, n_{il} is the number of units which assume the modality x_i and belong to the *l-th* group.
We can also construct this table for a discrete quantitative variable, while for continuous quantitative variables we should introduce a categorization. Usually a categorization in a large number of classes (i.e. 15-20 classes) has a little influence on the results of the analysis. Note that this table indicates the effect of clustering for each variable: for one variable a good partition is obtained when each modality corresponds to only one cluster. This case corresponds to a well-separated partition if the variable is qualitative, ordinal or nominal. Using the classification tables, one for each variable X_j considered, we can individuate a shortcut to calculating the measure $\delta_j(g_l)$.

Qualitative Nominal Variable
To calculate $\delta_j(g_l)$ for a nominal variable we have to consider:

1. the couples (a,b) formed by units of n_{il} ($X=x_i$ and $G=g_l$) compared to a unit h belonging to the units ($n_{i.} - n_{il}$) of the same row ($X=x_i$ and $G\neq g_l$) which give $p(a,b/h)=0$;
2. the couples formed by units of n_{il} compared to a unit of another row not belonging to the l-th cluster ($X\neq x_i$ and $G\neq g_l$), which gives $p(a,b/h)=1$;
3. the couples formed by one unit of n_{il} ($X=x_i$) and one unit of n_{kl} ($X=x_k$) compared to a unit not belonging to the *l-th* cluster but presenting the modality *i-th* or *k-th*, which are $(n_{il}*n_{kl})[(n_{i.}-n_{il})+(n_{k.}-n_{kl})]$, which give $p(a,b/h)=-1$;
4. the couples formed by one unit of n_{il} and one unit of n_{kl} compared to a unit not belonging to the *l-th* cluster and not presenting the modality *i-th* or *k-th*, which give $p(a,b/h)=1$;

Finally we obtain the sum of $p(a,b/h)$, with respect to a,b,h as

$$\gamma_l = \sum_{i=1}^{m} \frac{n_{il}(n_{il}-1)}{2} \cdot \left[(n - n_{.l}) - (n_{i.} - n_{il})\right] - \sum_{i=1}^{m-1} \sum_{k=i+1}^{m} n_{il} n_{kl} \left[(n_{i.} - n_{il}) + (n_{k.} - n_{kl})\right] +$$

$$+ \sum_{i=1}^{m-1} \sum_{k=i+1}^{m} n_{il} n_{kl} \left[(n - n_{.l}) - (n_{i.} - n_{il}) - (n_{k.} - n_{kl})\right] =$$

$$= \sum_{i=1}^{m} \frac{n_{il}(n_{il}-1)}{2} \cdot \left[n - n_{.l} - n_{i.} + n_{il}\right] + \sum_{i=1}^{m-1} \sum_{k=i+1}^{m} n_{il} n_{kl} \left[n - n_{.l} - 2(n_{i.} - n_{il} + n_{k.} - n_{kl})\right]$$

The number of comparisons considered is:

$$\tau_l = \sum_{i=1}^{m} \frac{n_{il}(n_{il}-1)}{2} \cdot (n - n_{.l}) + \sum_{i=1}^{m-1} \sum_{k=i+1}^{m} n_{il} n_{kl} (n - n_{.l})$$

Finally we obtain the isolation of the cluster g_l for one qualitative variable:

$$\delta_j(g_l) = \frac{\gamma_l}{\tau_l}$$

A similar procedure can be applied to calculate $\delta_j(g_l)$ for an ordinal or a quantitative variable. Having fixed the number of groups L, the classification algorithm starts from an initial partition and then tries to move the units from one group to another until the criterion (11) converges. Since the solution is highly dependent on the initial partition, we introduce a partially random choice of initial L couples in order to maximize the differences of the 'centers' of the initial groups. Then we assign each other unit to the centers considering the conditional proximity of the centers with respect to the unit. It is advisable to try several starts to avoid local optima. We then have the following steps:

1) evaluate the isolation of each cluster and calculate the criterion (11);
2) select the less isolated cluster, among the clusters not yet considered;
3) try to move a unit from the selected cluster to another cluster, with respect to the criterion (11);
4) go to 3) if there are other units in the selected cluster;
5) go to 2) if there are other clusters;
6) make another cycle if the criterion changed more than ε during the last cycle;

5. Application to real and simulated data

We generated simulated data for 3 clusters, 4 variables and 300 units, in the following way:

	X1	X2	X3	X4	Units
Cluster 1	Normal(10.;4.)	Poisson(0.5)	*a*	Uniform(0,1)	100
Cluster 2	Normal(5.;2.)	Poisson(5.0)	*b*	Uniform(0,1)	100
Cluster 3	Normal(0.;1.)	Poisson(2.0)	*c*	Uniform(0,1)	100

X1 and X4 are quantitative continuous variables, X2 is quantitative discrete and X3 is qualitative nominal. Applying MIXISO, the integer variable X2 was considered to be an ordinal variable. This usually is a good choice when the variable has an asymmetric distribution.
We compared MIXISO with the K-Means method obtaining the following results:

	% of correct classification		
Variables used	MIXISO	K-Means	Iso(G)
X1,X2	92.3	85.0	0.486
X1,X2,X3	98.7	97.3	0.626
X1,X2,X4	92.6	62.3	0.212
X1,X2,X3,X4	98.7	97.7	0.385

To apply K-Means, in such a way so as to give the same weight to each variable, we standardized all the variables. For the variable X3 we had to use 2 dummy variables. We chose the best result from several casual starts of the algorithm.
The reported results of MIXISO are the best values obtained after 10 casual starts of the algorithm, with the exception of the case (X1,X2,X4) which after 10 starts, obtained only 59.7% of correct classifications with Iso(G)=0.199.
The real data used, named "hepatitis-data", are available in the *www.ics.uci.edu/pub/machine-learning-database* and they are analyzed for example, in Diaconis & Efron (1983). They consist of 19 variables (6 quantitative, 13 dichotomous) observed on 155 units. The class variable assumes 2 categories: die (32 cases), live (123 cases). Some variables present a large number of missing values. Applying MIXISO to the 19 variables we obtained a classification accuracy of 84.5% while the K-Means method obtained 79.6%.

References

Chiodi M. (1990) A partition type method for clustering mixed data, *Italian Journal of Applied Statistics*, vol. 2, n. 2.

Diaconis P., Efron B. (1983). Computer-Intensive Methods in Statistics. *Scientific American*, Volume 248.

Di Ciaccio A. (1992) Simultaneous Clustering of Qualitative and Quantitative data with Missing Observations, *Italian Journal of Applied Statistics*, v. 4, n. 4.

Di Ciaccio A. (1999) MIXISO: a non-hierarchical clustering method for mixed-mode data, *CLADAG 1999, Book of short paper*, Rome 5-6 July 1999.

Everitt B.S., Merette, C. (1990) The clustering of mixed-mode data: a comparison of possible approaches, *Journal of Applied Statistics*, v. 17, n.3.

Gower J.C. (1971) A general coefficient of similarity and some of its properties, *Biometrics*, 27.

Jardine N., Sibson R. (1971) *Mathematical Taxonomy*, Wiley London.

Van Buuren S., Heiser W.J. (1989) Clustering N objects into K Groups under Optimal Scaling of variables, *Psychometrika*.

“Stable clusters”: a new Approach for Clustering Binary Variables

Raffaella Piccarreta
IMQ Università L. Bocconi, viale Isonzo 25, 20136 Milano,
e-mail: raffaella.piccarreta@uni-bocconi.it

Abstract: As pointed out by Openshaw (1980), monothetic divisive algorithms often need for some means of post-classification relocation. We introduce a new criterion for improving the results of a (particular) monothetic divisive algorithm for clustering individuals classified on the basis of Q binary variables. This is done by referring to an agglomerative algorithm having very interesting similarities with the monothetic divisive algorithm taken into account. The new algorithm, developed by combining information arising from the two procedures, is compared with a relocation algorithm proposed by Openshaw.

Keywords: Binary variables, Agglomerative algorithms, Monothetic divisive algorithms, Gini heterogeneity, Relocation methods.

1. Introduction

We consider the problem of clustering N individuals on the basis of Q binary variables, $Y_1, \ldots, Y_Q$, each measuring the presence (1) or the absence (0) of a certain characteristic. Let $\mathbf{y}_i$ and $\mathbf{y}_h$ denote the vectors characterizing the i-th and the h-th individual respectively: it is evident that if $\mathbf{y}_i = \mathbf{y}_h$, the two individuals will not be distinguishable, having the same characteristics. Let $C_{(1,G^*)}$, $C_{(2,G^*)}$, ..., $C_{(G^*,G^*)}$ be the groups constituted by undistinguishable individuals. This G^*-structure of the data is the most complicated one can define. On the other side, the simplest structure is the 1-structure corresponding to the situation when all the individuals constitute a single cluster, $C_{(1,1)}$. The aim of a clusterisation procedure is to individuate a “proper” G-structure of data, $C_{(1,G)}$, ..., $C_{(G,G)}$, which is an intermediate between the 1-structure and the G^*-structure. To do this, it is necessary to define a criterion to evaluate the quality of a certain grouping. The criterion depends, of course, on the nature of the involved variables.
In this paper we refer to a hierarchical divisive algorithm (in the following, DA) having as a starting point the 1-structure: at each step of the algorithm one cluster is split into two clusters and the process is continued until the G^*-structure is obtained. In particular, DA (for a more detailed description, see Piccarreta, 1998) is based on the concept of *total* and *within-groups* dispersion. More precisely, consider a group C: a proper measure of the mutability of the q-th variable within C, $V_q(C)$, is the *Gini heterogeneity*. The dispersion within C can be measured by $V(C) = \sum_q n(C)V_q(C)$, $n(C)$ indicating the number of cases in C.
Referring now to a G-structure, $C_{(1,G)}$, ..., $C_{(G,G)}$, a measure of the *within-groups*

dispersion is $V_G = \sum_g V_{g,G} = \sum_g \sum_q n(C_{(g,G)})V_q(C_{(g,G)})$. By construction, it is always $V_{G^*} \le V_G \le V_1$, V_1 being a measure of the *total* dispersion. A possible measure of the quality of the G-structure is then $\boldsymbol{M}_G = (V_1 - V_G)/V_1$. Moreover, if the groups are obtained following a hierarchical (divisive or agglomerative) procedure, it is $V_G \le V_{G-1}$, so that $\boldsymbol{M}_{G-1} \le \boldsymbol{M}_G$.

Referring to *DA*, each division step involves selecting a single variable, called the *splitting* variable, the values of which divide a cluster into two clusters, one constituted by individuals possessing the characteristic measured by the splitting variable, and one constituted by individuals not possessing the characteristic itself. We are thus considering a *monothetic* divisive algorithm. In particular, at the G-th step, starting from a $(G-1)$-structure, the cluster to be split and the splitting variable are selected, according to the $\boldsymbol{M}_G$ criterion, in such a way that the quantity $(\boldsymbol{M}_G - \boldsymbol{M}_{G-1})$ is maximum. $\boldsymbol{M}_G$ is thus maximized conditionally to the initial $(G-1)$-structure.

DA has the nice characteristic to lead to groups which are perfectly characterized, at least in terms of the splitting variables. Nevertheless, as noted by Openshaw (1980), monothetic divisive algorithms are characterized by a certain degree of "rigidity", due to the fact that few partitions of data are explored (this is due either to the hierarchy of the procedure and to the fact that the splits considered are only those induced by the categories (0/1) of the involved variables). This is the reason why Openshaw stresses the need for some means of post-classification relocation. In particular, Openshaw suggests applying the following *iterative subset reclustering algorithm* (in the following *RA*):

Step 1. Take a group in the classification.

Step 2. Reclassify this group, eventually with an iterative relocation stage, to identify two types of cases: those relating to the original step 1 group label and those which are placed in a new group.

Step 3. Repeat steps 1 and 2 until all groups have been processed.

Step 4. Pool all the cases belonging to the new groups and reclassify.

Step 5. Combine similar clusters and return to step 1.

This *expansion* and *contraction* process might give some indications of an appropriate number of groups and, moreover, might lead to a better overall classification.

In this paper we introduce a criterion, alternative to relocation, to improve the performance of the considered monothetic divisive algorithm, *DA*.

To do this, we refer to a hierarchical agglomerative algorithm (in the following *AA*), also based upon the $\boldsymbol{M}_G$ criterion. The starting point of *AA* is the G^*-structure: at each step two clusters are joined to form a single cluster. The process is continued until the 1-structure is obtained. In particular, starting from a G-structure, the clusters to be joined are selected in such a way that $(\boldsymbol{M}_G - \boldsymbol{M}_{G-1})$ is maximum. $\boldsymbol{M}_{G-1}$ is thus maximized conditionally to the initial G-structure.

This criterion consists in joining into the same cluster, $[C_{(g1,G)} \cup C_{(g2,G)}]$, the two clusters $C_{(g1,G)}$ and $C_{(g2,G)}$ assuring the quantity $V[C_{(g1,G)} \cup C_{(g2,G)}] - V(C_{(g1,G)}) - V(C_{(g2,G)})$ is minimum (this is the criterion the well known Ward's agglomerative algorithm is based upon).

Notice that *AA* and *DA* both attempt to maximize at each step the quantity $\boldsymbol{M}_G$, even if in two opposite "directions". *AA* is more flexible than *DA*, in that the selection of the groups to be joined is not related to considerations about the levels of a particular variable.

Our idea is to "combine" results obtained by applying *DA* and *AA* to determine a partition which is possibly better – in terms of the selected optimality criterion, but also from an interpretative point of view – than the original ones.

2. A new approach

Both algorithms in consideration being based upon the same optimality criterion, if G is the number of clusters adequate to describe the structure underlying data, then it should be $\boldsymbol{M}_{A,G} \cong \boldsymbol{M}_{D,G}$ (subscripts indicating the algorithm). One should then observe an increasing of $(\boldsymbol{M}_{A,H} - \boldsymbol{M}_{A,H-1})$ for $H < G$, indicating a further aggregation of clusters is not worthwhile, and a decreasing of $(\boldsymbol{M}_{D,H} - \boldsymbol{M}_{D,H-1})$ for $H > G$, indicating a further splitting of clusters is not worthwhile. Moreover, by construction, in *AA* the value of $\boldsymbol{M}_{A,H}$ is "under control" for high values of H. In *DD*, instead, the level of $\boldsymbol{M}_{D,H}$ is under control for small values of H. In this sense, a comparison of the results obtained by applying *AA* and *DA* can be useful to determine the number of clusters which is adequate to describe the structure of the data. Suppose now G is the number of clusters appearing adequate for both algorithms. Consider $A_{(1,G)}, \dots, A_{(G,G)}$ and $D_{(1,G)}, \dots, D_{(G,G)}$, the partitions obtained by applying *AA* and *DA* respectively (in the following, A-partition and D-partition). Let $n(A_{(h,G)} \cap D_{(k,G)})$ indicate the number of cases which are *simultaneously* placed in $A_{(h,G)}$, the h-th group of the A-partition, and in $D_{(k,G)}$, the k-th group of the D- partition: this quantity measures the degree of overlapping between $A_{(h,G)}$ and $D_{(k,G)}$. Moreover, the overlapping between the two partitions can be measured by $\sum_g n(A_{(g,G)} \cap D_{(g,G)})$, and the indices of the clusters of the two partitions are selected in such a way that the overlapping is maximum.

The individuals in $S_{(g,G)} = A_{(g,G)} \cap D_{(g,G)}$ are grouped into the g-th cluster by both algorithms. In this sense, these cases can be considered as probably constituting the "nucleus" of a cluster underlying the observed data. The G groups $S_{(1,G)}, S_{(2,G)}, \dots, S_{(G,G)}$, will be called *stable clusters*. Cases in the stable clusters are those mostly contributing (by construction) to the overlapping between the A-partition and the D-partition. Of course, not all the individuals placed in the stable clusters will be distinguishable one from another. In particular, let O_g denote the number of groups of undistinguishable individuals in $S_{(g,G)}$. Consider now that, being $O^* = \sum_g O_g$ the number of groups in the G^*-partition placed in the G stable clusters, there will be $F = G^* - O^*$ groups of undistinguishable individuals which are not members of any stable cluster. For the sake of simplicity, we will indicate these "unallocated" groups by $U_1, U_2, \dots, U_F$.

The starting point of our algorithm (in the following *SA*) is the $(G + F)$-structure constituted by the G stable clusters, $S_{(1,G)}$, ..., $S_{(G,G)}$, and by the F unallocated groups, U_1, ..., U_F. At the first step of the algorithm we reduce from the $(G + F)$-structure to a $[G + (F - 1)]$-structure by joining an *unallocated group*, say U_f, and a *stable cluster*, say $S_{(g,G)}$. The groups to be joined are selected in such a way that the quantity $V[U_f \cup S_{(g,G)}] - V(U_f) - V(S_{(g,G)})$ is minimum. The obtained partition is constituted by G updated "stable clusters", $S_{(1,G)}$, ..., $[U_f \cup S_{(g,G)}]$, ..., $S_{(G,G)}$ and $(F - 1)$ unallocated groups. The described procedure is applied again, and is continued until all the unallocated groups are allocated to a stable cluster. This procedure of aggregation of groups is hierarchical, but at each step only *stable-unstable* couples of groups are evaluated for joining. Moreover, notice that the number of clusters in the final partition, S-partition, obtained applying the algorithm is equal to G.

Before proceeding notice that in the case when, for example, it is $\boldsymbol{M}_{A,G} < \boldsymbol{M}_{D,G}$ or $\boldsymbol{M}_{A,G} > \boldsymbol{M}_{D,G}$ for all G, it is not worthwhile to apply *SA*. In fact, when this is the case, one would "merge" the results of a procedure performing uniformly better with the results performing uniformly worse. So, one could expect to obtain a final classification which is worse than the (the best of the) original ones.

As concerns a comparison between *SA* and the *RA* proposed by Openshaw, it is interesting to highlight that both algorithms are based upon the same intuition. Namely, in both cases the starting point is a first *expansion* of the D-partition, followed by a *contraction* phase (and by subsequent expansion phases in *RA*). In the first expansion phase a refinement of the D-partition is obtained. This is done by splitting each of the G clusters $D_{(1,G)}$, …, $D_{(G,G)}$, into two sub-clusters. The first sub-cluster can be viewed as the nucleus of the cluster. The second sub-cluster is constituted by a certain number of groups of undistinguishable individuals. The main difference between *SA* and *RA* is that in *RA* all of the groups "separated" by their original group are placed in an unique (residual) cluster, while in *SA* each unallocated group constitute a single cluster. In this sense, the partition obtained with *SA* in the first expansion phase is more refined than the one obtained by applying *RA*. Also, the contraction phase is very different. In fact, *SA* proceeds aggregating the unallocated groups to the nearest stable cluster in an agglomerative fashion. Thus, the stable clusters (eventually updated) are *never split* and are *never joined* one with another. *RA* proceeds, instead, by alternating expansion and contraction phases, disaggregating and aggregating "pieces" of the clusters obtained in the first phase to explore different partitions of the data structure in a non hierarchical fashion. In this sense, groups which we called "nuclei", obtained in the first expansion phase, can be split in subsequent phases and can be joined one with another, so that they are not "stable" at all.

Of course, there are no reasons to suppose one algorithm should perform better than the other, the performances of both depending upon the kind of data structure under examination. We are thus not affirming that *SA* is better than *RA*. Nevertheless, both algorithms are based upon the same intuition but follow a different strategy to visit partitions alternative to that obtained by applying *DA*. In this sense, it can be worthwhile to consider them as two alternative tools for breaking down the rigidity of the monothetic divisive algorithm.

3. An application to real data

In this section we briefly describe results obtained by applying *SA* to a real data set. We are interested in obtaining groups of firms according to their innovative activity in six technological classes (defined on the basis of the International Patent Office, see Malerba et. al., 1998, for details) related to properties of materials and to their processing and transformations: polymers (*P*), materials (*M*), chemical engineering (*C*), surfaces (*S*), materials processing (*MP*), environmental technology (*E*). At this aim we consider patent application to the European patent office (EPO) from firms of four European countries (Italy, France, Germany and UK) and from USA and Japan: the binary variables *P*, *M*, *C*, *S*, *MP* and *E*, assume values 0 or 1 according to the absence or presence of a firm in each of the six technological classes during the period 1988-1992 (attention is limited to firms patenting at least in one class).
The first problem to solve to apply *SA* is to select the adequate number of clusters. At this aim, in Figure 1the performances of *AA* and *DA* in terms of the $\boldsymbol{M}_G$ measure are reported.

Figure 1. *M_G measure for the agglomerative and the divisive algorithms.*

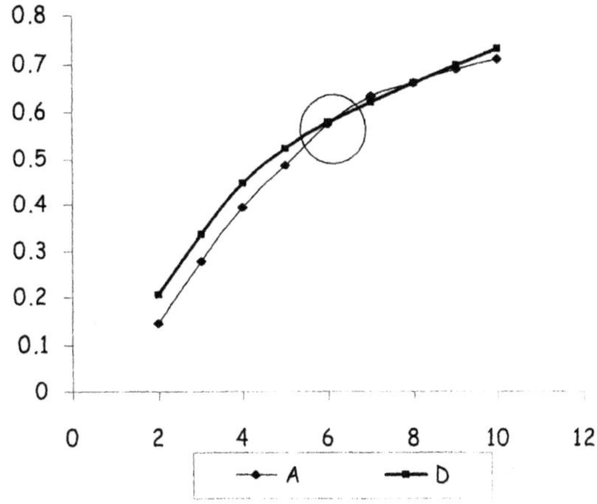

Notice that *DA* performs better than *AA* when $H < 6$, while for $H > 6$ an inversion of this tendency can be observed. A suitable value for *G* can then be 6 or 7.
In Table 1 results obtained by applying *AA* and *DA* to the considered data set are reported for $G = 6$. It is clear that none of the algorithms appears preferable with respect to the characterization of the clusters. It is interesting to notice that the obtained clusters are quite similar. In this sense, it seems sensible to combine information arising from the two partitions to obtain a possibly better characterization of the groups. In the first part of Table 2 a description of the stable clusters obtained by applying *SA* is reported. Notice that stable clusters have a simple interpretation: the second cluster is primarily constituted by firms patenting in all the technological classes, while the others contain firms patenting only in one class. The second part of Table 2 shows the final result of *SA*.

Table 1: *Description of clusters obtained by applying AA and DA* ($G = 6$).

	A-partition – $\boldsymbol{M}_{A,6} = 0.57$						
	$n(C_g)$	*C*	*E*	*P*	*MP*	*M*	*S*
1	547	38(*)	**100**	10	17	23	18
2	405	**70**	**44**	**90**	**71**	**97**	**65**
3	749	15	0	0	3	**66**	**50**
4	876	100	0	13	22	0	12
5	862	0	0	0	**100**	9	14
6	520	0	0	**100**	**36**	0	**21**
	D-partition – $\boldsymbol{M}_{D,6} = 0.58$						
	$n(C_g)$	*C*	*E*	*P*	*MP*	*M*	*S*
1	477	0	**44**	0	0	0	**60**
2	499	**54**	**33**	**60**	**100**	**55**	**100**
3	533	0	13	12	0	**100**	23
4	995	**100**	18	12	0	18	15
5	1152	19	7	19	**100**	12	0
6	348	0	4	**100**	0	0	15

(*) percentage of firms in the cluster patenting in *C* (chemical engineering). A similar interpretation holds for the other entries in the table.

Table 2: *Results obtained by applying SA*

	Description of the stable clusters						
	$n(S_g)$	*C*	*E*	*P*	*MP*	*M*	*S*
1	211	0(*)	**100**	0	0	0	9
2	220	**85**	**60**	**89**	**100**	**100**	**100**
3	409	0	0	0	0	**100**	**23**
4	682	**100**	0	8	0	0	8
5	743	0	0	0	**100**	6	0
6	333	0	0	**100**	0	0	**15**
	S-partition – $\boldsymbol{M}_{N,6} = 0.61$						
	$n(C_g)$	*C*	*E*	*P*	*MP*	*M*	*S*
1	467	34(*)	**100**	5	12	17	6
2	451	**88**	**51**	**76**	**88**	**75**	**80**
3	757	3	2	2	4	**64**	**58**
4	784	**100**	1	11	0	10	8
5	955	12	0	0	**100**	6	9
6	590	0	0	**100**	**36**	11	19

(*) See the note to Table 1.

Notice that applying *SA* an improvement of $\boldsymbol{M}_6$ can be observed. It is interesting to point out that for $G = 7$ it is $\boldsymbol{M}_{A,7} = 0.63$ and $\boldsymbol{M}_{D,7} = 0.62$. We think this result is quite encouraging, *SA* leading to a decrease of the within groups dispersion which

is achieved by *AA* and *DA* by "adding" one more cluster to the 6-structure.
Analyzing the *S*-partition, an improving appears also from the interpretative point of view. We think it is worthwhile to evidence that *AA* and *DA* still produced a "good" partition of data, either from an interpretative point of view and with respect to the M_G measure. In this sense, the improvement provided by *SA* is in our opinion valuable, even if it is not of very large size.
We also applied to the considered data set the *RA* (of course, we referred to the M_G criterion to guide the expansion and contraction phases in the algorithm). In the first part of Table 3 clusters obtained in the first expansion phase are illustrated. Notice there is a certain similarity between the first 6 groups and the stable clusters. The greatest difference regards the characterization of the first cluster, containing firms patenting only in the " surfaces" technological class (the first stable clusters contains firms patenting only in the "environmental technology" class). This is due to the fact that the first stable cluster is obtained by merging information arising from *AA* and *DA*, while the first cluster in (the first expansion phase of) *RA* is obtained by referring only to the *D*-partition.

Table 3: *Results obtained by applying RA*

Description of the clusters obtained in the first phase expansion							
	$n(S_g)$	*C*	*E*	*P*	*MP*	*M*	*S*
1	266	0	0	0	0	0	**100**
2	420	**56**	**31**	**72**	**100**	**47**	**100**
3	362	0	1	13	0	**100**	0
4	802	**100**	20	1	0	16	2
5	968	14	0	17	**100**	0	0
6	296	0	4	**100**	0	0	0
7	890	**35**	**46**	**27**	**29**	**49**	**44**
R-partition – $M_{R,6} = 0.53$							
	$n(C_g)$	*C*	*E*	*P*	*MP*	*M*	*S*
1	578	21	5	11	26	9	**100**
2	305	**70**	**42**	**100**	**97**	**66**	**100**
3	1266	**38**	**44**	8	25	**62**	16
4	571	**100**	0	0	0	0	0
5	692	0	0	0	**100**	0	0
6	592	17	1	**100**	32	13	0

(*) See the note to Table 1.

Analyzing the second part of Table 3, we observe that the *R*-partition shows a certain degree of similarity with the *S*-partition. In particular, clusters 4 and 5 have a very similar "structure" in both cases, even if it appears clearer in the *RA* result. So, on the one side *RA* individuates "pure" groups (clusters 4 and 5 are constituted by firms patenting *only* in *C* and in *MP* respectively) better than *SA*. On the other side, in the *R*-partition we observe that the less pure clusters are much more "confused" than their *SA* counterparts. This is the reason why a drastic worsening

of the M_G criterion can be noticed.
All the considerations made with reference to the 6-structures remain valid also when referring to the 7-structures. Moreover, *RA* gave the same indication provided by *AA* and *DA* (G=6 or G=7) with respect to the adequate number of clusters.
This application evidences the opportunity to refer either to *SA* and to *RA* to improve the performances of a monothetic divisive algorithm.

5. Conclusions

In this paper we introduce a new algorithm developed by combining information arising from the application of two hierarchical clustering algorithms which are based upon the same optimality criterion. The main aim of the algorithm is to improve results obtained by applying a monothetic divisive algorithm. This criterion is intended as alternative (or, better, complementary) to relocation criteria, such as for example, the one proposed by Openshaw (1980).
The performances of the new algorithm have been evaluated by referring to a real data set. The obtained structure is characterized by a level of within-clusters heterogeneity which is lower than that characterizing the original clusters. Moreover, the new algorithm performs better than the relocation algorithm.
In this sense, we think it is possible to refer to both kinds of methods to improve the performances of a monothetic divisive algorithm.

References

Agresti, A. (1990) *Categorical data Analysis*, Wiley, New York.

Bolasco, S. (1999) *Analisi multidimensionale dei dati*, Carocci: Roma.

Everitt, B.S. (1974) *Cluster Analysis*, London: Heinemann.

Kauffman L., Rousseeuw, P.J. (1990) *Finding Groups in Data*, Wiley, New York.

Malerba F., Breschi S., Lissoni, F. (1998) The knowledge proximity and technological diversification: an analysis of determinants of technological diversification in Europe, United States and Japan, *Paper prepared for the TSER Research Project Innovation System and European Integration, EU Commission DGXII.*

Openshaw, S. (1980) Monothetic divisive algorithms for classifying large data sets, *COMPSTAT 1980, Prooceedings in Computational Statistics*, pp. 419-245, Edinburgh.

Piccarreta, R. (1998a) Divisive Monothetic Algorithms for Clustering Binary Data: A Proposal, *Proceedings of the International Seminar on New Techniques and Methodologies for Statistics , Sorrento (NA)*, pp. 405-410.

Double k-means Clustering for Simultaneous Classification of Objects and Variables

Vichi Maurizio

Dep. Statistics, Probability and Applied Statistics
University "La Sapienza" Rome, Italy.
Email: Vichi@dmqte.unich.it

Abstract: In this paper a general model for simultaneous classification of objects and variables of a two mode data matrix is proposed. The model can identify both different classification structures of objects and variables (e.g., partitions, coverings) and different classification types (hard or fuzzy). The identification of the model is obtained by the numerical solution of a least-squares problem, which is typically NP-hard. An alternating least-squares algorithm is introduced to give an efficient solution to the fitting problem. The proposed classification model is named *double k-means* since its features recall for some extent the well-known clustering k-means methodology. Some real data sets are analyzed with the proposed methodology to show its features.

Keywords: Partition, Covering, Packing, Two Modes simultaneous classification.

1. Introduction

Large multivariate data sets are frequently observed which investigation needs the identification of strong syntheses both in the direction of objects and variables. A classification of objects and a classification of variables, or, more frequently, a factorial analysis of variables is often performed to detect the most relevant information in the data. This sequential approach may cause problems, as when principal component analysis (PCA) is followed by a clustering of objects based on the first relevant components. In fact, principal components may mask or obscure the clustering structure present in the data. This is due to the fact that PCA detects factors that explain the largest part of total variability in the data, but it does not necessarily maintain the relevant information for the classification. This paper deals with the problem of unifying the syntheses of objects and variables in a unique and simultaneous analysis, in order to avoid problems connected with a sequential approach of different multivariate techniques. A model for classifying objects and variables simultaneously is here proposed. It is named double k-means since it recalls, for some extent, the well-known methodology proposed by MacQueen (1967).

Some other tentative to propose techniques that simultaneously analyze both objects and variables have been given. De Sarbo, Howard and Jedidi (1991) developed MULTICLUS, a maximum likelihood based method for simultaneously performing multidimensional scaling and cluster analysis. De Soete and Carroll (1994) developed a procedure for clustering objects in a low dimensional subspace in order to represent the classification in a reduced factorial space.

This paper is organized as follows. In the remaining part of this section 1, the notation used in the paper is introduced. The model for classifying simultaneously objects and variables, allowing the fitting of different classification structures (partition, covering, packing) and classification types (hard, fuzzy) is discussed in section 2. Section 3 is devoted to the description of algorithms for classification problems considered in Section 2. Some examples are examined in section 4, in order to highlight the performance of the double *k*-means. Some conclusions are given in section 5.

Let $\mathbf{X}=[x_{ip}]$ be a $(n \times k)$ two-way two-mode (objects by variables) data matrix, regarding the *k*-variate profiles of *n* objects.
Without loss of generality we suppose that variables are standardized, in order to exclude, in the application of classifications, problems connected with different units of measurements of variables.
For the convenience of the reader the terminology used in this paper is listed here:

n, k	number of objects, and number of variables to be classified;
$I=\{o_1,\ldots,o_n\}$;	set of n objects to be classified;
$V=\{v_1,\ldots,v_k\}$	set of k variables to be classified;
$P=\{P_1,\ldots,P_c\}$	partition of I into c classes, where P_j is the jth class of P;
$Q=\{Q_1,\ldots,Q_m\}$	partition of V into m classes, where Q_l is the lth class of Q;
$\mathbf{U}=[u_{ij}]$	$n \times c$ membership function matrix, assuming values in [0, 1], specifying for each object o_i its membership to class P_j. Matrix **U**, in this case, identifies a fuzzy classification of objects. When values of **U** are 1 or 0, i.e., $u_{ij}=1$, object o_i belongs to P_j, while when $u_{ij}=0$, object o_i does not belong to P_j. In this last case matrix **U** is binary and identifies a hard classification of objects;
$\mathbf{V}=[v_{pl}]$	$k \times m$ membership function matrix, assuming values in [0, 1], specifying for each variable v_p its membership to class Q_l. Matrix **V**, in this case, identifies a fuzzy classification of variables. When values of **V** are 1 or 0, i.e., $v_{pl}=1$, variable v_p belongs to Q_l, while when $v_{pl}=0$ variable v_p does not belong to Q_l. In this last case matrix **V** is binary and identifies a hard classification of variables.
$\mathbf{C}=[c_{jl}]$	$c \times m$ *centroid* matrix specifying the mean of variable v_p in the class Q_l

The aim of this paper is to detect the best classification of objects (e.g., a partition P), specified by the membership matrix **U**, and the best classification of variables (e.g., a partition Q), specified by the membership matrix **V**. We are now in position to state the double *k*-means classification model.

2. Double *k*-means model

The double *k*-means model is formally specified as follows:

$$\mathbf{X} = \mathbf{UCV}' + \mathbf{E} \tag{1}$$

where matrix **E** is the error component matrix. The first term in model (1) pertains to the portion of information of **X** that can be explained by the simultaneous classification of objects and variables. Different classification structures can be defined imposing different constraints on the elements of **U** and **V**, as it will be shown in a moment.

The least-squares fitting problem connected with model (1) leads to the formulation of the following quadratic optimization problem with respect to variables u_{ij}, v_{pl} and c_{jl}, where, constraints can be set according to the *classification structure* (partition, covering or packing) and *type* of classification (hard or fuzzy) required:

$$\min \| \mathbf{X} - \mathbf{UCV}' \|^2 \qquad [P1]$$

subject to

	$u_{ij} \in \{0, 1\}$	$i=1,\ldots,n; j=1,\ldots,c;$	(hard object classification)	(2)
or	$u_{ij} \in [0, 1]$	$i=1,\ldots,n; j=1,\ldots,c;$	(fuzzy object classification)	(2')
	$\sum_{j=1}^{c} u_{ij} = 1$	$i=1,\ldots,n;$	(object partitioning)	(3)
or	$\sum_{j=1}^{c} u_{ij} \geq 1$	$i=1,\ldots,n;$	(object covering)	(3')
or	$\sum_{j=1}^{c} u_{ij} \leq 1$	$i=1,\ldots,n;$	(object packing)	(3'')
	$v_{pl} \in \{0, 1\}$	$p=1,\ldots,k; l=1,\ldots,m;$	(hard variable classification)	(4)
or	$v_{pl} \in [0, 1]$	$p=1,\ldots,k; l=1,\ldots,m;$	(fuzzy variable classification)	(4')
	$\sum_{l=1}^{m} v_{pl} =0\ 1$	$p=1,\ldots,k;$	(variable partitioning)	(5)
or	$\sum_{l=1}^{m} v_{pl} \geq 1$	$p=1,\ldots,k;$	(variable covering)	(5')
or	$\sum_{l=1}^{m} v_{pl} \leq 1$	$p=1,\ldots,k;$	(variable packing)	(5'')

Several classification problems are defined by combining constraints in the program [P1]; some relevant cases are listed below:

(i) [P1] subject to (2), (3), (4) and (5) identifies a hard partition of both objects and variables;
(ii) [P1] subject to (2'), (3), (4') and (5) identifies a fuzzy partition both of objects and variables;
(iii) [P1] subject to (2), (3), (4') and (5) identifies a hard partition of objects and a fuzzy partition of variables;
(iv) [P1] subject to (2), (3), (4') and (5') identifies a hard partition of objects and a fuzzy covering of variables;
(v) [P1] subject to (2'), (3'), (4') and (5') identifies a fuzzy covering both of objects and variables;
(vi) [P1] subject to (2'), (3), (4) and (5) identifies a fuzzy partition of the objects and a hard partition of the variables;
(vii) [P1] subject to (2'), (3'), (4) and (5) identifies a fuzzy covering of the objects and a hard partition of the variables.

Case (i) is the logical extension of the k-means methodology for hard partitioning of both objects and variables. Problems (ii) and (v) are fully fuzzy in the two modes, identifying partitions and coverings, respectively. These are the two most flexible models, in the sense that can better fit the data, and they should be first adapted to data in order to understand whether a covering or a partition better summarizes the taxonomic information in **X**.
Models (iii) and (iv) (models (vi) and (vii)) are useful to represent objects (variables) into a reduced dimensional space, highlighting the classification of objects (variables) detected by the double k-means. In fact, the fuzzy partition or covering of variables (or objects), specified by matrix **V** (or **U**), can be used as a factor loading matrix for weighting the original variables (or objects). Thus, derivational modes (i.e., virtual objects or factors) **XV** (or **X'U**) are those linear combination of variables (or objects) that better contribute to identify the classification of objects (or variables). Note that these factor loading matrices are not column-wise orthogonal.

Remark 1: When problem (i) is considered and matrix **V** is fixed equal to the identity matrix $\mathbf{I}_k$ of order k the double k-means reduces to the k-means algorithm, applied on matrix **X** for classifying objects according to variables. In fact, in this case constraints (4) and (5) are not active.
Moreover, when problem (i) is considered and matrix **U** is fixed equal to the identity matrix $\mathbf{I}_n$ of order n the double k-means reduces to the k-means algorithm, applied on matrix **X** for classifying variables according to objects. In this case constraints (2) and (3) are not active.

Let us suppose that matrices **V** and **U** are estimated. Problem [P1] reduces to the unconstrained LS regression problem,

$$\min \| \mathbf{X} - \hat{\mathbf{U}}\mathbf{C}\hat{\mathbf{V}}' \|^2,$$

which solution is,

$$\mathbf{C} = (\mathbf{U}'\mathbf{U})^{-1}\mathbf{U}'\mathbf{X}\mathbf{V}(\mathbf{V}'\mathbf{V})^{-1}, \tag{6}$$

where $(\mathbf{U}'\mathbf{U})^{-1}$ and $(\mathbf{V}'\mathbf{V})^{-1}$ are the generalized inverses of matrices **U'U** and **V'V**, which may be singular. Thus, when considering (6), problem [P1] is reduced to:

$$\min \| \mathbf{X} - \mathbf{U}(\mathbf{U}'\mathbf{U})^{-1}\mathbf{U}'\mathbf{X}\mathbf{V}(\mathbf{V}'\mathbf{V})^{-1}\mathbf{V}' \|^2,$$

subject to some constraints among: (2) – (5'').

3. Algorithms for the Double *k*-means

When double *k*-means involves continuous variables u_{ij} and v_{pl} (i.e., as in problems: [P1'] subject to (2'); (3), or (3'), or (3''); (4'); (5), or (5') or (5'')), and only fuzzy solutions are required, double *k*-means can be solved directly using a sequential quadratic programming algorithm (SQP). Comparative studies of non-linear programming algorithms indicate that SQP performs very well in terms of successful solutions with a superlinear rate of convergence (Powell 1983).

When double *k*-means involves only binary variables u_{ij} and v_{pl} and a hard partition of objects and variables is required (i.e., as in problem [P1'] subject to (2), (3), (4) and (5)), double *k*-means can be solved using an alternating least-squares (ALS) algorithm. It sequentially and recursively solves two assignment problems, as shown below in table 1. The ALS algorithm, at each step, does not increase and usually decreases the objective function, thus converging to at least a local optimum.

When the double *k*-means involves a mixed classification problem, i.e., both a fuzzy and a hard classification are required (e.g., as in cases (iii) and (iv)), the algorithm for solving this problem is an ALS procedure, that uses for the hard classification one of the two steps in table 1, depending on whether the partition of objects or variables is required, while for the fuzzy classification step, a SQP algorithm is directly employed.

Remark 2: In any case, the double *k*-means algorithm needs an initial matrix **U** and **V**, which can be randomly chosen. Using a large number of initial matrices **U** and **V** and retaining the best solution of the double *k*-means algorithm, the chance to obtain the optimal solution is increased. The solutions of the ordinary *k*-means algorithm, obtained by partitioning objects and variables independently from each other, can be also used as initial solutions for the double *k*-means.

Table 1: ALS algorithm for double *k*-means when hard partitions of objects and variables are required.

Step *a* given $\hat{\mathbf{V}}$, new values $\mathbf{U}=[u_{ij}]$ are given for $(i=1,\ldots,n)$:

$u_{ij}=1$, if $F(\hat{\mathbf{V}},[u_{ij}])=\min\{F(\hat{\mathbf{V}},[u_{iv}]) : v=1,\ldots,c\}$;
$u_{iv}=0$ otherwise;

Step *b* given $\hat{\mathbf{U}}$, new values $\mathbf{V}=[v_{pl}]$ are given for $(p=1,\ldots,k)$:

$v_{pl}=1$, if $F([v_{pl}], \hat{\mathbf{U}})=\min\{F([v_{pv}], \hat{\mathbf{U}}) : v=1,\ldots,d\}$;
$v_{pv}=0$ otherwise;

---

4. Macroeconomic performance of industrialized countries

The proposed methodology has been used to analyze the average macroeconomic performances of the G7 most industrialized countries: France (*FRA*), Germany (*GER*), Great Britain (*GB*), Italy (*ITA*), United States of America (*USA*), Japan (*JAP*), Canada (*CAN*), plus Spain (*SPA*), according to 7 variables: Gross Domestic Product (*GDP*), Inflation (*INF*), Budget deficit/gdp (*DEF*)*, Public debt/gdp (*DEB*)*, Long term interest rate (*INT*), Trade balance/gdp (*TRB*)*, unemployment rate (*UNE*) - most of these considered in the Maastrich treaty - in the period 1980-1990. This data set has been also studied by Rizzi & Vichi (1995), using a factorial three-way technique, to observe the convergence of the countries' trajectories.

In the present analysis, variables were standardized in order to allow the cross-sectional comparison. On the 8 × 7 column standardized data matrix, the double *k*-means algorithm has been applied, by fixing the number of classes $c=3$ and $m=2$ for countries and variables, respectively. Then, hard partitions for both countries and variables were required. The initial partition of double *k*-means was obtained by the best ordinary *k*-means partitions of objects and variables among 1000 random initial estimates. This multistart strategy increases the chance to obtain the global minimum solution.

The ALS algorithm, described in Table 1, converged after two iterations (convergence constant = 10^{-4}), with objective function equals to 3.5282. The best countries' partition obtained by the double *k*-means algorithm is,

P={{**ITA, SPA**}, {**FRA, GB, USA, CAN**}, {**GER, JAP**)},

while the best variables' partition is,

---

* both aggregates measured in domestic currency and at current price

$$Q=\{\{\mathbf{GDP, DEF, DEB, TRB}\}, \{\mathbf{INF, INT, UNE}\}\}.$$

The double *k*-means solution can be plotted into a factorial plane given by the objects' centroids, i.e. $\mathbf{X}\hat{\mathbf{V}}(\hat{\mathbf{V}}'\hat{\mathbf{V}})^{-1}$, where $\hat{\mathbf{V}}$ is the optimal (hard) membership function matrix for variables given by the double *k*-means.

Double *k*-means was applied again on the same data set, but requiring (as in case (iii)), both a hard partition of the countries and a fuzzy partition of the economic variables. The problem was solved with an ALS algorithm that alternated between two steps:

1. solving the problem [P1'] subject to (2), (3), (4') and (5) by using a SQP algorithm;
2. solving step *b* of Table 1, which assigns each object to the closest centroid.

Figure 1: *Double k-means model* (i), *where hard partitions of both countries and economic variables are required. Countries are plotted on the centroids* $\mathbf{X}\hat{\mathbf{V}}(\hat{\mathbf{V}}'\hat{\mathbf{V}})^{-1}$

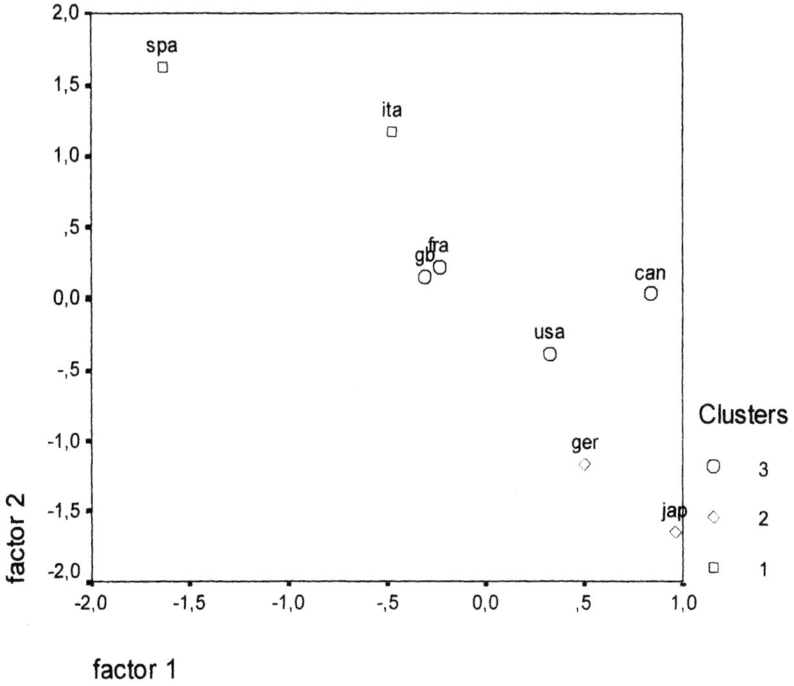

Even this time the initial solutions for the double *k*-means were given by the best *k*-means partitions of objects and variables, among 1000 iterations of the ordinary *k*-means algorithm performed on random initial estimates. The best hard partition of objects coincides with the one previously obtained by solving the fully hard partitioning double *k*-means. The algorithm converged after two iterations. The sequential quadratic programming algorithm used for identifying the fuzzy partition of the variables converged after 166 and 31 function evaluations, respectively (requiring the termination tolerance for variables and objective function to be 10^{-4} and for constraint violation to be equal to 10^{-6}). The optimal value of the objective function is 2.9028. The final fuzzy membership function matrix **V** is,

	GDP	*INF*	*DEF*	*DEB*	*INT*	*TRB*	*UNE*
V'=	0.8235	0.0004	0.9881	0.4465	0.0143	0.9997	0.0707
	0.1765	0.9996	0.0119	0.5535	0.9857	0.0003	0.9293

A plot of the countries partition on the factors defined by **XV** allows to show the classification of the countries (Figure 2). From the comparison between Figure 1 and 2 it can be observed that model (iii) (Figure 2) increases the interpretability of results since classes of countries appear better separated.

Figure 2: *Double k-means model* (iii), *where a hard partition of the countries and a fuzzy partition of the economic variables is required. Countries are plotted on the factors* **XV**.

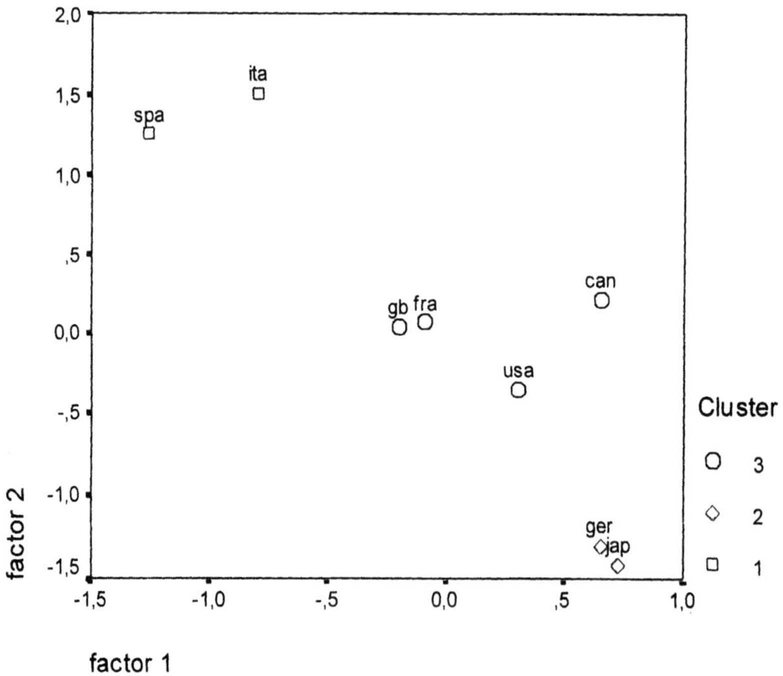

The double *k*-means algorithm was again applied on the 8×7 data matrix, requiring a fuzzy partition of both countries and variables.

The sequential quadratic programming algorithm used to solve model (ii), converged after 1231 function evaluations (even this time the termination tolerance for variables and objective function was set to be 10^{-4} and for constraint violation was fixed equal to 10^{-6}). The optimal value of the objective function was 2.60946. The final fuzzy membership function matrix **U** was,

	FRA	*GER*	*GB*	*ITA*	*SPA*	*USA*	*JAP*	*CAN*
	0.3241	0.0748	0.3684	0.5163	0.7458	0.1714	0.0356	0.0265
U'=	0.4414	0.8675	0.4954	0.0002	0.2541	0.5241	0.9028	0.1270
	0.2345	0.0577	0.1362	0.4835	0.0001	0.3045	0.0615	0.8464

Note that when a fuzzy partition of objects is required, Canada tends to form a singleton class. A relevant membership of USA and Great Britain and France to the class mainly formed by Japan and Germany is also observed. Italy is still forming a class with Spain but it has a relevant membership to the class formed by Canada.
The final fuzzy membership function matrix **V** was,

	GDP	*INF*	*DEF*	*DEB*	*INT*	*TRB*	*UNE*
V'=	0.7775	0.0024	0.8048	0.4510	0.0194	0.8833	0.0383
	0.2225	0.9976	0.1952	0.5490	0.9806	0.1167	0.9617

This matrix is similar to the one obtained when a hard partition of the objects was required.

5. Discussion

Double *k*-means is a new clustering methodology that allows to classify, simultaneously, objects and variables of a two mode data matrix. Different clustering structures (partition, covering packing), and classification types (hard or fuzzy) can be fitted to the data.
The identification of the model is statistically formalized as a least-squares problem, through a quadratic objective function subject to a set of constraints in accordance with the required classification structure (partition, covering, packing) and classification type (hard or fuzzy).
The least-square constrained problems with continuous variables can be solved using a sequential quadratic programming algorithm, that in the examples analyzed resulted very efficient (few seconds were always sufficient to obtain the final solution). When a hard partition of objects and/or variables is required, an alternating least-squares algorithm has been proposed.
The double *k*-means methodology can be used to select those variables (objects) that most contribute to identify the best classification of objects (variables). This model selection capability is obtained when a hard partition of objects (variables) and a fuzzy partition of variables (objects) are required in the double *k*-means

(e.g., as in problems (iii), (iv), (vi) and (vii)). The fuzzy membership function matrix **V** (**U**) can be considered as a factor loading matrix.

References

DeSarbo, W., S., Howard, D., J., and Jedidi, K. (1991). MULTICLUS: A new method for simultaneous performing multidimensional scaling and clustering, *Psychometrika*, 56, 121-136.

De Soete G., and Carroll, J. D. (1994). *K*-means clustering in a low-dimensional Euclidean space, in: Diday et al. (eds), *New approaches in classification and data analysis*, Springer, Heidelberg, 212-219.

MacQueen, J. (1967). Some Methods for Classification and Analysis of Multivariate Observations, in: *Proceedings of the Fifth Berkeley Symposium on Mathematical Statistics and probability, Volume 1 Statistics*, Eds, L.M. Le Cam and J. Neyman, Berkeley CA: University of California Press, 281-297.

Powell, M. J.D. (1983). Variable Metric Methods for Constrained Optimization, in: *Mathematical Programming: the State of the Art*, Eds. A. Bachem, M. Grötshel and B. Korte, Berlin: Springer-Verlag, 288-311.

Rizzi, A., and Vichi, M. (1995). Three-way Data Set Analysis, in: A. Rizzi (ed.) *Some relations between matrices and structures of multidimensional data analysis*, 93-166.

Categorical FDA Under Prospective Sampling Scheme: a Proposal for Variable Selection

Francesco Palumbo
Dipartimento di Istituzioni Economiche e Finanziarie
Università di Macerata, Via Crescimbeni 14, I-62100 Macerata
email: palumbo@unimc.it

Abstract: Given a population described by p explanatory and one dependent categorical variables, we assume that the dependent variable defines a partition of the population into g groups.
Discriminant Analysis studies the relation between the p explanatory variables and the dependent variable finding the subset of variables that has the most predictive power. Generally, in categorical discriminant analysis, the *a priori* probabilities associated to the g groups are assumed known. In this paper we summarise some suitable approaches under the hypothesis of unknown group *a priori* probabilities and we propose a new variables selection algorithm.

Keywords: Variable selection; Categorical Variables; Factorial Discriminant Analysis.

1. Introduction and motivation

This paper intends to summarise some results on the Factorial Discriminant Analysis (FDA) in case of categorical predictors and proposes new tools for variable selection under specific sampling hypotheses.
Let Ω be a population described by $p+1$ categorical variables, where $X_1,\ldots,X_p$ are said to be *explanatory* variables and the last is called *dependent* variable and denoted by Y. We indicate by g the number of categories of Y, that defines a partition of the population Ω into g groups $(\omega_1, \omega_2, \ldots, \omega_g)$ defined *a priori*.
We assume that E and E^* are two independent samples of Ω having respectively n *and* n^* elements on which all $p+1$ variables have been recorded. The sample E is called *training set*, whereas the sample E^* is called *test set*.
Discriminant analysis aims to define a parametric or non-parametric classification model, on the basis of E, capable of assigning statistical units of $E^* \subset \Omega$, to the g classes. The classification rule performance is measured by the ratio between the number of statistical units correctly classified and the sample size. It can be easily shown that the Correct Classification Ratio (CCR) is strictly dependent on the number of variables involved into the classification rule. Introducing more variables the CCR increases, reaches a maximum and decreases (Hand, 1981). The crucial aspect is variable selection.
Tackling the problem from a geometrical point of view, FDA aims to describe the g groups on the basis of a new set of m *explanatory* variables, with $m<p$, obtained by a linear combination of the p original variables. FDA permits to graphically represent, into the reduced subspace, the g groups, as well as the n and n^*

statistical units. Of course, we want that points belonging to the same group should be close and, at same time, that groups appear as much as possible separate. Moreover, making reference to the new set of m variables, it is also possible to define a geometrical or a probabilistic decision rule to classify a new set of statistical units.

Usually, in descriptive FDA on categorical data, no assumptions on the variable distributions are formulated, whereas when the FDA has a decisional aim, it is assumed that the *a priori* probability associated to the g groups is known and that groups are *i.i.d.* subpopulations. This hypothesis is quite realistic in many conditions and defines the so called *retrospective* sampling scheme. Holding this hypothesis, the number of observations in each group is fixed *a priori*, whereas the *explanatory* variable marginal distributions are assumed unknown. We refer to Celeux and Nakache (1994) and to McLachlan (1992) for an exhaustive treatment of the topic.

In this paper we will refer to the less frequent, but no so unrealistic, case of unknown groups *a priori* probability, that it is equivalent to assume a *prospective* sampling scheme. These conditions are typical in the satellite images recognition, and in general in all conditions where we do not know much about $\mathbf{\Omega}$.

Under this hypothesis the Non-Symmetrical Factorial Discriminant Analysis (NS-FDA) can be consistently adopted as a method for the definition of the *canonical variables* or discriminant factors (Palumbo, 1995; Verde and Palumbo, 1996).

In the following, we will briefly summarise the properties of the NS-FDA and we remark its consistency under the *prospective* sampling scheme (Palumbo and Siciliano, 1999). A new variable selection method, based on an asymptotic parametric test, will be proposed in the section three.

2. Categorical FDA under the prospective sampling scheme

In a previous paper Verde and Palumbo (1996) defined the NS-FDA as a special case of the Non-Symmetrical Correspondence Analysis, proposed by Lauro and D'Ambra (1984), under the hypothesis of independence among explanatory variables. There are two possibilities to satisfy the independence hypothesis: *i*) perform a stepwise selection procedure that selects the more predictive *explanatory* variables *iff* they are independent among them; *ii*) define *a priori* a orthogonal system of *explanatory* variables, like the design of experiments.

Let $\mathbf{X}$ be the matrix formed by the juxtaposition of the p sub-matrices $X_1, X_2, \cdots\cdots X_p$ where X_i refers to the ith predictor with binary coding of the predictor categories (where $1 \leq i \leq p$). We denote by J^+ the total sum of the number of categories of all p predictors. The notation $\mathbf{Y}$ refers to the binary coding matrix associated to the dependent variable Y

The NS-FDA solves the following equation:

$$(1/n)\left(\mathbf{Y}'\mathbf{X}\Delta_{\mathbf{X}}^{-1}\mathbf{X}'\mathbf{Y} - \frac{p}{n}\mathbf{Y}'\mathbf{1}\mathbf{1}'\mathbf{Y}\right)\mathbf{v}_m = \mu_m \mathbf{v}_m \tag{1}$$

where **1** is a $(n,1)$ unit vector, Δ_X is the (J^+,J^+) diagonal matrix obtained by the column marginals of **X**, and $1 \le m \le (g-1)$. Symbols μ_m and $\mathbf{v}_m$ are respectively the generic eigenvalue and the generic eigenvector associated to the left hand side of the equation (1), and are determined under the usual ortho-normality constraints.
In case of only one *explanatory* variable the equation (1) decomposes the numerator of Goodman-Kruskal's τ_{YX_i} index. (1954)

In case of two or more independent variables X_i the use of the diagonal matrix Δ_X imposes the independence condition among the explanatory variables. In this case, the trace of the left hand side in (1) corresponds to the quantity:

$$\mathrm{tr}\left(\mathbf{Y}'\mathbf{X}\Delta_{\mathbf{X}}^{-1}\mathbf{X}'\mathbf{Y} - \frac{p}{n}\mathbf{Y}'\mathbf{1}\mathbf{1}'\mathbf{Y}\right) = \mathrm{tr}\left(\mathbf{Y}'\mathbf{Y} - \frac{1}{n}\mathbf{Y}'\mathbf{1}\mathbf{1}'\mathbf{Y}\right)\sum_i \tau_{YX_i} \tag{2}$$

More precisely, in case of p independent *explanatory* variables, NS-FDA decomposes the sum of p numerators of the Goodman-Kruskal's τ_{YX_p}. Let us remark that the denominator of the τ does not vary in function of the independent variables, so that it can be assumed as constant in the NS-FDA context.

The maximisation of the $\sum_i \tau_{YX_i}$ criterion in the right hand side in (2) is equivalent to the First Order Independence Model (FIOM) (Goldestein and Dillon, 1978). Both of them maximise a monotonic transformation of the distance between the unconditional and conditional group probability distribution. The FIOM maximises the logarithmic transformation, whereas NS-FDA considers the square function. Square and logarithmic functions are well approximated by each other in $\Re^+$.
On the basis of these results, Verde and Palumbo (1996) proposed a forward variable selection approach based on the τ index. The CATANOVA test (Margolin and Light, 1974) permits to approximate the τ index to an exact distribution and to sort the independent variables on the basis of their explicative power. Variables enter the analysis following the order of the significance of the test. The rule stops when the CCR begins to decrease. The unbiased CCR is estimated on the set E^* or, equivalently, *jackknifing* or *bootstrapping* the sample E (Efron, 1986).
This variable selection approach presents two main drawbacks: it only works under the assumption of independence among the X's and is cumbersome to compute.

3. A proposal for the NS-FDA variable selection

In this section we propose a new approach for variable selection, based on an asymptotic test.
Let us define the following notation:

$$\mathbf{T} = \mathbf{Y}'\mathbf{Y} - \frac{1}{n}\mathbf{Y}'\mathbf{1}\mathbf{1}'\mathbf{Y} \qquad \text{and} \qquad \mathbf{B} = \mathbf{Y}'\mathbf{X}\Delta_{\mathbf{X}}^{-1}\mathbf{X}'\mathbf{Y} - \frac{p}{n}\mathbf{Y}'\mathbf{1}\mathbf{1}'\mathbf{Y}.$$

where **T** and **B** are respectively the total variance matrix and the between variance

matrix. According to the usual variance decomposition, we denote with **W** the within variance matrix corresponding to the quantity **W** = (**T** - **B**)
Under the independence hypothesis among explanatory variables and taking into account the Multinomial approximation to the Normal distribution when $n \to \infty$ (Mardia et. al., 1979), it can be shown that the statistics expressed by the ratio in (3) follows the Wilks' Λ distribution under the null hypothesis H_0: $\mathbf{B} = \mathbf{0}$.

$$|\mathbf{T}-\mathbf{B}|/|\mathbf{T}| \sim \Lambda \tag{3}$$

Let us make the two following considerations: *i*) there exists a relationship between the Wilks'Λ in the above expression and the eigenvalues $\{\lambda_1, \lambda_2, \cdots, \lambda_{g-1}\}$ of the matrix $\mathbf{W}^{-1}\mathbf{B}$, in fact $\Lambda = \prod_{i=1}^{g-1} \frac{1}{1+\lambda_i}$ (Rencher, 1998); *ii*) the same result holds also if we consider the matrix $\mathbf{T}^{-1}\mathbf{B}$ instead of the matrix $\mathbf{W}^{-1}\mathbf{B}$. Notice that the matrix **B** represent the NS-FDA equation and that the matrix **T** is not affected by the *explanatory* variables.
Holding these results we propose to adopt the Wilks' Λ for the selection of the most discriminant subset of independent variables.
The Wilks' Λ-distribution is indexed by three parameters: the dimension d=g, where g are the *a priori* defined groups by the variable Y; the degrees of freedom $\nu_Y = (J^+ - p)$ and $\nu_n = (n-g-1)$.
In case of normally distributed independent variables, Wilks' Λ can be approximated by a F distribution.

$F = \frac{1-\Lambda^{1/t}}{\Lambda^{1/t}} \frac{wt - 1/2(d\nu_Y - 2)}{d\nu_Y}$, where $w = \nu_n + \nu_Y - 1/2(d + \nu_Y + 1)$ and the quantity $t = \sqrt{(d^2\nu_Y^2 - 4)/(p^2 + \nu_Y^2 - 5)}$.

An easier approximation of Λ distribution is the Bartlett's approximation to a χ^2 distribution on $d\nu_Y$ degrees of freedom expressed in the formula (4) (Krzanowski, 1988):

$$\chi^2_{(g-1)\nu_Y} = -\left[\nu_n - \frac{1}{2}(g - \nu_Y)\right] \log\left(\Lambda_{g,\nu_n,\nu_Y}\right) \tag{4}$$

The results can be extended to the case of categorical data assuming the usual asymptotic normal theory, so that applying the central limit theorem the multinomial distribution can be approximated by a normal distribution when n increases.
The proof of the above relationship is quite known in the literature in case of continuous variables. In the next section we prove that it also holds in case of categorical variables and in the special case in which a prospective sampling scheme is adopted. This is empirically proved in the next subsection.
On the basis of this result, we propose a *sequential* variable selection procedure based on the Wilks' Λ approximation via F or χ^2 distribution. The latter is made up of four steps: *i*) independent variables are sorted on the basis of their predictive power; *ii*) the variable associated to the maximum criterion is chosen as first; *iii*)

new variables are iteratively added on the significance of the proposed asymptotic test; *iv*) the process stops when no more variables induce a significant test. Consistently with the independence assumption, no account is taken of the inter-relationships among *X*'s.

3.1. Empirical proof

Five independent multinomial random variables (X_1, X_2, X_3, X_4, X_5) with 4, 3, 5, 2, 3 categories respectively were generated. From each of them were extracted 1000 samples consisting of 300 observations. An equal number of samples was extracted by the multinomial random variable *Y* with 4 categories.

Figure 1: *Approximation of the proposed statistics to a chi-square distribution for different combinations of explicative variables*

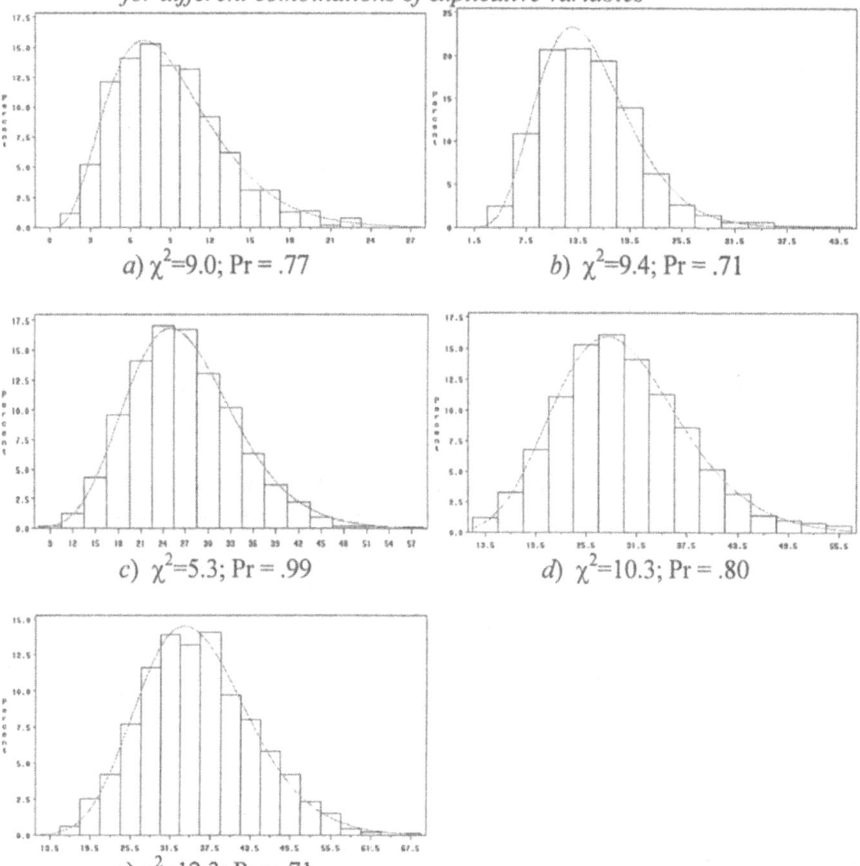

a) χ^2=9.0; Pr = .77

b) χ^2=9.4; Pr = .71

c) χ^2=5.3; Pr = .99

d) χ^2=10.3; Pr = .80

e) χ^2=12.3; Pr = .71

The five plots in Figure 1 prove the approximation of the Wilks' Λ to the chi-square distribution, in case of categorical data under the independence assumption.
More precisely, the image *a* refers to the approximation for the first multinomial

random variable the image b considers the first and the second at same time and so on. The number of categories of the above mentioned random variables does not affect the proof. Further empirical proofs demonstrated that these results hold for any combination of multinomial random variables, under the independence assumption. The relevance of the showed result consists in the fact that, varying the number of *explanatory* variables, the proposed statistics holds its property.

4. A real data set application

In this section we show and discuss some results accomplished on a real data set. The data set refers to a survey committed by a car manufacturer, that wanted to predict how its clients would choose among four different models on the basis of a set of categorical variables. The four categories of the dependent variable Y correspond to the models offered, that in the following are indicated as Model "A", Model "B", etc.
The data where collected by a mailed questionnaire to a list of people who bought a car of the manufacturer in the past years. A share of them answered to the questionnaire. Because of the missing questionnaires and the lack of information related to whole population, it should be reasonable to assume that the a priori probabilities associated to the classes A, B, C and D are unknown.

Table 1: *Distribution of the dependent variable "Model currently owned"*

Model	n_i	%	% cum.
A	114	27.2	27.2
B	122	29.1	56.3
C	117	27.9	84.2
D	66	15.8	100.0
Tot.	**419**	**100.0**	

In the table 2 are collected the questions and the relative frequencies associated to each modalities. Listed variables are also defined on the basis of short labels (**V1, V2**, ..., **V11**) that are used to identify the variables in the output analysis. Notice that: only a subset of answers has been selected; questionnaires having missing answers have not been included in the analysis.
The analysis is performed taking into account the chi-square approximation and the procedure selected the following variables into the given order at the 95% significance level: **V10, V1, V2, V8**. At each procedure step we computed the CCR using the Leave One Out (LOO) algorithm and obtained the following unbiased estimations: 39.5%, 47.2% 49.7%, 49.6%. The values in the previous list refer to the model with one variable, two variables, etc. .
We note that the 4th variable selection does not contribute to the CCR, but the procedure selected it all the same. This is a consequence of the independence assumption that has unfavourable effects on the analysis. Tests on the interdependence among predictors can prevent this drawback.

Table 2: *List of available explicative variables and their modalities.*

Question	***Modalities***	***%***
V1 Which kind of gear-box do you prefer?	4-gear	17.1
	5-gear (overdrive)	68.4
	5-gear (sport)	5.9
	Automatic	8.7
V2 Do you use the car exclusively for...	personal matters	30.7
	private and business affairs	63.5
	business matters	5.9
V3 Did you accomplish a test before buying the car?	yes	36.8
	no	63.2
V4 Did publicity influence decision?	yes	93.9
	no	6.1
V5 Did specialised magazines influence decision ?	yes	44.7
	no	55.3
V6 Did other circumstances influence decision?	yes	43.6
	no	56.4
V7 How would you describe your driving-manners ?	economical	4.4
	normal, not too powerful	50.6
	powerful, but not extreme	35.4
	speedy, sometimes extreme	9.6
V8 What is your preferred velocity on highways?	between 60 - 80 km/h	8.2
	between 81 - 95 km/h	35.8
	between 96 - 110 km/h	45.2
	above 110 km/h	10.8
V9 Sex of car-owner	Male	95.6
	Female	4.4
V10 Kind of professional activity	self employed	36.3
	free lanced	20.1
	employee / workman	43.6
V11 Size of household	one/two-person household	38.2
	three persons or more	61.8

The selected variables are involved in the NS-FDA and the result is shown in Figure 2.

Figure 2: *Graphical representation on the first factorial plan (98% of the total inertia)*

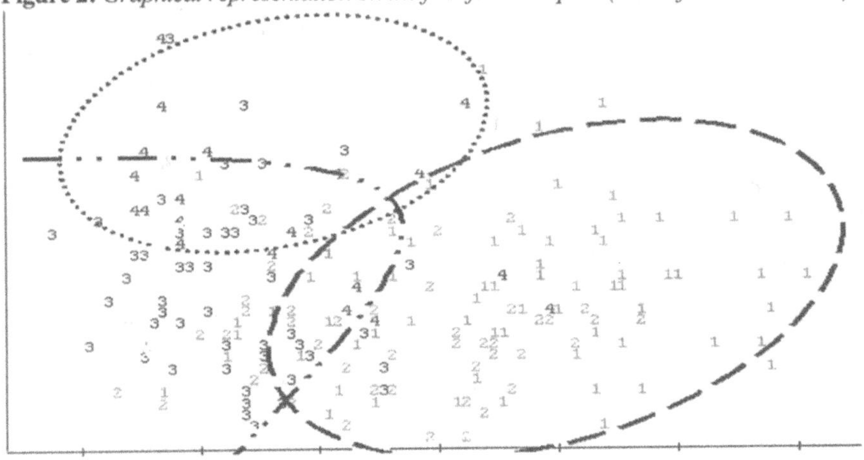

The Figure shows that there is an overlapping of the points belonging to the different four groups. The factorial axes from one to tree have been selected for the classification algorithm. Table 3 contains the redistribution matrix obtained with the LOO algorithm and the unbiased estimation of the CCR (49.6%).

Table 3: *Redistribution matrix and CCR*

	Model	Predicted A	B	C	D	
Actual	A	71 (**53.7%**)	24	18	1	114
	B	43	31 (**43.0%**)	44	4	122
	C	13	7	82 (**48.0%**)	15	117
	D	5	10	27	24 (**54.4%**)	66
		132	72	171	44	419 (**49.6%**)

5. Conclusion and perspective

The proposed procedure represents a consistent variable selection approach under the prospective sampling scheme. In fact, there is a coherence between the sampling schema with both the variable selection and the factorial approach.
When the independence hypothesis among *explanatory* variables is not satisfied, the criterion value tends to be a biased estimation and induces over parameterised models. The problem can be fixed testing the interdependence of the last select predictor with respect to each predictor already selected.

References:

Celeux, G., J.P. Nachake (1994) *Analyse discriminante sur variables qualitative.* Polytechnica, Paris.

Efron, B., (1986) How biased is the Apparent Error Rate of a Prediction Rule? *Journal of American Statistical Association,* **81**, pp. 461-470.

Goldstein, M., W.R.Dillon (1978) *Discrete Discriminant Analysis.* Wiley, N.Y.

Goodman, L.A., W.H. Kruskal (1954) Measure of Association for Cross-classifications, *Journal of American Statistical Association,* **49**, pp. 732-764.

Hand, D.J. (1981) *Discrimination and Classification.* John Wiley & Sons, N.Y.

Krzanowski, W.J. (1988) *Principles of Multivariate Analysis*, Oxford Science , Oxford.

Lauro, N.C., L.D'Ambra (1984) L'analyse non simmetrique des correspondances, in *Data Analysis and Informatics.* Ed. E. Diday et al., Elsevier Publishers, North-Holland.

Mardia, K, J.Kent and J.Bibby (1979), *Multivariate Analysis*, Academic Press, London.

Margolin, B.H., R.J. Light (1974) An Analysis of Variance for Categorical Data. *Journal of American Statistical Association,* **66**, pp. 534-544.

McLachlan G.J. (1992), *Discriminant Analysis and Statistical Pattern Recognition*, John Wiley & Sons, N.Y.

Palumbo, F. (1995) *Selezione e quantificazione dei predittori qualitativi nell'analisi discriminante*, Ph.D. thesis, DMS Univ. di Napoli.

Palumbo, F., R. Siciliano (1999) Factorial Discriminant Analysis and Probabilistic Models. *Metron*, **LVI**, pp. 185-198.

Rencher, A.C. (1998) *Multivariate Statistical Inference and Applications*, John Wiley & Sons, Inc.

Siciliano, R. (1990) Asymptotic Distribution of Eigenvalues and Statistical Tests in Non-Symmetric Correspondence Analysis. *Statistica Applicata*, **2**, pp. 259-276

Verde, R., F. Palumbo (1996) Non-Symmetrical Factorial Discriminant Analysis on Qualitative Predictors. In *XXXVIII Riunione Scientifica SIS*, Rimini (in italian).

The Effect of Telephone Survey Design on Discriminant Analysis

Alessandra Petrucci
Dipartimento di Statistica "G. Parenti"
Università di Firenze, Viale G.B. Morgagni 59, 50134, Firenze, Italy

Monica Pratesi
Dipartimento di Matematica, Statistica, Informatica ed Applicazioni
Università di Bergamo, Piazza Rosate 2, 24129, Bergamo, Italy

Abstract: In this paper the effect of telephone survey sample design on discriminant analysis is studied. The population is divided into two groups and a superpopulation model with multivariate normal distribution is assumed for the variables. The learning sample is composed of the randomly generated telephone numbers which correspond to the eligible population units (RDD). The rate of misclassification of the discriminant function and the effect of sample design on this rate of misclassification are discussed.

Keywords: Discriminant Analysis, RDD surveys, Misclassification rate

1. Introduction

Recently, many surveys have been designed adopting the telephone as a tool for contacting and selecting the sample of respondents. Especially in marketing research, the collected data have often been analyzed by discriminant analysis (Albaum and Backer, 1976; Crask and Perreault, 1977; Hair et al., 1992). One of the goals of discriminant analysis is the classification of individuals in the population into mutually exclusive and exhaustive groups on the basis of a set of characteristic variables. Variable values are collected on a sample of individuals called the "learning sample" and the results of the analysis are presented together with the estimation of the probability of incorrect classification of the individuals into groups (misclassification rate). In practice, the learning sample is often the result of the Random Digit Dialling (RDD) selection method. While the effect of a complex design on the misclassification probabilities has been recently studied (Leu and Tsui, 1997), the effect of the RDD selection has not yet been investigated. In this paper we consider the inverse sampling scheme associated with the random generation of telephone numbers and study the effect of this design on the misclassification rate, assuming a superpopulation model.
The sampling selection is described in Section 2 and the superpopulation model is outlined in Section 3. Design based estimators of the superpopulation parameters

are presented and the effect of the design on the misclassification rate is examined in Section 4. The results of the simulation study are reported in Section 5 and some concluding remarks are drawn in Section 6.

2. Random Digit Dialling Selection of the Learning Sample

In many countries, telephone companies assign to each subscriber a telephone number (TN) made up of two groups of digits: the first group is an area code and the second is the subscriber code. The random generation of a TN, given that the area code is fixed, is the random generation of the digits corresponding to the subscriber code. If the subscriber code has l digits the area code contains up to 10^l subscribers.

Assume that the population of N individuals is reachable by telephone and divided into two groups g=1,2 consisting of N_1 and N_2 individuals. If the learning sample is selected by a random generation process of telephone numbers (TNs), the fixed number of individuals to be selected from each population group, respectively n_1 and n_2, indicate the fixed number of successes to be reached in the process of generating TNs. The probability of finding an individual in group 1 is $\pi_1 = 10^{-l} \cdot N_1$, while $\pi_2 = 10^{-l} \cdot N_2$ is the probability of finding an individual in group 2. The total number of needed trials is $m > n_1 + n_2$.

Suppose that associated with each unit i in group g there is a discriminator x_{gi}, a $p \times 1$ vector, to be measured in the survey, and a $q \times 1$ vector z_{gi} of known values that could be a group variable for stratified sampling of TNs. In RDD surveys the role of z_{gi} can be linked to the structure of the generated number. The auxiliary information z can be association of the subscriber code with an area code. In stratified RDD designs, the sampling design depend on z_g but not on x_g. In the following we study the effect of an RDD design without considering the strata of TNs. The sampling design does not depend on x_g.

3. The Superpopulation Model

The vectors x_{gi} and z_{gi} can be considered as the realized outcome of a vector random variable $(X_{gi}, Z_{gi})^T$ with $i = 1, \cdots, N_g$ and $g = 1,2$. As in Leu and Tsui (1997), assume that the random vectors $(X_{gi}, Z_{gi})^T$ are independent and the superpopulation model, i.e., the joint distribution of the vector random variable, is the multivariate normal distribution:

$$\begin{pmatrix} X_{gi} \\ Z_{gi} \end{pmatrix} \sim N_{p,q}\left[\begin{pmatrix} \mu_g \\ \mu_z \end{pmatrix}, \Omega\right] = N_{p,q}\left[\begin{pmatrix} \mu_g \\ \mu_z \end{pmatrix}, \begin{pmatrix} \Omega_{11} & \Omega_{12} \\ \Omega_{21} & \Omega_{22} \end{pmatrix}\right]. \tag{1}$$

Ω_{22} is assumed to be positive definite, the two populations have a common covariance matrix (Ω), and the means of the random vectors Z_{gi} are equal to μ_z. In other words, the only discriminator is the random vector X_{gi}.

Our interest is on the effect of the RDD selection on the application to the learning sample of Fisher's linear discriminant function:

$$L(x_0) = (x_0 - (\mu_1 + \mu_2)/2)^T \Omega_{11}^{-1} (\mu_1 + \mu_2). \tag{2}$$

As is well known, the classification rule is to allocate x_0 to group 1 if $L(x_0) > 0$ and to allocate x_0 to group 2 otherwise. Under the superpopulation model the distribution of $L(x_0)$ is normal: $L \sim N(\Delta^2/2, \Delta^2)$, when x_0 comes from group 1 and $L \sim N(-\Delta^2/2, \Delta^2)$ when x_0 comes from group 2, where $\Delta^2 = (\mu_1 + \mu_2)^T \Omega_{11}^{-1} (\mu_1 + \mu_2)$ is the Mahalanobis distance. As a consequence, when the superpopulation parameters are known, $p(2|1) = P(L(x_0) < 0 | \text{group } 1) = \Phi(-\Delta/2)$ is the probability of misclassifying a unit from group 1 into group 2, where $\Phi(\cdot)$ is the cumulative distribution function of the standard normal distribution. And, similarly, the probability of misclassifying a unit from group 2 into group 1 is $p(1|2) = P(L(x_0) > 0 | \text{group } 2) = \Phi(\Delta/2)$.

Obviously, in most applications, the superpopulation parameters Ω_{11}, μ_1 and μ_2 are not known and are to be estimated by the learning sample.

4. Sampling Effects on Rates of Misclassification

The RDD learning sample provide data to estimate both the parameters Ω_{11}, μ_1 and μ_2 and the parameters N_1, N_2, π_1 and π_2 which are unknown in many applications.

The group size N_g (g = 1, 2) can be estimated without bias by $\hat{N}_g = \hat{\pi}_g \cdot 10^l$, where $\hat{\pi}_g = (n_g - 1)/(m-1)$ and m denotes the number of generated random TNs for calculating n_g eligible units (Haldane, 1945; Johnson et al., 1997).

The estimators of μ_1 and μ_2 are, respectively, $\hat{\mu}_1 = \sum_{s_1} \omega_i x_{1i}, \hat{\mu}_2 = \sum_{s_2} \omega_j x_{2j}$, with $\omega_i = \hat{N}_1^{-1} \hat{\pi}_1^{-1}, \omega_j = \hat{N}_2^{-1} \hat{\pi}_2^{-1}$. If N_1, N_2, and π_g are known, $\hat{\mu}_1$ and $\hat{\mu}_2$ are Horvitz and Thompson estimators and they are unbiased for μ_1 and μ_2, otherwise they are design biased ratio estimators.

The superpopulation parameter Ω_{11} can be estimated by a pooled covariance matrix estimator:

$$\hat{\Omega}_{\pi\varpi} = \frac{\hat{N}_1}{\hat{N}}\left[(\hat{N}_1 - 1)\hat{\pi}_1\right]^{-1}\left[X_1^T X_1 - (\hat{N}_1\pi_1)^{-1} t_1 t_1^{\,T}\right] + \\ + \frac{\hat{N}_2}{\hat{N}}\left[(\hat{N}_2 - 1)\hat{\pi}_2\right]^{-1}\left[X_2^T X_2 - (\hat{N}_2\pi_2)^{-1} t_2 t_2^{\,T}\right] \tag{3}$$

where X_g is the matrix $n_g \times p$ containing the values of the discriminators in the sample selected from group g, and $t_g = \left(\sum_{s_g} x_{1i}, \sum_{s_g} x_{2i}\right)^T$ is the vector of sample totals (Särndal et al., 1992). The pooled estimator is approximately design-unbiased for the finite population covariance only if in (3) N_1, N_2, N and π_g are known.

The previous estimators allow for the following sample-linear discriminant function:

$$W(x_0) = (x_0 - (\hat{\mu}_{1,\pi\omega} + \hat{\mu}_{2,\pi\omega})/2)^T \hat{\Omega}_{\pi\omega}^{-1} (\hat{\mu}_{1,\pi\omega} + \hat{\mu}_{2,\pi\omega}), \tag{4}$$

where x_0 is a vector of values of the discriminator of a new unit. As in Leu and Tsui (1997), "the classification rule is to allocate x_0 to group 1 if $W(x_0) > k$ and to allocate x_0 to group 2 otherwise, where k is a constant depending on prior information (misclassification cost and prior probabilities). In the following for simplicity k=0, i.e., equal prior probability and equal cost".

The sampling distribution of $W(x_0)$ can be far from the normality of $L(x_0)$. The effect of the inverse design on the distribution of $W(x_0)$ is likely to be transmitted to the misclassification rates $p_w(2|1)$ and $p_w(1|2)$.

In order to appreciate the effect of the design on the misclassification rate, one may compare the estimated misclassification rates with the rates obtained applying $L(x_0)$ to a simple random learning sample from the population. The design effect on the probabilities of misclassification can be expressed by the ratio $D = \frac{(p_w(2|1) + p_w(1|2))/2}{\Phi(\Delta/2)}$, where the average misclassification rate is compared with the true misclassification probability under the superpopulation model. If D is either grater than or less than 1, the design is presumed not to have a negligible effect on the misclassification probabilities, depending on the values of $\hat{\mu}_{1,\pi\varpi}, \hat{\mu}_{2,\pi\varpi}, \hat{\Omega}_{\pi\varpi}, \hat{N}_1$ and $\hat{N}_2$ obtained from the learning data.

5. Some Simulation Results

To estimate the probabilities of misclassification a simulation study was carried out following these steps:

1. two finite populations each having 1000 units were created. For each population unit, the vector x was generated from the same multivariate normal superpopulation model chosen by Leu and Tsui (1997), $\mu_1 = (1,2)$, $\mu_2 = (5,4)$ and common covariance matrix $\Omega = \begin{pmatrix} 8 & 2 \\ 2 & 10 \end{pmatrix}$, using the random number generator of GAUSS rel. 3.1. The Mahalanobis distance, Δ^2 is 2.105.

2. For several telephone coverage levels 100 independent samples of size 100 ($n_1 = n_2$) were selected from each population. For each sample the design based estimators of parameters Ω_{11}, μ_1, μ_2, N_1, N_2, π_1 and π_2 were calculated. In Table 1 the estimated values of μ_1, μ_2, N_1, N_2 have been reported only for two couples of π_1 and π_2 telephone coverage rates.

Table 1: *Design estimation of telephone coverage and population size for two couples of π_1 and π_2*

	π_1	π_2	π_1	π_2
	0.2	*0.8*	*0.4*	*0.6*
$\hat{E}(\hat{\pi}_g)$	0.21	0.79	0.39	0.63
$\sqrt{M\hat{S}E(\hat{\pi}_g)}$	0.017	0.035	0.027	0.037
$\hat{E}(\hat{\mu}_1)$	(1.14, 2.11)	(0.52, 1.77)	(1.25, 2.25)	(0.91, 2.11)
$\sqrt{M\hat{S}E(\hat{\mu}_1)}$	(3.61, 4.13)	(3.96, 4.19)	(3.94, 4.15)	(3.62, 4.18)
$\hat{E}(\hat{\mu}_2)$	(4.78, 3.94)	(4.83, 3.86)	(5.30, 4.27)	(4.55, 3.80)
$\sqrt{M\hat{S}E(\hat{\mu}_2)}$	(4.12, 4.17)	(4.08, 4.43)	(3.50, 4.02)	(3.76, 3.80)
$\hat{E}(\hat{N}_g)$	201.13	798.11	397.49	599.07
$\sqrt{M\hat{S}E(\hat{N}_g)}$	17.33	34.19	28.88	36.68

3. As shown in Section 3, the misclassification probabilities $p(2|1)$ and $p(1|2)$ depend on the value of Δ. An estimate of Δ is given by $\hat{\Delta}_\pi = (\hat{\mu}_{1,\pi\omega} + \hat{\mu}_{2,\pi\omega})^T \hat{\Omega}_{\pi\omega}^{-1} (\hat{\mu}_{1,\pi\omega} + \hat{\mu}_{2,\pi\omega})$. The conditional probabilities of misclassification, given the values of $\hat{\mu}_{1,\pi\varpi}, \hat{\mu}_{2,\pi\varpi}, \hat{\Omega}_{\pi\varpi}, \hat{N}_1$ and $\hat{N}_2$ obtained

from the learning data, are estimated by the proportion of units in the original population that the classification rule $W(x_0)$ classifies incorrectly in each learning sample ($p_w(2|1)$ and $p_w(1|2)$).

4. Given the Mahalonobis distance, the true rates of misclassification of the Fisher discriminant function are equal to $p(2|1) = p(1|2) = \Phi(-\Delta/2) = \Phi(-0.7274) = 0.234$. A good sample of Fisher's discriminant function should have nearly the sample properties of $L(x_0)$ and, as a result, the misclassification probabilities $p_w(2|1)$ and $p_w(1|2)$ should be nearly 0.234. This allows us to judge the design effect on misclassification probabilities both on the basis of the closeness of the probability to 0.234, and of the closeness of $p_w(2|1)$ to $p_w(1|2)$. The design effect on the probabilities of misclassification can be estimated by the ratio $D = \dfrac{(p(2|1) + p(1|2))/2}{0.234}$.
In Table 2 the misclassification probabilities for several telephone coverage rates (π_1 and π_2) are reported for the two groups. The rows indexed with the symbol *G1* contain the $p_w(2|1)$ while those indexed with *G2* contain the $p_w(1|2)$. The numerator of D is in bold type. The results of the design effect D are shown in Table 3.

Table 2: *Achieved misclassification based on 100 independent learning samples*

$\pi_1 \backslash \pi_2$		**0.2**	**0.3**	**0.4**	**0.5**	**0.6**	**0.7**	**0.8**	**0.9**
	G1	0.29	0.18	0.34	0.33	0.33	0.29	0.34	0.21
0.2	*G2*	0.24	0.26	0.28	0.37	0.23	0.31	0.21	0.31
	Average	***0.27***	***0.22***	***0.31***	***0.35***	***0.28***	***0.30***	***0.28***	***0.26***
	G1	0.22	0.25	0.21	0.22	0.21	0.26	0.32	0.28
0.4	*G2*	0.26	0.26	0.32	0.32	0.31	0.27	0.19	0.29
	Average	***0.24***	***0.26***	***0.27***	***0.27***	***0.26***	***0.27***	***0.26***	***0.29***
	G1	0.37	0.30	0.22	0.29	0.26	0.22	0.27	0.27
0.6	*G2*	0.25	0.27	0.25	0.29	0.37	0.30	0.28	0.28
	Average	***0.31***	***0.29***	***0.24***	***0.29***	***0.32***	***0.26***	***0.28***	***0.26***
	G1	0.33	0.28	0.30	0.22	0.23	0.27	0.24	0.33
0.8	*G2*	0.24	0.29	0.21	0.29	0.21	0.24	0.26	0.31
	Average	***0.29***	***0.29***	***0.26***	***0.26***	***0.22***	***0.26***	***0.25***	***0.32***

Table 3: *Design effect (D) on misclassification probability*

$\pi_1 \backslash \pi_2$	**0.2**	**0.3**	**0.4**	**0.5**	**0.6**	**0.7**	**0.8**	**0.9**
0.2	1.13	0.94	1.32	1.49	1.20	1.28	1.78	1.11
0.4	1.02	1.09	1.13	1.15	1.11	1.13	1.08	1.22
0.6	1.32	1.22	1.00	1.23	1.34	1.11	1.17	1.09
0.8	1.22	1.22	1.09	1.09	0.94	1.09	1.07	1.37

6. Concluding Remarks

From the simulation results some conclusions can be drawn:

- the RDD design effect on misclassification probability is not negligible (D is mostly higher than 1). The RDD selection of the learning sample seems to increase the probability of misclassifying a population unit;

- the achieved level of the probability of misclassification seems not to depend on the telephone coverage rate of the population. We obtained nearly the same values of D for each possible pair π_1 and π_2 in the two populations;

- the importance of the effect of the RDD design on the misclassification probability can be appreciated by comparing it with the design effect of other sampling designs. Our results can be directly compared with those obtained for the same population with different complex sampling designs by Leu and Tsui (1997; p. 285). In each case, the effect of the RDD design (D_{RDD}) is greater than that obtained for several sampling designs by Leu and Tsui (simple random sampling of size 100: D=1.01; proportional allocation: D=1.01; increasing allocation: D=1.04; u-shaped allocation: D=1.03). The most reliable comparison, the one with simple random sampling of the same size, indicates that the inverse design produces a greater design effect (D_{RDD}=1.13 vs D=1.01, D_{RDD} =1.34 vs D=1.01).

We remark that the results heavily relies on the covariance matrix estimator (3) and future studies should be directed to explore the analytic properties of the estimator under the superpopulation model. Further investigation is also required to evaluate the goodness of the discriminant function $W(x_0)$ when comparing the misclassification probabilities in different telephone survey conditions: different sample sizes and more complex RDD designs (Waksberg, 1978; Lepkowsky, 1988).

References

Albaum G., Baker K. (1976), The sampling problem in validation of multiple discriminant analysis, *Journal of the Market Research Society*, 18.

Crask M. R., Perreault W. D. Jr. (1977), Validation of discriminant analysis in marketing research, *Journal of marketing research*, n.14, pp. 60-68.

Hair J. F. Jr., Anderson R. E., Tatham R. L., Black W. C. (1992), *Multivariate Data Analysis with Readings*, 3rd ed., Macmillan, New York.

Haldane F. R. S. (1945), On a method of estimating frequencies, *Biometrika*, 33, pp. 222-225.

Johnson N. L., Kotz S. and Balakrishnan N. (1997), *Discrete Multivariate Distributions*, Wiley, New York.

Leu, C. H. and Tsui, K. W. (1997). Discriminant analysis of survey data, *Journal of Statistical Planning and Inference*, 60, 273-290.

Lepkowsky, J. M. (1988). Telephone sampling methods in the United States, in Groves R. M. et al. (eds), *Telephone Survey Methodology*, John Wiley & Sons, New York, 161-173.

Särndal C. E., Swensson B, Wretman J. (1992), *Model Assisted Survey Sampling*, Springer-Verlag, New York.

Skinner, C. J., Holt D., Smith T. M. F. (eds) (1989) *An Analysis of Complex Surveys*. Wiley, Chichester.

Waksberg J. (1978), Sampling Methods for Random Digit Dialling, *Journal of the American Statistical Association*, 73, pp. 40-46.

A Dissimilarity Measure between Probabilistic Symbolic Objects

Laura Bocci
Dipartimento di Statistica, Probabilità e Statistiche Applicate
Università degli Studi di Roma "La Sapienza", 00185 Roma, Italia
e-mail: bocci@pow2.sta.uniroma1.it

Abstract: This paper presents an approach to calculate the dissimilarity between probabilistic symbolic objects. The proposed dissimilarity measure is based on both a comparison function and an aggregation function. Comparison function is a proximity coefficient based on statistical information given by each probabilistic elementary event. The aggregation function is a proximity index, related to Minkowski measure, which aggregates the results given by comparison functions.

Keywords: Symbolic objects, probability, dissimilarity measure.

1. Introduction

The definition of a symbolic object involves two dual concepts of intent and extent: a symbolic object can be defined as the description (*intent*), according to M observed variables Y_i, of a set of objects which belong to its *extent*. Therefore a symbolic object is defined by its intent and by a way of finding its extent.
Probabilistic symbolic objects (Diday, 1995) are represented in their intent by a logical conjunction of "internal modal elementary events". Each elementary event e_i is characterized by the set of values that it assumes for a variable Y_i. The intent of a probabilistic elementary event represents the set of weighted values that the variable can take on it, where a probabilistic weighting system is adopted.
Since the dissimilarity between two symbolic object is evaluated comparing their intents, only the intent of a symbolic object will be considered without mentioning its extent.
The purpose of this paper is to present an approach to calculate the dissimilarity between probabilistic symbolic objects (PSO). Our approach is based on two functions. First, given two PSO, the comparison function evaluates the dissimilarity between the intents of a couple of probabilistic elementary events (PEE) comparing the set of values that they assume for the same observed variable. Then the aggregation function, which is related to Minkowski measure, combines the dissimilarity measure computed with respect to each variable.
The comparison between PEE is based on the *potential energy* of a PEE, which is a form of statistical information given by it, and on two new kind of PEE which will be called, respectively, *Cartesian join probabilistic elementary event* and *Cartesian meet probabilistic elementary event*.

2. Probabilistic Symbolic Objects

Let consider a set of symbolic objects E which are described by M variables Y_i. The intent of a probabilistic symbolic object (PSO) s, which is made up of M probabilistic elementary events (PEE) e_i, is defined as

$$s = \bigwedge_{i=1}^{M} e_i \text{ where } e_i = \left[Y_i = \left\{ c_{ij} \ , \ {}_s p_{ij} \right\}_{j=1 \ldots k_i} \right] \tag{1}$$

The intent of a PEE e_i describes the set of values that it assumes for an observed statistical variable Y_i which is a quantitative or qualitative variable defined by associating to the set of observed "values", which it can take on e_i, a weighting system.
In the present context, only quantitative variable with a frequency or a probabilistic weighting system will be considered.
Therefore in (1) the statistical variable Y_i assumes on e_i "value" c_{ij} with frequency or probability ${}_s p_{ij}$ such that $0 \leq {}_s p_{ij} \leq 1$ for all $j = 1, \ldots, k_i$ and $\sum_{j=1}^{k_i} {}_s p_{ij} = 1$. The set $\left\{ c_{ij} \ j = 1 \ldots k_i \right\}$ form a partition of the observation set V_i^s of the variable Y_i on the PEE e_i. If Y_i is a continuous quantitative variable then c_{ij} is an interval and therefore e_i can be represented by a histogram, while if Y_i is a discrete quantitative variable then $c_{ij} \subset \Re$ is a discrete set of real values and therefore e_i can be represented by a diagram.

2.1 Potential energy of a probabilistic symbolic object

Let consider a PSO s as defined in (1). Let us define its *potential energy* as

$$E(s) = \prod_{i=1}^{M} E(e_i) \tag{2}$$

where $E(e_i)$ is the *potential energy* of PEE e_i defined as

$$E(e_i) = \mu(V_i^s) \sum_{j=1}^{k_i} \frac{1}{\mu(c_{ij}) \big/ \mu(V_i^s)} \left({}_s p_{ij} \right)^2 = \left[\mu(V_i^s) \right]^2 \sum_{j=1}^{k_i} \frac{1}{\mu(c_{ij})} \left({}_s p_{ij} \right)^2 \tag{3}$$

In the last expression we define $\mu(V_i^s)$ and $\mu(c_{ij})$ as the length of V_i^s and the length of interval c_{ij} if Y_i is a continuous quantitative variable, otherwise they are the number of possible values included in the set V_i^s and in c_{ij}, respectively, if Y_i is a discrete quantitative variable.
The potential energy $E(e_i)$ is a positive quantity that allows us to have some statistical information given by a probabilistic elementary event. It is related to the concept of homogeneity introduced first by Gini (Leti, 1979), but the difference between them is that in $E(e_i)$ the squared frequency/probability $({}_sp_{ij})^2$ is weighted. Since subsets c_{ij} $(j = 1,\ldots,k_i)$, of the partition of V_i^s, have not the same length or number of elements, these weights take into account their relative "size" (length or number of elements) with respect to the "size" of V_i^s itself.
If the variable Y_i is uniformly distributed on V_i^s then the potential energy of the PEE e_i is the same as the description potential defined by De Carvalho (De Carvalho, 1996) for a Boolean elementary event $e_i = [Y_i = V_i^s]$, therefore $E(e_i) = \mu(V_i^s)$.

2.2 Cartesian probability space model

Let us define a new mathematical model called *Cartesian probability space model* $(U^{(M)},\oplus,\otimes)$ where $U^{(M)}$ is a M-dimensional frequency or probability space on the variables, and $\oplus$ and $\otimes$ are two operators, the Cartesian join and the Cartesian meet, defined on the variable frequency/probability space.
Let consider a PSO s whose intent is defined in (1): its intent can be viewed as a pattern of the M-dimensional frequency/probability space on the variables $U^{(M)} = U_1 \times \cdots \times U_i \times \cdots \times U_M$ where U_i is the frequency/probability space on variable Y_i. Assuming this, we will say that a PSO is an element of $U^{(M)}$.
Let consider two PSO a and b whose intents, elements of $U^{(M)}$, are

$$
\begin{aligned}
a &= \bigwedge_{i=1}^{M} a_i = \bigwedge_{i=1}^{M} \left[Y_i = \left\{ a_{ij}\ ,\ {}_ap_{ij} \right\}_{j=1\ldots k_i} \right] \\
b &= \bigwedge_{i=1}^{M} b_i = \bigwedge_{i=1}^{M} \left[Y_i = \left\{ b_{i\ell}\ ,\ {}_bp_{i\ell} \right\}_{\ell=1\ldots h_i} \right]
\end{aligned}
\tag{4}
$$

Let denote the observation set of Y_i on a_i and b_i as, respectively, V_i^a and V_i^b. V_i^a is partitioned on k_i subsets a_{ij}, while V_i^b is partitioned on h_i subsets $b_{i\ell}$, where $k_i \neq h_i$.
The intent of *Cartesian join probabilistic symbolic object* $a \oplus b$ is the logical conjunction of the intents of *M Cartesian join probabilistic elementary events*

$a_i \oplus b_i$. The intent of *Cartesian meet probabilistic symbolic object* $a \otimes b$ is the logical conjunction of the intents of *M Cartesian meet probabilistic elementary events* $a_i \otimes b_i$. Both the intents of $a \oplus b$ and $a \otimes b$ are elements of $U^{(M)}$.

To apply Cartesian join and Cartesian meet operators to a couple of PEE a_i and b_i it is necessary to make some algebra on them. First two new observation sets V_i^* and S_i and suitable partitions of them have to be defined.

V_i^* is the observation set of the statistical variable Y_i on $a_i \oplus b_i$: it is the Cartesian join of observation sets V_i^a and V_i^b, defined by Ichino & Yaguchi (Ichino & Yaguchi, 1994) as

$$V_i^* = \begin{cases} \left[\min(V_i^{a\ell}, V_i^{b\ell}), \max(V_i^{au}, V_i^{bu})\right] & \textit{if} \quad Y_i \text{ continuous variable} \\ V_i^a \cup V_i^b & \textit{if} \quad Y_i \text{ discrete variable} \end{cases}$$

where, in the case Y_i is a continuous quantitative variable, V_i^a and V_i^b are two intervals such that $V_i^a = [V_i^{a\ell}, V_i^{au}]$ and $V_i^b = [V_i^{b\ell}, V_i^{bu}]$ with $V_i^{a\ell}$, $V_i^{b\ell}$ and V_i^{au}, V_i^{bu} the lower and upper bounds of V_i^a and V_i^b respectively. Let consider a partition $\{c_{ij}^*, j = 1, \ldots, T_i\}$ of V_i^*.

S_i is the observation set of the statistical variable Y_i on $a_i \otimes b_i$: it is called by Ichino & Yaguchi (Ichino & Yaguchi, 1994) the Cartesian meet of observation sets V_i^a and V_i^b and it is defined as $S_i = V_i^a \cap V_i^b$ where Y_i can be either a discrete or continuous quantitative variable. It should be noted that when $V_i^a \cap V_i^b \neq \varnothing$ therefore $S_i \subseteq V_i^*$. This means that some subsets of partition of V_i^* are the same subsets of partition of S_i: for this reason we use a unique notation c_{ij}^* for the subsets of both partitions. Then a partition of S_i will be $\{c_{ij}^*, j = 1, \ldots, G_i\}$.

Now we have to re-define the frequency/probability distribution of Y_i on V_i^a and V_i^b with respect to partition $\{c_{ij}^*, j = 1, \ldots, T_i\}$ of V_i^*: this means that new frequencies/probabilities ${}_a\hat{p}_{ij}$ and ${}_b\hat{p}_{ij}$, with which the statistical variable Y_i takes values c_{ij}^* on a_i and b_i, respectively, have to be calculated from the original ones defined in (4). In the case that $c_{ij}^* \not\subset V_i^a$ or $c_{ij}^* \not\subset V_i^b$ a probability ${}_a\hat{p}_{ij} = 0$ or ${}_b\hat{p}_{ij} = 0$ will be considered. Therefore the two PEE a_i and b_i become

$$a_i = \left[Y_i = \left\{c_{ij}^*, \; {}_a\hat{p}_{ij}\right\}_{j=1,\ldots,T_i}\right] \text{ and } b_i = \left[Y_i = \left\{c_{ij}^*, \; {}_b\hat{p}_{ij}\right\}_{j=1,\ldots,T_i}\right] \tag{5}$$

Now taking into account the relation (5) it is possible to define the Cartesian join

and the Cartesian meet probabilistic elementary events.
The intent of the *Cartesian join probabilistic elementary event* $a_i \oplus b_i$ is

$$a_i \oplus b_i = \left[Y_i = \left\{ c_{ij}^* , p_{ij}^* \right\}_{j=1\ldots T_i} \right] = \left[Y_i = \left\{ c_{ij}^* , p_{ij}^* = \frac{1}{2}\, {}_a\hat{p}_{ij} + \frac{1}{2}\, {}_b\hat{p}_{ij} \right\}_{j=1\ldots T_i} \right]$$

The intent of the *Cartesian meet probabilistic elementary event* $a_i \otimes b_i$ is

$$a_i \otimes b_i = \left[Y_i = \left\{ c_{ij}^* , \tilde{p}_{ij} \right\}_{j=1\ldots G_i} \right] = \left[Y_i = \left\{ c_{ij}^* , \tilde{p}_{ij} = \frac{{}_a\hat{p}_{ij}\ {}_b\hat{p}_{ij}}{\sum_{j=1}^{G_i} {}_a\hat{p}_{ij}\ {}_b\hat{p}_{ij}} \right\}_{j=1\ldots G_i} \right]$$

Here the numerator of frequency/probability $\tilde{p}_{ij}$, with which the statistical variable Y_i takes value c_{ij}^* on $a_i \otimes b_i$, is calculated in the following way

$${}_a\hat{p}_{ij}\ {}_b\hat{p}_{ij} = \Pr ob\{(Y_i = c_{ij}^* \mid a_i) \cap (Y_i = c_{ij}^* \mid b_i)\} = \Pr ob(Y_i = c_{ij}^* \mid a_i) \Pr ob(Y_i = c_{ij}^* \mid b_i)$$

where it is assumed that the two PEE a_i and b_i are independent.
When $V_i^a \cap V_i^b = \varnothing$ then $S_i = \varnothing$ therefore we define $a_i \otimes b_i = [Y_i = \varnothing]$: its potential energy is $E(a_i \otimes b_i) = 0$ and $\mu(S_i) = 0$.

3. Dissimilarity Measure

The dissimilarity measure between PSO is calculated using a comparison function d_e, applied on each pair of PEE intents attached to the same variable, and an aggregation function which aggregates the M comparisons d_e.
Let consider two PSO a and b whose intents are defined in (4). The aggregation function, related to Minkowski measure, is a mapping defined as

$$d(a,b) = \underbrace{\left\{ \sum_{i=1}^{M} \underbrace{[d_e(a_i,b_i)]^r}_{\text{comparison}} \right\}^{1/r}}_{\text{aggregation}} \qquad (6)$$

where $r \geq 1$ and the comparison function d_e between a_i and b_i is a mapping
$d_e : U_i \times U_i \rightarrow [0,1]$

defined as

$$d_e(a_i,b_i)=\underbrace{\left[1-\frac{E(a_i\otimes b_i)}{A+B+C}\right]}_{Q_1}\times\underbrace{\frac{\left|\overline{R}_i^a-\overline{R}_i^b\right|}{\left|V_i^*\right|}}_{Q_2} \tag{7}$$

where:

$$A=\mu(V_i^a)^2\sum_{V_i^a\backslash V_i^b}\frac{1}{\mu(c_{ij}^*)}\,({}_a\hat{p}_{ij})^2 \quad \text{and} \quad B=\mu(V_i^b)^2\sum_{V_i^b\backslash V_i^a}\frac{1}{\mu(c_{ij}^*)}\,({}_b\hat{p}_{ij})^2$$

$$C=\mu(S_i)^2\sum_{S_i}\frac{1}{\mu(c_{ij}^*)}\left[\max\left(p_{ij}^*,\tilde{p}_{ij}\right)\right]^2$$

$\overline{R}_i^a$ and $\overline{R}_i^b$ are the means of the interquartile range of a_i and b_i, respectively, and $\left|V_i^*\right|$ is the length of the set V_i^*.

Proposition 2. The comparison function d_e defined in (7) is a dissimilarity measure for all $i=1,\ldots,M$.

Proof. To be a dissimilarity measure the comparison function d_e has to satisfy the similarity, straightness and null diagonal properties. It is immediate to see that it satisfies the first two:

- similarity: for all a_i and b_i in U_i $d_e(a_i,b_i)=d_e(b_i,a_i)$
- straightness: for all a_i and b_i in U_i $d_e(a_i,b_i)\geq d_e(a_i,a_i)$

To verify the null diagonal property it should be noted that if $a_i=b_i$ then $a_i\oplus b_i=a_i\otimes b_i$ and thus $p_{ij}^*=\tilde{p}_{ij}$ for all $j=1,\ldots,T_i$ since $T_i=G_i$. This means that $A=B=0$ and $C=E(a_i\otimes b_i)$ and therefore the term Q_1 on the right of (7) is equal to zero. Also the term Q_2 on the right of (7) is equal to 0 since a_i and b_i have the same interquartile range and thus $\overline{R}_i^a=\overline{R}_i^b$. So it is possible to conclude that if $a_i=b_i$ then $d_e(a_i,b_i)=0$. □

The comparison function d_e is the product of two quantity Q_1 and Q_2. Q_1 is based on the *potential energy* of each PEE and their corresponding *Cartesian join* and *Cartesian meet probabilistic elementary events*. This term can be interpreted as an extension of Jaccard dissimilarity since A is the potential energy of a_i on that part of its support which a_i does not share with b_i, B is the potential energy of b_i on that part of its support which b_i does not share with a_i and C is the statistical information given both by a_i and b_i. Q_2 is based on the distance between the mean of each PEE interquartile range.

The coefficient d_e, which takes values in [0,1], allows us to evaluate the

dissimilarity between PEE even if a_i and b_i do not share any part of their support and therefore S_i is the empty set. In this situation Q_1 is equal to 1 since $E(a_i \otimes b_i) = 0$, while Q_2 gives us information about the distance between the position of masses of a_i and b_i.
If the statistical variable Y_i is a continuous quantitative variable and it is uniformly distributed on both the observation sets V_i^a and V_i^b on PEE a_i and b_i, respectively, then

$$d_e(a_i,b_i) = \underbrace{\left[1 - \frac{\mu(S_i)}{\mu(V_i^*)}\right]}_{Q_1} \times \underbrace{\frac{\left|\frac{V_i^{a\ell} + V_i^{au}}{2} - \frac{V_i^{b\ell} + V_i^{bu}}{2}\right|}{\left|V_i^*\right|}}_{Q_2}$$

This means that in this particular situation the dissimilarity between PEE is equal to a specific dissimilarity between Boolean elementary events $a_i = [Y_i = V_i^a]$ and $b_i = [Y_i = V_i^b]$. Indeed Q_1 and Q_2 are the De Carvalho dissimilarity (De Carvalho, 1996) with $\gamma = 0$ and $\gamma = 0.5$, respectively.
Since each comparison function d_e for all $i = 1,\dots,M$ is a dissimilarity measure, also $d(a,b)$, defined in (6), is a dissimilarity measure.

4. An Application

We present an example in order to illustrate the usefulness of the dissimilarity measure proposed. The data set used for this example consists of Fisher's Iris data. From this set of "classic" data described by four quantitative variables, we have first constructed seven probabilistic symbolic objects. Then to this set of symbolic objects we applied two different kinds of clustering algorithm by using dissimilarity measure defined in (6): the single linkage and the complete linkage method. When the dendrograms (figure 1) obtained by each of the clustering method applied are cut at an appropriate level, we get three clusters: {a, b} {c, d} and {e, f, g} by the complete linkage algorithm and {a, b} {c} and {d, e, f, g} by the single linkage algorithm. Although it can be observed that these results are slightly different, the clusters obtained represents the flowers structure in three different species. The difference depends on the fact that the extent of PSO d is mainly characterized by flowers belonging to two different species.

5. Conclusions

The proposed methodology, which takes into account either the statistical

Figure 1: *Dendrograms obtained by single linkage (A) and complete linkage (B) algorithm*

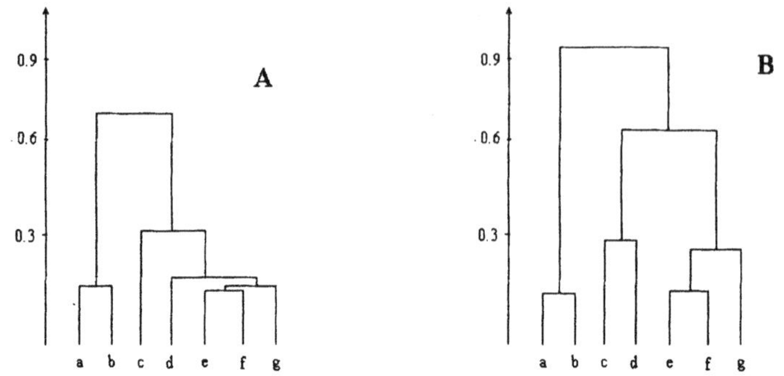

information of each PEE and its position in the plane, represents a new approach to calculate the dissimilarity between particular kind of symbolic objects, the probabilistic ones, for which not many proximity measures can be find in literature. Also from the concept of potential energy and from the evaluation of dissimilarity between PEE, we have seen that a Boolean symbolic object can be considered as a particular kind of probabilistic symbolic object. According to this the dissimilarity measure proposed can be viewed as a generalization of dissimilarity measures for both Boolean and probabilistic symbolic objects.
Further developments of this work should be intended to consider PEE characterized by the set of values that it assumes for either quantitative or qualitative variable. Some other generalizations of the method have also been undertaken: to represent actual knowledge the intent of a probabilistic symbolic object must take into account different kind of constraints given by logical dependencies between variables.

References

De Carvalho F.A.T. (1996) Histogrammes et Indices de Proximité en Analyse des Données Symboliques, in: *Actes de l'Ecole d'Ete sur Analyse des Données Symbolique*, Lise - Ceremade, Université Paris - IX Dauphine, Paris.

Diday E.(1995) Probabilist, possibilist and belief objects for knowledge analysis, *Annals of Operations Research*, 55, 227-276.

Diday E. (1998) Symbolic Data Analysis: a Mathematical Framework and Tool for Data Mining, in: *Advances in Data Science and Classification*, Rizzi, A. & Vichi, M. & Bock H.H. (Eds.), Springer Verlag, 409-416.

Ichino M., Yaguchi H. (1994) Generalized Minkowski Metrics for Mixed Feature-Type Data Analysis, *IEEE Transactions on Systems, Man, and Cybernetics*, 24, 4, 698-708.

Leti G. (1979) *Distanze e indici statistici*, La Goliardica Editrice, Roma.

Rizzi A. (1998) Metriche nell'analisi dei dati simbolici, *Statistica*, 4, 577-588.

Recent Experimentation on Euclidean Approximations of Biased Euclidean Distances[1]

Sergio Camiz
Dipartimento di Matematica "Guido Castelnuovo",
Università di Roma "La Sapienza", E-mail: camiz@mat.uniroma1.it

Georges Le Calvé
Laboratoire de Statistique - Université de Rennes II

Abstract: Given a set of 16 points on a grid, a set of randomly biased distances matrices is built and ten methods for their Euclidean approximantion are compared, to identify which minimize the stress. The Principal Coordinates Analysis of Torgerson's (1958) matrix of biased distances, limited to positive eigenvalues, proved to be more effective than methods based on monotonous transformations, aiming at getting the corresponding Torgerson's (1958) matrix positive semidefinite prior to *PCoA*. Its behaviour resulted close to Kruskal Non-Metric Multidimensional Scaling and Bennani Dosse (1998) Optimal Scaling, with the advantage of the identification *a posteriori* of the suitable dimension.

Keyword: Dissimilarities, Distances, Eigenanalysis, Euclidean approximations, Multidimensional scaling, Non-metric multidimensional scaling, Optimal scaling.

1. Introduction

Data representing relations among units, as deviations or dissimilarities, are conveniently represented as distances in the graphical space. When data are not originally Euclidean distances, a bias is introduced at representation time, in order to adjust to the Euclidean nature of representation space.

The problem of *Multidimensional Scaling* (*MDS*), the representation on a Euclidean space of non-Euclidean distances or dissimilarities among n points $D = (d_{ij})$, was long investigated, considering two main directions for its solution. The first consists in minimizing the introduced bias, as expressed by a suitable loss function. It is the domain of *Optimal Scaling* (*OS*), where as loss functions are considered either the *stress*, the sum of squared errors between original data and Euclidean distances, computed using the coordinates in the representation space

[1] The first author was granted by CNR contributions n. 97.05087.CT12, 97.03844.CT15 and 98.01545.CT10. The second author was supported, as visiting professor, by Roma University La Sapienza in 1998. Thanks are due to M. Bennani Dosse, for kindly providing the computations concerning his optimal scaling method and to M. Vichi, for his kind suggestions.

$$\sum_{ij}\left(d_{ij}-\sqrt{\sum_{\alpha=1}^{q}(u_{\alpha i}-u_{\alpha j})^2}\right)^2=\text{minimum}\ ,\ q<n. \qquad (1)$$

or the *sstress*, the squared errors among their squares

$$\sum_{ij}\left(d_{ij}^{\,2}-\sum_{\alpha=1}^{q}(u_{\alpha i}-u_{\alpha j})^2\right)^2=\text{minimum}\ ,\ q<n. \qquad (2)$$

In both cases, some distances may be weakly and some strongly modified, in order to obtain the *best possible Euclidean approximation.* This may be found in any *a priori* fixed dimensional space, through numerical approximation techniques. It is known (Le Calvé, 1976) that only in this way a global solution may be found, but up to now even the best existing algorithms do not ensure to find such global optima. In this respect, the method proposed by Bennani Dosse (1998) seems promising. Using these methods, it must be noticed that no relations exist among the solutions in different dimensions. Often, this representation causes some exchange in the relations between some deviations, in particular their order.
The second direction considers that the important information does not concern the absolute magnitude of deviations but their relations. In particular, attention may be concentrated on the order of the original values. It is the domain of *Non-Metric Multidimensional Scaling* (*NMMDS*), where a monotonous transformation of D is searched, able to get Euclidean the transformed distances Δ. This may be obtained either through Kruskal's (1964a, 1964b) algorithm, whose solution dimension may be chosen *a priori*, or through monotone functions, as described in the following. For these, Gower's (1966) *Principal Coordinate Analysis* (*PCoA*) on Torgerson's (1958) $T^{(2)}(\Delta)$ matrix, given by

$$t_{ij}^{(2)}=\tfrac{1}{2}\left(\delta_{i\cdot}^{\,2}+\delta_{j\cdot}^{\,2}-\delta_{\cdot\cdot}^{\,2}-\delta_{ij}^{\,2}\right) \qquad (3)$$

is a technique commonly used to reduce the dimensionality of the solution, dual to principal components analysis. With this aim, monotonous transformations are useful to get positive semi-definite (*psd*) $T^{(2)}(\Delta)$, since otherwise interpretation problems were supposed to raise. Despite the fact that none of these techniques would lead to *the* best solution in the sense of the stress, several advantages are in their favour, in particular i) if the solution has some optimal properties, it is a *global* optimum that is found, and ii) the solutions in different dimensions are encapsulated, so that the one in $k + p$ dimensions is the sum of the solution in k plus the best one in p in its orthogonal space. This, false in both *OS* and Kruskal solution, allows additive decomposition of distances, according to their principal components, that may be represented on the different orthogonal factors spanning the representation space. As a drawback, the reduction of dimensionality through *PCoA* projections of the deviations may lose the original monotonicity of the transformations.
In order to suggest a possible choice among directions and methods one should behave as follows, according to his aims:

1) If a *global minimum bias* solution is searched in fixed dimensions, *optimal scaling* techniques may be used. In the following, the method proposed by Bennani Dosse (1998) will be taken into account, as a reference for evaluating the quality of the others. It consists of the search of the global optimum of the smoothed stress function, obtained by the original stress through integral Gauss transformation (Coleman and Wu, 1996).
2) If the monotonicity of the solution in a reduced dimensional space is a primary goal, Kruskal's algorithm is the only possible, since the monotonicity of the transformation is strictly observed, at least in the full dimensional solution. This is not true when a reduced dimensional solution is searched by using *PCoA* on a matrix of transformed distances Δ, since projections are not monotonous transformations.
3) If a hierarchy of encapsulated solutions in several dimensions is searched and attention is drawn to distances decomposition according to principal axes, one may search, among the methods that may be submitted to *PCoA*, the one giving both minimum bias and maximum agreement to the true dimension of the problem.
In this paper we concentrate particularly on this last point, and we aim at testing the qualities of the different methods of transformation proposed in literature. Since the monotonicity of the solution is not ensured in reduced dimensions, we wanted to check Mardia (1978) argument in favour of Gower (1966) *MDS* solution, namely the *PCoA* of Torgerson's (1958) $T^{(2)}(D)$ matrix, limited to the real eigenvectors corresponding to positive eigenvalues. It was recently reexamined by Camiz and Tulli (1996), and Camiz (1999) tried to give some meaning to its imaginary part, showing that it contains information on the composition of the introduced bias.
A composite solution is suggested in literature (Gower, 1966; Seber, 1984), consisting of the use of a metric approximation in a reduced dimensional space, to be used as starting configuration for a following Kruskal algorithm. Compared with a random starting configuration, it may lead to a faster computation and maybe approach a better optimum (Kruskal, 1964a,b), albeit losing monotonicity. In our work, this method will be used to check to what extent the Mardia *MDS* solution may be further improved.

2. Monotone transformations

We shall consider the following six monotonous transformations:
2.1 Lingoes additive constant in d^2. Lingoes (1971) proposes to find a constant c such that $|c|$ be minimum and Δ, defined by $\delta_{ij}^2 = d_{ij}^2 + c$, be Euclidean. The problem is solved by $c = |\lambda_n|$, λ_n being the minimum negative eigenvalue of $T^{(2)}(\Delta)$. It is shown that this solution is in a *(n-2)*-dimensional space.
2.2 Cailliez additive constants in d. Cailliez (1983) found the solution to the old problem to find a c so that $\delta_{ij} = d_{ij} + c$ be Euclidean under the constraint that $|c|$ be minimum. Given the matrix T whose elements are

$$t_{ij} = \frac{1}{2} \left(d_{i\cdot} + d_{j\cdot} - d_{\cdot\cdot} - d_{ij} \right) \tag{4}$$

and the non-symmetric matrix B given by

$$B = \begin{pmatrix} 0 & 2T^{(2)} \\ -I & -4T \end{pmatrix} \tag{5}$$

the solution is given by B's largest eigenvalue λ_1. Cailliez shows in addition that $c' = |\lambda_n|$ is also a solution of the problem

2.2' find c such that $\delta_{ij} = |d_{ij} - c|$ be Euclidean with $|c|$ minimum.

Both transformations 2.2 et 2.2' find their rationale in rounding or measure errors.

2.3 Non-Metric Full Multidimensional Scaling. Vichi (1999) deals with the problem of finding the linear transformation $\delta_{ij} = bd_{ij} + a$ that minimizes the stress, under the condition that $T^{(2)}(\Delta)$ is *psd*. He shows that the solution is given by

$$b = \frac{c\ \Sigma_{ij} d_{ij} + \Sigma_{ij} d^2_{ij}}{c^2 n(n-1) + \Sigma_{ij} d^2_{ij} + 2c\ \Sigma_{ij} d_{ij}},\ a = c\,b, \tag{6}$$

where c is Cailliez (1983) λ_1 : in practice, it is a homothety of Cailliez solution.

2.4 Star distances. Le Calvé (1985) showed that any distance is a sum of a Euclidean and a star distances, the latter c_{ij} defined as $c_{ij} = c_i + c_j$, $c_{ii} = 0$. The problem to find a vector $\boldsymbol{c}$ such that $\delta_{ij}^2 = d_{ij}^2 + c_i + c_j$ allows infinitely many solutions. The most obvious is the half of Cailliez (1983) additive constant, that uniformly adjusts all interpoint distances, so that var($\boldsymbol{c}$) is constrained to be minimum. On the opposite, Bénasséni (1994) proposes to least modify only those distances responsible of non-Euclideanity. This means to impose the constraint that var($\boldsymbol{c}$) be maximum. We chose the solution proposed by Béninel *et al.* (1994), corresponding to var($\boldsymbol{c}$) = 1. They showed that, if $T^{(2)}$ matrix has only one negative eigenvalue, the solution is given by the square coordinates of points on its corresponding eigenvector, suitably normalized. In absence of a suitable algorithm for an optimal computation, we developed a method converging to a solution, by iterating on the orthogonal of the real part of $T^{(2)}$. We have no evidence of optimality, proved only in Béninel *et al.* (1994) case.

2.5 Euclidean index. Joly and Le Calvé (1986) showed that, for any dissimilarity, there exists a *Euclidean index* ε, such that $\delta_{ij} = d_{ij}^{\beta}$ is Euclidean for all $0 < \beta \leq \varepsilon \leq 1$. The space of representation is maximal (n-1) for $\beta < \varepsilon$. No analytical solution is known for this problem. We developed a convergent algorithm, based on iterative quadratic interpolation: once that two previous approximations give the last eigenvalue opposite in sign, we estimated a solution in the inner of their interval.

2.6 Mardia Principal Coordinates Analysis limited to real part. This is a metric *MDS*, since no monotonous transformation is performed. Since Torgerson's matrix contains the scalar products among the vectors connecting the centroid to all points, Gower's (1966) *PCoA* is a principal components analysis of the latter. If $T^{(2)}$ is not *psd*, Mardia (1978) strongly suggests to take into account the Euclidean approximation given by the subspace spanned by the real eigenvectors, corresponding to its positive eigenvalues, and shows the optimality of the solution considering the corresponding scalar products. Camiz and Tulli (1996) checked its consistency in respect to Lingoes (1971) additive constant, and Camiz (1999) discusses some possible interpretation of the imaginary part. This is analogous to relativity theory, where distance is not Euclidean, since the time component is negative. Here, we may state that every dissimilarity, in particular every distance, may be represented as the difference between two Euclidean squared distances. Their relative importance may be evaluated in terms of the share of the sums of corresponding eigenvalues to the Torgerson's matrix trace. In particular, the sum of positive eigenvalues overestimates the trace. In addition, the classical inspection of contribution of units to the negative eigenvectors composition is a helpful tool that may be used in practice, to identify which distances are furthest from Euclideanity.

3. An experiment: methods and results

In order to compare the different behaviour of these transformations, a 16-points plane grid was taken. In the experiment, still in progress on few hundreds of matrices, distances have been biased, by introducing a uniformly distributed random error. In this way, we are allowed to search for an underlying two-dimensional Euclidean structure. To build each matrix from the grid distances, an error percentage *p* was fixed and each Euclidean distance e_{ij} was biased according to the formula

$$d_{ij} = e_{ij}\,(\,1 + r\,p\,) \tag{7}$$

where *r* is a uniform random variate, whose range is (-.5, +.5). Here the results concerning only three such matrices will be discussed, two obtained with 10% error percentage and one with 20%.

The biased matrices, *D*, were transformed according to the said six transformations, giving Δ, stresses between *D* and Δ were computed, and the matrix $T^{(2)}(\Delta)$ was submitted to *PCoA*, to estimate the dimension of the solution. $T^{(2)}(D)$ was also submitted to *PCoA* to get Mardia *MDS* solution. In order to approach a minimum stress solution, we considered three other methods: i) we further investigated Mardia solution, by searching among all possible reduced dimensional subspaces the one where the stress of the points configuration is minimum; ii) Mardia two-dimensional solution was used as a starting configuration for Kruskal algorithm; and iii) Bennani Dosse *OS* in two dimensions was attempted. This allowed to compare how far from an optimal solution are those obtained by monotonous transformations. In order to compare the stresses of the different solutions, they were also normalized, through a homotethy of ratio $\|D\| / \|\Delta\|$. In this way the distances among points in the different solutions were comparable. All methods, but *OS*, were tested through a Fortran program written for the purpose; *OS* solutions were kindly given by M. Bennani Dosse and used by the program to compute the stresses. All computations were performed in double precision.

Table 1: *Comparison of stress (1st column) and normalized (2nd column) stress of the different solutions for three differently biased matrices. In parentheses the seed used for the error randomization routine.*

Method	*error 10% (23)*		*error 10% (45)*		*error 20% (12)*	
	stress	*normalized stress*	*stress*	*normalized stress*	*stress*	*normalized stress*
Lingoes	19.5090	10.150	17.9065	9.56	28.7132	12.85
Cailliez λ_1	40.3896	9.83	29.6135	8.01	65.9990	12.76
Cailliez λ_n	407.15529	31.203	406.16235	31.407	401.51686	30.15
Vichi *NMFMDS*	9.810	9.83	8.001	8.01	12.7151	12.76
Star distance	16.4472	8.74	14.3096	7.68	26.4154	12.45
Euclidean index	35.8204	10.486	31.2475	8.63	41.2422	13.24
Mardia *MDS*	5.361	3.84	4.286	3.00	7.535	5.3
Mardia best subspace	2.141	2.13	2.169	1.89	3.918	3.9
Kruskal 2 dimensions	65.8809	2.17	65.8064	1.91	64.9101	4.0
Bennani 2 dimensions	1.969	1.96	1.664	1.66	3.537	3.5

In table 1 both non-normalized and normalized stresses of used methods on three differently biased matrices are shown. Looking at the table, it is evident the interest in normalizing the solutions, since so doing the stress is always dramatically reduced: exceptions are both Vichi and Bennani Dosse methods, since stress minimization is part of the algorithm, and Mardia best subspace solution. Observing the columns containing normalized stresses, three different groups of solutions may be identified: the Cailliez λ_n solution has a stress around three-times bigger than the second group, in which the stress is improving from power method, to Lingoes, to both Cailliez λ_1 and Vichi *NMFMDS*, and to star distance. Vichi method actually minimizes the stress, that, normalized, is identical to that of Cailliez λ_1 solution, since solutions are homothetical. The group of minimum stress, around three-times smaller than the second, contains Mardia *MDS*, Kruskal algorithm on Mardia 2-dimensional solution, Mardia's subspace solution, and the best, Bennani Dosse *OS*.
In table 2 are shown both the suggested dimension and the explained variance of every method succeptible of *PCoA*. The suggestion is based on the big gap between the first few eigenvectors and all the remaining. This rule was not applied to Mardia best solutions, so that the actual dimensions are displayed, regardless of the suggestion: it must be noted that in case of 10% bias the best solutions are in three-dimensions, whereas for 20% bias the best solution is four-dimensional. Most solutions agree with the two-dimensions suggestion, keeping nearly the same amount of variance explained by the first two axes. Only Cailliez λ_n and Euclidean index fail: in Cailliez λ_n solution, 13 among 14 non-zero eigenvectors keep nearly the same importance, within 6 and 9% each, whereas the Euclidean index solution gives a first axis summarizing 50% of total variance and other two nearly equal, summarizing other 30-40%. Concerning explained variance, one may consider it as a measure of the relative importance of the chosen solution to the full-rank one: Mardia solutions are the highest, a sign of quality of the solution given in a correct number of dimensions; in this case, the amount of explained variance may overcome 100%, since the real solution overestimates original data, as Camiz (1999) discusses in detail. The other solutions improve from Lingoes,

Table 2: *Comparison of likely dimensions and corresponding percentage of explained variance of the solutions for three differently biased matrices. In parentheses the seed used for the error randomization routine.*

Method	*error 10% (23)*	*error 10% (45)*	*error 20% (12)*
Lingoes	2 71.21	2 74.14	2 59.63
Cailliez λ_1	2 70.63	2 77.37	2 58.63
Cailliez λ_n	13 99.79	13 99.88	13 99.51
Vichi *NMFMDS*	2 70.63	2 77.37	2 58.63
Star distance	2 75.34	2 78.98	2 62.28
Euclidean index	3 83.76	3 87.12	3 78.10
Mardia *MDS*	2 99.91	2 101.34	2 97.60
Mardia best solution	3 100.24	3 102.90	4 99.29

to the identical Cailliez λ_1 and Vichi, to star distance and Euclidean index, ranging from 58% to 79%.

4. Discussion and conclusions

The experimentation proposed in this paper aims at identifying the best methods for a Euclidean approximation of non-Euclidean distances, based on minimization of the normalized stress, and to the best reproduction of the original pattern of a two-dimensional grid. The idea of biased Euclidean distances allows to check for a two-dimensional solution.

The first few examples of a larger experimentation in progress agree in showing that the Cailliez λ_n solution is the further from both the biased data reality and the original non-biased pattern; the solution based on the Euclidean index seems competitive in terms of both stress and explained variance, but does not help in identifying a correct pattern, due to a suggested three-dimensional solution. Based on the same principle of adding constants to data, the following group of methods of medium adjustment does not show big differences among each other. The systematic improvement of the solutions from Lingoes to Cailliez λ_1 and Vichi to the star distance, is due to a finer tuning of the solution model, in particular in the latter case.

It is most interesting the Mardia *MDS* solution, since it approaches best of all others the minimum stress values of Bennani Dosse *OS*. In particular, its best solution in a reduced dimensional subspace is competitive with the improvement solution in two dimensions through Kruskal algorithm, with the advantage of both its speed and its easy identification of the true Euclidean pattern underlying the biased distances. Instead of being a method refused, because of the existence of negative eigenvalues, the new interpretation of the eigenvectors corresponding to the negative eigenvalues proposed by Camiz (1999) allows, in addition, to identify the units whose distances are most responsible of the deviation from Euclideanity, a precious information for the investigation relative to the reasons of such deviation and, if possible, its correction.

References

Bénasséni, J. (1984) Partial Additive Constant, *J. Stat. Comp. Simul.*, 49, 179-193.

Béninel, F., Qannari, A., Qannari, E.M. (1994) Distances à centre additives, *Revue de Recherche Opérationnelle (RAIRO)*, 4.

Bennani Dosse, M. (1998) Approximation quadratique d'une dissimilarité - Optimisation globale. *Actes des XXXèmes Journées de Statistique* de l'ASU-SFdS, vol. II, 37-38.

Cailliez, F. (1983) The Analytical Solution of the Additive Constant Problem, *Psychometrika*, 48, 2, 305-308.

Camiz, S. (1999) Comparison of Euclidean Approximatins of non-Euclidean Distances. in: *Classification and Data Analysis - Theory and Application*, M. Vichi and O. Opitz (eds.), Springer, Berlin, Studies in Classification, Data Analysis, and Knowledge Organization, pp. 139-146.

Camiz, S., Tulli, V. (1996) Relations between Negative Eigenvalues, Additive Constant and Bias of Euclidean Representation when Distances are not Euclidean. in: *Compstat 1996 short communications,* A. Prat and E. Ripoll (eds.), Barcelona, pp. 17-18.

Coleman, T, Wu, Z. (1996) Parallel Continuation-based Global Optimization for Molecular Conformation and Protein Folding, *Journal of Global Optimisation*, 8, 49-65.

Gower, J.C. (1966) Some Distance Properties of Latent Root and Vector Methods used in Multivariate Analysis, *Biometrika*, 53, 325-338.

Joly, S., Le Calvé, G. (1986) Étude des puissances d'une distance, *Statistique et Analyse des données*, 11, 3, 30-50.

Kruskal, J.B. (1964a) Multidimensional scaling by optimizing goodness of fit to a nonmetric hypothesis, *Psychometrika*, 29, 1-27.

Kruskal, J.B. (1964b) Nonmetric multidimensional scaling: a numerical method, *Psychometrika*, 29, 115-129.

Le Calvé, G. (1976) *Quelques remarques sur certains aspects de l'analyse factorielle*, Lab. Analyse des Données, Université de Rennes II, Cahier n. 2.

Lingoes, J.C. (1971) Some Boundary Conditions for a Monotone Analysis of Symmetric Matrices, *Psychometrika*, 36, 195-203.

Mardia, K.V. (1978) Some Properties of Classical Multi-Dimensional Scaling, *Communications in Statistics - Theory and Methods*, Series A, 7, 13, 1233-1241.

Seber, G.A.F. (1984) *Multivariate Observations*, J. Wiley, New York.

Torgerson, W.S. (1958) *Theory and Methods of Scaling*, J.Wiley, NewYork.

Vichi, M. (1999) Non-Metric Full Dimensional Scaling. in: *Classification and Data Analysis - Theory and Application*, M. Vichi and O. Opitz (eds.), Springer, Berlin, Studies in Classification, Data Analysis, and Knowledge Organization, pp. 139-146. pp. 163-170.

Comparing Capital Structure through Similarity Analysis: Evidence about two Industrial Districts[1]

Fabrizio Cipollini
Dipartimento di Statistica *G. Parenti*, viale Morgagni 59, 50134, Firenze
e-mail: cipollin@ds.unifi.it

Piero Ganugi
Università Cattolica-Dises, Piacenza
e-mail: piero.ganugi@tin.it

Abstract: We compare fundamental aspects of Capital Structure of Companies located in two different industrial districts. Starting from the Aggregate Balance Sheets of both districts, on whom ratios of Capital Structure are calculated, the emerging differences and similarities are investigated by means of a statistical analysis of similarity. The problem is worked out through three different approaches: a modified version of a simple Gini's index, a generalised distance between densities and the overlapping area between them. All these measures are used in non parametric sense: Gini's does not assume any particular distribution; we compute the other measures by using kernel estimates of the involved densities. Data are two 4-years panels of 843 and 187 Balance Sheets of Industrial textile Companies of Prato and Biella.

Keywords: Aggregate Balance Sheet, Gini's index of dissimilarity, Kernel density estimation, Index ratios, Measures of similarity.

1. Introduction

Industrial districts are characterised by a strong peculiarity of the production process which is the key of the high product differentiation by whom old and new markets are maintained and conquered. Then, even when characterised by the same industry, one industrial district is never the larger or smaller scale of another one. According to this, aim of the paper is to analyse in which measure the peculiarity of each district is reflected by Capital Structure of the Balance Sheets of its firms.

The paper has the following structure: in section 2 we present the data; in section 3 we describe the statistical tools used in the analysis; in the 4-th we show the results; in section 5 the resume.

[1] The whole work derive from a perfect collaboration of the authors. However subsection 3.3 is edited by Piero Ganugi, subsections 3.1, 3.2, 3.4 by Fabrizio Cipollini and the remaining by the two authors. We thanks Pietro Rigo and the referee for helpful comments.
The research has been financed by SIEL-Osservatorio dei Bilanci- Provincia of Prato.

2. The data

The analysis is developed on a 4-years panel Balance Sheets of the textile districts of Biella and Prato. The Companies Accounts are drawn by the *Cerved-Infocamere* files and represent the universe of the companies active in the two districts since 1993 to 1996: 187 for Biella and 843 for Prato.

We have considered main ratios of Capital Structure, constrained by two facts:

1) the ratios must be those of Aggregate Balance Sheets obtained on the basis of *Cerved* file (Ganugi-Gozzi 1997);
2) the ratios must satisfy some properties: in particular they must be definite[2] and monotonous[3] (Foster 1986).

Following these criteria we have this list of 6 ratios of Capital Structure:

Table 1: *The analysed ratios of Capital Structure*

x_5	= Manufacturing Equipments / Total Assets
x_8	= Circulating Medium / Total Assets
x_{21}	= Total Debt / Total Assets
x_{26}	= Capital and Reserves / Total Assets
x_{31}	= TFR (Retirement Fund) / Total Assets
x_{37}	= (Added Value – General Expenditures) / Total Assets

3. The methodology

The Aggregate ratios of Balance Sheets of Biella and Prato are the benchmark of our analysis. It may be worth to remember that the ratios reckoned up Aggregate Balance Sheet are ratios between sums. According to this feature, we may interpret these as weighted means of the same indexes with their denominators as weights. Consequently, ratios extracted by Aggregate Balance Sheets are sufficiently robust to outliers. Obviously the robustness of these means fails if the sector is characterised by high concentration.

---

[2] Because they are liable to have zero items at the denominator.

[3] This peculiarity is typical of the ratios whose denominator can be negative and it is particularly annoying for the interpretation of the ratio. To explain this aspect let $I = N/D$ a balance sheet ratio composed by the items N and D, where D can result both positive and negative (we exclude $D = 0$ on the basis of the previous note). To make easier the interpretation, let us suppose the ratio I negatively correlated with the health of the company: greater it is, worse it is the company with respect to the ratio. For simplicity let us suppose N equal to a fixed quantity.
Then, starting from *great* values of D, and choosing progressively smaller values of the same D, the ratio I enlarges tending to $+\infty$ for D tending to 0^+: the economic interpretation is that when D decreases the health of the company decreases consequently. If D diminish further on assuming small but negative values, I is negative but tends to enlarge: the health of the company improves.
This asymmetry of interpretation is particularly troublesome when automatic procedures are used in treating the data. A possible remedy is then to consider the reciprocal of the ratio, $I_1 = D/N$, but this is possible only when N is always positive, i.e. if N does not show the same problems of D. In the opposite case, it is necessary to give up the ratio, at least with the methodology of this paper.

The values of the ratios obtained by Aggregate Balance Sheets synthesize the aspects of Capital Structure about whom the two districts are different or, at the opposite, much alike. We investigate the different degrees of likeness of the six ratios considered by means of a statistical analysis of similarity. This analysis is performed with three different methods, illustrated in the two following subsections, to verify the stability of the results.

3.1. A modified Gini's index

The first level of analysis is a modified version of the Gini's index of dissimilarity (Leti 1983). Indicating the two data sets as *A* and *B*, our version of the index is

$$\Delta = \frac{1}{\sigma} \sum_{i=1}^{N} \left| F_{Ai} - F_{Bi} \right| \left(x_{i+1} - x_i \right) \tag{1}$$

where $x = (x_1, \ldots, x_N)$ is the ascending sorted union of the two datasets, respectively of N_A and N_B observations ($N = N_A + N_B$), F_{Ai} and F_{Bi} are the empirical cdf's corresponding to x_i, σ is the empirical standard deviation of x (we remember that the variance σ^2 *poole* those of *A* and *B*: $\sigma^2 = \dfrac{N_A \sigma_A^2 + N_B \sigma_B^2}{N_A + N_B}$).

The aim of this our change is to curb the impact of the different scale of ratios on the magnitude of the index[4]. However, even if useful, the transformation is not neutral respect to outliers, because the empirical standard deviation at the denominator is very sensitive to these observations.

3.2. A generalised affinity approach

The second approach we have chosen to measure the degree of dissimilarity between distributions has been started by Bhattacharyya and improved until now by Krzanowski-Marriot (1995). In this domain we have chosen L_2 distance between the square root of the densities (called Matusita's or Hellinger's distance), given by

$$\Delta_2 = \left[\int \left[f_A(x)^{1/2} - f_B(x)^{1/2} \right]^2 dx \right]^{1/2} = (2 - 2\rho)^{1/2} \tag{2}$$

where $\rho = \int \left[f_A(x) f_B(x) \right]^{1/2} dx$ is the *affinity* between the densities $f_A(x)$ and $f_B(x)$ relative to *A* and *B* (we remember that $\rho \in [0,1]$ by the Schwartz inequality).

However, not all the analysed ratios are absolutely continuos variables: some of these have points (essentially 0) with non-zero mass. Then we have considered a natural generalisation of ρ for mixture distributions, given by

[4] In the Gini's version of the index, the right side is not divided by σ. With our modification, if we linearly transform the variable x as $y = a+bx$ (a and b are constants), then $\Delta(a+bx) = \Delta(x)$ (the symbol $\Delta(.)$ indicates the index Δ computed on the variable between parenthesis).

$$\rho = (\alpha_A \alpha_B)^{1/2} \rho^{ac} + [(1-\alpha_A)(1-\alpha_B)]^{1/2} \rho^d \quad (3)$$

where $\rho^{ac} = \int [f_A^{ac}(x) f_B^{ac}(x)]^{1/2} dx$ is the affinity between the absolutely continuous components of the two distributions, $\rho^d = \sum [f_A^d(x) f_B^d(x)]^{1/2}$ is the affinity between the discrete components, α_A and α_B are the weights of the absolutely continuous components in the mixture representation of the two distributions[5].

3.3. The overlapping area

Finally, as a third method, for each ratio we have calculated the overlapping area (OA) between the different couples of distributions of the two districts:

$$OA = \int \min\{f_A(x), f_B(x)\} dx . \quad (4)$$

This is not a very usual measure of similarity but it enjoys an intuitive appealing because the more two distributions incline to overlap, the more they are similar: when the overlapping area is equal to 1, the distributions are identical. Furthermore, since in general $|a-b| = a+b-2\min\{a,b\}$, OA is lied to the L_1 distance between the two densities $f_A(x)$ and $f_B(x)$ by the relation[6]

$$\Delta_3 = \int |f_A(x) - f_B(x)| dx = 2(1 - OA) \quad (5)$$

Then we considered Δ_3 as measure of dissimilarity lied to the overlapping area. By the same motivations of the previous subsection, to compute this index we used again a decomposition in absolutely continuous and discrete components:

$$OA = \int \min\{\alpha_A f_A^{ac}(x), \alpha_B f_B^{ac}(x)\} dx + \sum \min\{(1-\alpha_A) f_A^d(x), (1-\alpha_B) f_B^d(x)\}. \quad (6)$$

---

[5] This generalisation can be easily deduced. Let $F_i = \alpha_i F_i^{ac} + (1-\alpha_i) F_i^d$ (i = A,B) the probability measure of the variables in the two sets A and B, where F_i^{ac} and F_i^d are respectively dominated by Lebesgue (m) and counting measure (c). Then, because F_i^{ac} and F_i^d are dominated by the measure $m+c$, we can define

$$\rho = \int \left[\left[\alpha_A f_A^{ac} + (1-\alpha_A) f_A^d\right] \left[\alpha_B f_B^{ac} + (1-\alpha_B) f_B^d\right] \right]^{1/2} d(m+c)$$

where f_i^{ac} and f_i^d are the densities with respect to Lebesgue and counting measure. Developing the previous formula we obtain (4).

[6] We thanks the referee for pointing us this.

3.4. The density estimation

In the formulas of the two previous sub-sections, the densities $f_A(x)$ and $f_B(x)$ are obviously unknown and must be estimated from the data.
About the study of the distributions, the prevailing procedure used by the scholars has been to remove outliers and, after transforming the ratios if necessary, to fit the same with simple and known distributions, as for example the normal (Frecka-Hopwood 1983, Watson 1990), the Stable Paretian (So 1987), the Pearson systems or the Ramberg-Schmeiser systems of distributions (Lau-Lau-Gribbin 1995).
However, the nature of Balance Sheet data does not support, as general rule, the choice of simple and known distributions, first of all the normal, for this kind of data. Among the 33 ratios obtained on the basis of our panels and analysed in Cipollini-Ganugi (1999), only 4 exhibit skewness lower than 1 and curtosis a bit different from 3. The non normality of Balance Sheet ratios looks then as a structural feature of ratios also in industrial districts.
In addition to this we have to face the issue of outliers. A strategy commonly used to force the fitting of known distributions to the data, lies in the exclusion of the firms which are outlier. If this strategy can be unavoidable when the aim of the research is the distress analysis developed by means of parametric discriminant models, it does not suit our case. Our aim is in fact the analysis of the distributions obtained by two different universes by comparing them.

By virtue of these considerations, we chose to estimate the density functions of the ratios using a nonparametric kernel approach, because of its simplicity and easiness of implementation. A kernel estimate of a density function $f(x)$ based on the data $(x_1,\ldots,x_n)$ is given by

$$\hat{f}(x) = \frac{1}{nh}\sum_{i=1}^{n} K\left(\frac{x - x_i}{h}\right) \tag{7}$$

where $K(.)$ is the kernel function and the parameter $h > 0$ is the bandwidth (Wand-Jones 1995, Devroye-Gyorfi 1985).
The main issue with this method is the selection of h. We made many experiments, with both simulated and our real data, to choose the kernel and the bandwidth computation method, trying to verify the sensitivity to various options. As expected, density estimates turned out to be substantially insensitive to the kind of kernel (even if this is important in the estimation of derivatives), whereas the value of h demonstrates a bigger influence. We chose the Gaussian kernel, computing h with both Least Squares Cross-Validation (LSCV) method and Oversmoothed bandwidth Selection (OS) method (Wand-Jones 1995, ch. 3) to compare the obtained h and the consequent degree of smoothing (sometimes LSCV gave us undersmoothed estimates)[7].

[7] In our simulations, adaptive kernel estimates showed often better performances than fixed kernel, especially with very asymmetric distributions (as often happens with balance sheet ratios).

However we point out that the distributions of each ratio in the two districts have usually very similar characteristics, both in range and shape. Consequently the mentioned choices are not so important as in other contexts, because being interested only to distance between these estimates, overmoothing or undersmoothing affects usually both distribution in similar manner, so that results based on different options are usually very close.
Finally, for the ratios with points of non-zero mass (x_5 and x_{31} in our case - see sections 2, 3.2 and 3.3) we estimated the weights of the two components of the distribution (discrete and absolutely continuous) by their empirical relative frequency.

4. The results

In table 2 we show the ratios obtained from aggregate Balance Sheets of the two textile districts. From this table we evidence the following results:

1) remarkable differences for Leverage, Net Capital (with progressive slight reduction) and TFR on Total Assets (respectively x_{21}, x_{26} and x_{31} ratios) in all the four years;
2) during the same period a sharp reduction of the differences between the two districts about the structure of Total Assets, i.e in Circulating Medium/Total Assets (x_8) and in Manufacturing Equipments/ Total Assets (x_5);
3) an high similarity in the ratio Added Value/Total Assets (x_{37}), so that the two districts show substantially the same ability in creating income[8].

We have reported the dissimilarity analysis in table 3[9].
This analysis corroborates only partially evidence 1). In fact, during the four years the three indexes increase conspicuously about Leverage (x_{21}) and Net Capital/Total Assets (x_{26}).
At the contrary, for 2) it is fully confirmed the reduction of difference about the structure of Assets, i.e Circulating Medium/Total Assets (x_8) and Manufacturing Equipments/Total Assets (x_5), whereas the dissimilarity for x_{31} remains substantially constant.
About Added Value/ Total Assets (x_{37}), Gini's give different answers with respect to the other measures: the ratio decreases according to the first and increases slightly according to density based indexes. The Aggregate Balance Sheets agree with this second kind of measures. It is worth recording that the ratio considered here does not have points with non-zero mass. Then the different behaviour of the dissimilarity measures computed on it is not due to this feature.

However we don't used this approach, following the advices of a referee that tell us about terrible performances of this kind of kernel in the applications.

[8] We note however that it is trivial a certain difference in the ratio Added Value/Revenues given the different degree of decentralization of industrial structure in the two districts. Because of this we not reported the analysis of this ratio.

[9] All these methods are implemented with code written in GAUSS language. The involved integrals are computed using the *intquadl* procedure.

Table 2: *Ratios from Aggregate Balance Sheets*

	Prato				*Biella*			
Ratios	*1993*	*1994*	*1995*	*1996*	*1993*	*1994*	*1995*	*1996*
x_5	0.17	0.16	0.17	0.18	0.23	0.20	0.20	0.20
x_8	0.77	0.79	0.75	0.73	0.68	0.71	0.72	0.71
x_{21}	0.69	0.71	0.70	0.69	0.64	0.66	0.66	0.64
x_{26}	0.24	0.23	0.24	0.26	0.28	0.26	0.27	0.28
x_{31}	0.038	0.035	0.037	0.037	0.053	0.049	0.049	0.050
x_{37}	0.35	0.33	0.34	0.29	0.33	0.32	0.34	0.30

Table 3: *Dissimilarity indexes*

	Modified Gini's index				***Generalized L_2 distance*** *(affinity based measure)*				***Generalized L_1 distance*** *(overlapping area based measure)*			
Ratios	*1993*	*1994*	*1995*	*1996*	*1993*	*1994*	*1995*	*1996*	*1993*	*1994*	*1995*	*1996*
x_5	0.31	0.24	0.21	0.25	0.29	0.29	0.27	0.27	0.50	0.51	0.48	0.48
x_8	0.36	0.30	0.21	0.26	0.30	0.28	0.23	0.22	0.49	0.50	0.41	0.38
x_{21}	0.25	0.24	0.25	0.30	0.31	0.18	0.42	0.36	0.30	0.27	0.48	0.42
x_{26}	0.21	0.18	0.22	0.29	0.21	0.25	0.21	0.27	0.25	0.35	0.29	0.36
x_{31}	0.24	0.23	0.20	0.21	0.31	0.30	0.32	0.33	0.46	0.43	0.46	0.47
x_{37}	0.19	0.15	0.16	0.16	0.19	0.19	0.22	0.20	0.29	0.28	0.34	0.33

5. Conclusions

We have analysed the Capital Structure of the companies of two Italian Industrial Districts. Data are two 4-years panels of Balance Sheets of Textile Companies in Prato and Biella.

The benchmark of the analysis is Aggregate Balance Sheet whose results are compared with three different approaches to verify the stability of the results: a modified version of a Gini's index, a generalised measure of affinity between density estimates and the overlapping area between them.

The analysis of similarity does not confirm the constancy of the dissimilarities between the two districts about Leverage and Capital Structure evidenced by Aggregate Balance Sheets pointing out an increase of the difference in the years.

At the contrary, Similarity analysis and Aggregate Balance Sheets overlap in showing a remarkable reduction of difference in the structure of Total Assets, i.e Manufacturing Equipments /Total Assets and Circulating Medium /Total Assets.

About the capacity of each district in creating Added Value for unit of Assets, Gini's give different answers with respect to the other measures. The result supplied by Aggregate Balance Sheets agree with the density based measures.

Based on the evidence supplied in the paper we can remark that:

- differences among ratios must be evaluated not only on the basis of synthetic measures of their distributions but also by means of a more analytical approach, of whom similarity analysis is a good example;
- the choice of the measure of similarity is not completely neutral for the evaluation of differences among ratios, even if usually these measures agree.

References

Cipollini, F., Ganugi, P. (1999). La selezione degli indicatori di bilancio per la classificazione delle società tessili mediante l'analisi della transvariazione, in Ganugi, P., ed., *Ricerche quantitative per la politica economica nei distretti industriali,* Franco Angeli.

Devroye, L., L. Gyorfi (1985). *Nonparametric Density Estimation: the L_1 View.* Wiley, N.Y.

Ferrero, G., Dezzani, F., Pisoni, P., Puddu, L. (1994). *Le analisi di bilancio*, Giuffrè Editore, Milano.

Foster, G. (1986), *Financial Statement Analysis*, 2-nd ed., Prentice Hall, New York.

Frecka, T. J., Hopwood, W. S. (1983). The effects of Outlier on the Cross-Sectional Distributional Properties of Financial Ratios, *The Accounting Review*, vol. 78, 115-128.

Ganugi, P., Gozzi, G.(1997). La previsione delle insolvenze in presenza di limitata informazione contabile, Atti del convegno SIS Torino 1997 Statistica per le imprese, 155-162.

Krzanowski, W.J. and Marriott, F. H. C. (1994). Multivariate Analysis, Edward Arnold, London.

Lau, H.S., Lau, A.H.L., Gribbin, D.W. (1995). On Modelling Cross Sectional Distributions of Financial Ratios, *Journal of Business Finance and Accounting*, vol. 22, 521-549.

Leti, G. (1983). *Statistica descrittiva.* Il Mulino, Bologna.

Marbach, G. (1989). *Previsioni e misure di efficienza aziendale. Una mentalità statistica per l'impresa*, Nuova Italia, Roma.

Silverman, B.W. (1986). *Density estimation for statistics and Data Analysis.* Chapman and Hall, London.

So, J. (1987). Some empirical evidence on outliers and the non normal distribution of financial ratios, *Journal of Business Finance and Accounting,* vol.14, 483-495.

Wand, M.C., Jones, M.P. (1995). *Kernel Smoothing.* Chapman and Hall, London.

Watson, C. (1990). Multivariate distributional properties, outliers and transformations of financial ratios, *The Accounting Review*, vol. 85, 682-696.

The Geometric Approach to the Comparison of Multivariate Time Trajectories

Renato Coppi, Pierpaolo D'Urso
Dipartimento di Statistica, Probabilità e Statistiche Applicate,
Università di Roma "*La Sapienza*"

Abstract: To compare time trajectories different approaches might be envisaged. In this paper, considering the *geometric* approach, several dissimilarity measures between time trajectories are taken into account. An empirical comparison of the dissimilarity measures is also shown.

Keywords: Time Array, Time Trajectories, Geometric Approach, Dissimilarity Measures.

1. Three-way time array and time trajectories

Let $\mathbf{X} \equiv \{x_{ijt}\}$, $i = 1, I$ (objects), $j = 1, J$ (quantitative variables), $t = 1, T$ (times), be a (*three-way*) *time array*. It is possible to study the *time array* $\mathbf{X}$ depending on how the indices i, j, t are aggregated. Let $\mathbf{X}(I,J,T)$ be the *time array* $\mathbf{X}$ where the three indices are not aggregated. If we combine the indices of sets I and T in such a way as to obtain a two way matrix, where the rows are defined by combination of I and T and the columns are defined by J, we have the bidimensional form $\mathbf{X}(IT,J)$ of $\mathbf{X}$. Likewise, we obtain the other bidimensional forms of $\mathbf{X}$: $\mathbf{X}(JT,I)$ and $\mathbf{X}(IJ,T)$. In this paper, $\mathbf{X}(IT,J)$ is considered.

Let $\mathbf{X}_t$ be the $(I{\times}J)$-matrix at time t: $\mathbf{X}_t \equiv (\mathbf{x}_{1t},...\mathbf{x}_{jt},...\mathbf{x}_{Jt})$, with $\mathbf{x}_{jt} = (x_{1jt},...x_{ijt},...x_{Ijt})'$; let $\mathbf{X}_i$ be the $(T{\times}J)$-matrix, for object i: $\mathbf{X}_i \equiv (\mathbf{x}_{i1},..., \mathbf{x}_{it},..., \mathbf{x}_{iT})$, with $\mathbf{x}_{it} = (x_{i1t},...,x_{ijt},...,x_{iJt})'$. Then, it is possible to analyze the bidimensional form $\mathbf{X}(IT,J)$ of $\mathbf{X}$ in two ways: $\mathbf{X}(IT,J) \equiv \{\mathbf{X}_t\}_{t=1,T}$ and $\mathbf{X}(IT,J) \equiv \{\mathbf{X}_i\}_{i=1,I}$.

Let R^{J+1} be the vectorial space, where the axes are referred to the J variables and time. In this space we represent each object i by means of a vector, for each t: $\mathbf{u}'_{it} = (x_{i1t},..., x_{ijt},..., x_{iJt}, t)$. Fixed t, the scatter of the points (vectors), $N_I(t) \equiv \{\mathbf{u}_{it}\}_{i=1,I}$, represents the matrix $\mathbf{X}_t$. For each t, the scatters $N_I(t)$ are placed on hyperplanes parallel to the sub-space R^J. Fixed i, $N_T(i) \equiv \{\mathbf{u}_{it}\}_{t=1,T}$ represents $\mathbf{X}_i$. This scatter describes the *time trajectory* of object i across the time and $\{N_T(i) \equiv \{\mathbf{u}_{it}\}_{t=1,T}\}_{i=1,I}$ represents the set of the time trajectories. Each time trajectory $N_T(i)$ crosses the T hyperplanes parallel to R^J.

2. The geometric approach

We classify *time trajectories* in several typologies: *quantitative*, *qualitative* or *mixed* (with quantitative and/or qualitative variables); *univariate* or *multivariate* (with one or more variables); *discrete* or *continuous* (with discrete or continuous variables); *short* or *long* (with few or many times); *synchronous* or *asynchronous* (with values of variables observed at the same or different times); with *constant* or *variable time support* (with equidistant or non equidistant time intervals).
To compare time trajectories there are different approaches: *inferential approaches* and *exploratory approaches*, e.g. the *geometric approach*.
In this paper, we analyze the *dissimilarity measures*, relevant to the *geometric approach*, for *quantitative* (single numerical values) *discrete synchronous multivariate time trajectories* with *constant time support*. We introduce an *Euclidean metric E* in the space R^{J+1} such that: $d(\mathbf{x}_{it},\mathbf{x}_{it})=0$, $d(\mathbf{x}_{it},\mathbf{x}_{lt})=d(\mathbf{x}_{lt},\mathbf{x}_{it})$, $d(\mathbf{x}_{it},\mathbf{x}_{lt})\geq 0$, $d(\mathbf{x}_{it},\mathbf{x}_{lt})+d(\mathbf{x}_{ht},\mathbf{x}_{lt})\geq d(\mathbf{x}_{it},\mathbf{x}_{ht})$ $(i,l,h=1,I,\ i\neq l\neq h)$, where $d(\mathbf{x}_{it},\mathbf{x}_{lt})\equiv\|\mathbf{x}_{it}-\mathbf{x}_{lt}\|$ indicates the Euclidean distance between objects i and l, at time t. Before computing these distances, the variables have to be normalized or standardized.
In the geometric approach, in order to compare time trajectories we consider the geometric-structural characteristics of the time trajectories: *location* (*position*), *slope*, *concavity* and *convexity*. In particular, we distinguish:

1. *instantaneous* (*cross sectional*) *dissimilarity measures*,
2. *evolutive* (*longitudinal*) *dissimilarity measures*,
3. *mixed dissimilarity measures*.

1. To define the *instantaneous dissimilarity*, the *location* (*position*) of time trajectories is considered (Carlier 1986, 1991; D'Urso, Vichi, 1998, if $w_t=1$):

$$d_1^2(i,l)=\sum_{t=1}^{T}\|\mathbf{x}_{it}-\mathbf{x}_{lt}\|^2 w_t\ ,$$

where w_t is a weight at time t.
2. *Slope*, *concavity* and *convexity* of the time trajectories are taken into account in order to define *evolutive* (*longitudinal*) *dissimilarity measures*.
An evolutive dissimilarity is (Carlier 1986, 1991; D'Urso, Vichi, 1998, if $w_{t-1,t}$=1):

$$d_2^2(i,l)=\sum_{t=2}^{T}\|\mathbf{v}_{it-1,t}-\mathbf{v}_{lt-1,t}\|^2 w_{t-1,t}\,,$$

where $w_{t-1,t}$ is a weight assigned to time interval $[t-1,t]$, $\mathbf{v}_{it-1,t}=(\mathbf{x}_{it}-\mathbf{x}_{it-1})$ and $\mathbf{v}_{lt-1,t}=(\mathbf{x}_{lt}-\mathbf{x}_{lt-1})$ are the *velocities* of i-th and l-th time trajectory[1]. In R^2, the

---

[1] In fact the velocity definition of the i-th time trajectory, referred to time interval $[t-1,t]$, is $\mathbf{v}_{it-1,t}=\dfrac{(\mathbf{x}_{it}-\mathbf{x}_{it-1})}{t-(t-1)}=(\mathbf{x}_{it}-\mathbf{x}_{it-1})$ (D'Urso, Vichi, 1998).

velocity of each segment of the trajectory is the *slope* of the straight line passing through it: if the velocity is negative (positive) the slope will be negative (positive) and the angle made by each segment of the trajectory with the positive direction of the t-axis will be obtuse (acute).
Another evolutive dissimilarity measure, which is a weighted Euclidean version of a squared distance proposed by D'Urso and Vichi (1998), is:

$$d_3^2(i,l) = \sum_{t=3}^{T} \left\| \mathbf{a}_{it-2,t} - \mathbf{a}_{lt-2,t} \right\|^2 w_{t-2,t},$$

where $w_{t-2,t}$ is the weight for interval $[t-2,t]$, $\mathbf{a}_{it-2,t} = \frac{1}{2}(\mathbf{v}_{it-1,t} - \mathbf{v}_{it-2,t-1})$ and $\mathbf{a}_{lt-2,t} = \frac{1}{2}(\mathbf{v}_{lt-1,t} - \mathbf{v}_{lt-2,t-1})$ are the *accelerations* of i-th and l-th time trajectories[2]. Geometrically, the acceleration of each pair of segments of trajectory represents its convexity or concavity: if the acceleration is positive (negative) the trajectory of the two segments is convex (concave).
3. The *mixed dissimilarities* are a compromise between location and evolution of the time trajectories. The following mixed dissimilarity is considered (D'Urso, Vichi, 1998, if $w_t = w_{t-1,t} = w_{t-2,t} = 1$):

$$d_{m_1}(i,l) = \sum_{r=1}^{3} \alpha_r d_r(i,l),$$

where α_r, $r = 1,2,3$, are suitable positive weights, indicating the contribution of each trajectory's characteristic to determining $d_{m_1}(i,l)$. If $d_3(i,l) = 0$, the mixed dissimilarity proposed by Carlier (1986, 1991), called $d_{m_2}(i,l)$, is found.
There are different criteria to compute the weights. It is possible to define the weights subjectively (Carlier, 1986), but other methods could be considered. A procedure has been suggested by D'Urso and Vichi (1998) and other methods by Gordon (1990) in a different context. These can be taken into account in the present study, when an appraisal of the dissimilarities by an assessor is available. These methods are illustrated in the following sections.
3.1 *Max Variance*: considering $d_{m_1}(i,l)$, the weights α_r are determined so that the variance of the mixed dissimilarity is maximized as shown below. Let $\mathbf{D}_{m_1} = [d_{m_1}(i,l)]$ be the $(I \times I)$-matrix of mixed dissimilarities between i-th and l-th time trajectories; let $\mathbf{D}_r = [d_r(i,l)]$ be the $(I \times I)$-matrices obtained by considering the dissimilarities between i-th and l-th trajectories for the three, $r = 1,2,3$, features of

---

[2] This result is due to the acceleration definition of the i-th time trajectory, for time interval $[t-2,t]$, that is $\mathbf{a}_{it-2,t} = \dfrac{(\mathbf{v}_{it-1,t} - \mathbf{v}_{it-2,t-1})}{t-(t-2)} = \frac{1}{2}(\mathbf{v}_{it-1,t} - \mathbf{v}_{it-2,t-1})$ (D'Urso, Vichi, 1998).

each trajectory. The matrices $\mathbf{D}_{m_1}$ and $\mathbf{D}_r$ may be represented as vectors defined by the elements of the triangle below (above) the diagonal of the matrices: $\overline{vec}(\mathbf{D}_{m_1})$ and $\overline{vec}(\mathbf{D}_r)$; therefore $\overline{vec}(\mathbf{D}_{m_1})=\sum_{r=1}^{3}\alpha_r\overline{vec}(\mathbf{D}_r)$. To compute the weights of the mixed dissimilarities, $\overline{vec}(\mathbf{D}_{m_1})$, the following problem, with respect to $\alpha=(\alpha_1,\alpha_2,\alpha_3)'$, has to be solved:

$$\begin{cases} max\ \ var\,(\overline{vec}\,(\mathbf{D}_{m_1})) \\ \alpha'\alpha = 1. \end{cases}$$

3.2 *Least Squares Multiple Linear Regression*: if it is available $\delta(i,l)$, denoting a measure of the dissimilarity between i-th and l-th time trajectories, provided by an assessor, where $\delta(i,l)=\delta(l,i), 1\le l<i\le I$, it is possible to set: $\delta(i,l)\cong c d_{m_1}(i,l)$. This model can be fitted by the least squares multiple linear regression method (without intercept) (Gordon, 1990).

3.3 *Least Squares Monotone Regression*: it is supposed that $\delta(i,l)$ and $d_{m_1}(i,l)$ are monotonically related: $\delta(i,l)\cong M(d_{m_1}(i,l))$ (M denotes a monotone function). This model is fitted using Kruskal's method of least squares monotone regression, with primary treatment of ties (Gordon, 1990). Values $\hat{d}_{m_1}(i,l)$ are sought that minimize:

$$\left[\frac{\sum(d_{m_1}(i,l)-\hat{d}_{m_1}(i,l))^2}{\sum(d_{m_1}(i,l))^2}\right]^{1/2} \quad \text{or, as we alternatively suggest,} \quad \left[\frac{\sum(d_{m_1}(i,l)-\hat{d}_{m_1}(i,l))^2}{\sum(d_{m_1}(i,l)-\bar{d}_{m_1})^2}\right]^{1/2}$$

subject to: $\delta(i,l)<\delta(l,m)\rightarrow\hat{d}_{m_1}(i,l)\le\hat{d}_{m_1}(l,m)$, where the summations are over all relevant pairs of subscripts and $\bar{d}_{m_1}$ is the mean of the dissimilarities $d_{m_1}(i,l)$, $1\le l<i\le I$. In this two criteria the only difference regards the normalizing factor.

To define a mixed dissimilarity another procedure has been proposed (Saporta, Lavallard, 1996). Various problems have been pointed out by Saporta and Lavallard concerning $d_{m_2}(i,l)$. The measure $d_{m_2}(i,l)$ must be calculated considering T position-coordinates and (T-1) evolution (velocities)-coordinates, while each trajectory is defined by only T coordinates. Furthermore, the position and evolution (velocity) affect in a different way the $d_{m_2}(i,l)$. Due to this problem a new dissimilarity, which is based on T coordinates only, has been advocated by Saporta and Lavallard. These coordinates are chosen in order to obtain the same effect of position and evolution (velocity) in the dissimilarity. Firstly two definitions are considered.

Definition 1: A configuration is a set of T binary numbers $\{a_1,\dots,a_T\}$ such that $a_1=1$ (the first coordinate is always a *position-coordinate*) and, $\forall t=2,T$, $a_t=1$ ($a_t=0$) if a *position-coordinate* (*evolution-coordinate*) at time t is considered.

Definition 2: For the configuration $\{a_1,...,a_T\}$, I_T represents the *total inertia* which is decomposable as follows: $I_T = I_{T_1} + I_{T_2}$, where $I_{T_1} = \sum_{i=1}^{I} \sum_{\{t:\, a_t=1\}} \| \mathbf{x}_{it} - \bar{\mathbf{x}}_t \|^2 p_i$, $I_{T_2} = \sum_{i=1}^{I} \sum_{\{t:\, a_t=0\}} \| (\mathbf{x}_{it} - \mathbf{x}_{it-1}) - (\bar{\mathbf{x}}_t - \bar{\mathbf{x}}_{t-1}) \|^2 p_i$ (p_i=weight for i).

The procedure consists of two steps:

1. the *optimal configuration* $\{a_1,..., a_T\}$ is determined by considering the criterion:

$$\min_{\{a_1,...,a_T\}} \left| I_{T_1} - I_{T_2} \right|;$$

2. the mixed dissimilarity between time trajectories is calculated as follows:

$$d^2_{m_3}(i,l) = \sum_{\{t:\, a_t=1\}} \| \mathbf{x}_{it} - \mathbf{x}_{lt} \|^2 + \sum_{\{t:\, a_t=0\}} \| (\mathbf{x}_{it} - \mathbf{x}_{it-1}) - (\mathbf{x}_{lt} - \mathbf{x}_{lt-1}) \|^2 .$$

Since $d_{m_1}(i,l)$ suffers from the same above mentioned problems, we propose an extension of the Saporta and Lavallard procedure in which the three dissimilarities (location, velocity and acceleration dissimilarity) are considered simultaneously:

$$d^2_{m_4}(i,l) = \sum_{\{t:\, a_t=2\}} \| \mathbf{x}_{it} - \mathbf{x}_{lt} \|^2 + \sum_{\{t:\, a_t=1\}} \| \mathbf{v}_{it} - \mathbf{v}_{lt} \|^2 + \sum_{\{t:\, a_t=0\}} \| \mathbf{a}_{it} - \mathbf{a}_{lt} \|^2 ,$$

where a_t=2 if a position-coordinate at time t is considered and so forth. The optimal configuration $\{a_1,...a_T\}$ is determined by minimizing the three way squared distance (Joly, Le Calvé, 1995) among the following types of inertia: $I_{T_1} = \sum_{i=1}^{I} \sum_{\{t:\, a_t=2\}} \| \mathbf{x}_{it} - \bar{\mathbf{x}}_t \|^2 p_i$, defined considering the position of the objects with respect to the objects' barycentre, $I_{T_2} = \sum_{i=1}^{I} \sum_{\{t:\, a_t=1\}} \| \mathbf{v}_{it} - \bar{\mathbf{v}}_t \|^2 p_i$, defined taking into account the difference between the objects and the barycentre velocity, $I_{T_3} = \sum_{i=1}^{I} \sum_{\{t:\, a_t=0\}} \| \mathbf{a}_{it} - \bar{\mathbf{a}}_t \|^2 p_i$, defined taking into account the difference between the objects and the barycentre acceleration. Therefore, the dissimilarity $d^2_{m_4}(i,l)$ is determined on the basis of the criterion:

$$\min_{\{a_1,...,a_T\}} d^2(I_{T_1}, I_{T_2}, I_{T_3})$$

where $d^2(I_{T_1}, I_{T_2}, I_{T_3}) = d^2(I_{T_1}, I_{T_2}) + d^2(I_{T_1}, I_{T_3}) + d^2(I_{T_2}, I_{T_3})$ is the three way squared distance among I_{T_1}, I_{T_2} and I_{T_3}.

A dissimilarity measure between time trajectories, which takes into account the geometric structure of the trajectories, is suggested by D'Urso and Gastaldi (1999). This dissimilarity is based on the *polygonal representation* of each univariate time trajectory. In particular, the polygonal representation of a univariate time trajectory is:

$$x_{it'} = \sum_{t=1}^{T} b_{it} \left| t' - t \right| \quad t' = 1,T$$

where b_{it} (i=1,I; t=1,T) are the *polygonal coefficients*.

The solutions are: $b_{i1} = \frac{1}{2}(x_{i2} - x_{i1} + A)$, $b_{it} = \frac{1}{2}[(x_{it+1} - x_{it}) - (x_{it} - x_{it-1})]$ (t=1,T-2) and $b_{iT} = \frac{1}{2}(A - x_{iT} + x_{iT-1})$, where $A = \frac{x_{i1} + x_{iT}}{T-1}$. The polygonal coefficients b_{it} represent the semi-difference of the velocities of the univariate time trajectories in the different time intervals.

Thus, the dissimilarity between the *polygonal coefficients* of each multivariate time trajectory is examined in the following way. Let

$$\mathbf{X}_i = \mathbf{C}\,\mathbf{B}_i$$

be the polygonal representation of the i-th multivariate time trajectory, where

$$\mathbf{C} = \begin{pmatrix} 0 & |1-2| & \cdots & |1-T| \\ |2-1| & 0 & \cdots & |2-T| \\ \vdots & \vdots & \ddots & \vdots \\ |T-1| & |T-2| & \cdots & 0 \end{pmatrix} = \begin{pmatrix} 0 & 1 & \cdots & T-1 \\ 1 & 0 & \cdots & T-2 \\ \vdots & \vdots & \ddots & \vdots \\ T-1 & T-2 & \cdots & 0 \end{pmatrix}$$

is an invertible matrix and

$$\mathbf{B}_i = \begin{pmatrix} b_{i11} & \cdots & b_{ij1} & \cdots & b_{iJ1} \\ \vdots & & \vdots & & \vdots \\ b_{i1t} & \cdots & b_{ijt} & \cdots & b_{iJt} \\ \vdots & & \vdots & & \vdots \\ b_{i1T} & \cdots & b_{ijT} & \cdots & b_{iJT} \end{pmatrix} = \begin{pmatrix} \mathbf{b}'_{i1} \\ \vdots \\ \mathbf{b}'_{it} \\ \vdots \\ \mathbf{b}'_{iT} \end{pmatrix}$$

is the matrix of the coefficients of the polygonal representation of the i-th trajectory. Therefore, the polygonal coefficients can be obtained as follows:

$$\mathbf{B}_i = \mathbf{C}^{-1}\mathbf{X}_i$$

and then the dissimilarity based on the polygonal representation is provided by:

$$d_4^2(i,l) = \sum_{t=1}^{T} \| \mathbf{b}_{it} - \mathbf{b}_{lt} \|^2 .$$

This dissimilarity allows us to quantify the divergence between the evolutive aspects of the time trajectories which are embodied in the polygonal coefficients. Finally, we would suggest a dissimilarity which considers the *area* between each pair of time trajectories. In the multivariate case, this dissimilarity summarizes the areas between trajectories represented onto each of the J coordinate bidimensional sub-spaces, whose axes refer to the time and each one of the J variables.

3. An illustrative example for comparative purposes

The various dissimilarities have been applied to the same time array in order to compare their performances. In order to check these performances the consumption behaviour (described by 2 variables: daily number of food items bought and average daily expediture) of 10 individuals in a supermarket at 5 different times has been observed. In this connection, we have calculated the following dissimilarities between all pairs of time trajectories: $\mathbf{D}_1$ (location dissimilarity), $\mathbf{D}_2$ (velocity dissimilarity), $\mathbf{D}_3$ (acceleration dissimilarity), $\mathbf{D}_4$ (mixed (location + velocity) dissimilarity), $\mathbf{D}_5$ (mixed (location + velocity + acceleration) dissimilarity), $\mathbf{D}_6$ (compromise dissimilarity proposed by Saporta and Lavallard), $\mathbf{D}_7$ (compromise dissimilarity proposed by Coppi and D'Urso), $\mathbf{D}_8$ (polygonal dissimilarity). These dissimilarity matrices have been compared by considering the following dissimilarity measure between matrices:

$$\mathbf{D}^* = \| \mathbf{D}_r - \mathbf{D}_s \| \quad r,s = 1,\ldots,8, \quad r \neq s .$$

The dissimilarity matrix $\mathbf{D}^*$, obtained in the experiment, is shown in Table 1.

Table 1: *Dissimilarity Matrix* $\mathbf{D}^*$

	$\mathbf{D}_1$	$\mathbf{D}_2$	$\mathbf{D}_3$	$\mathbf{D}_4$	$\mathbf{D}_5$	$\mathbf{D}_6$	$\mathbf{D}_7$	$\mathbf{D}_8$
$\mathbf{D}_1$		4.01	3.7	24.97	14.88	30.01	17.94	20.78
$\mathbf{D}_2$			1.9	25.89	15.56	30.99	18.71	21.64
$\mathbf{D}_3$				26.21	15.78	31.32	18.99	21.93
$\mathbf{D}_4$					10.92	5.75	7.26	5.55
$\mathbf{D}_5$						16.37	4.15	6.19
$\mathbf{D}_6$							12.57	11.05
$\mathbf{D}_7$								4.24
$\mathbf{D}_8$								

The dendrogram obtained by the average linkage method is shown in Figure 1.

Figure 1: *Dendrogram*

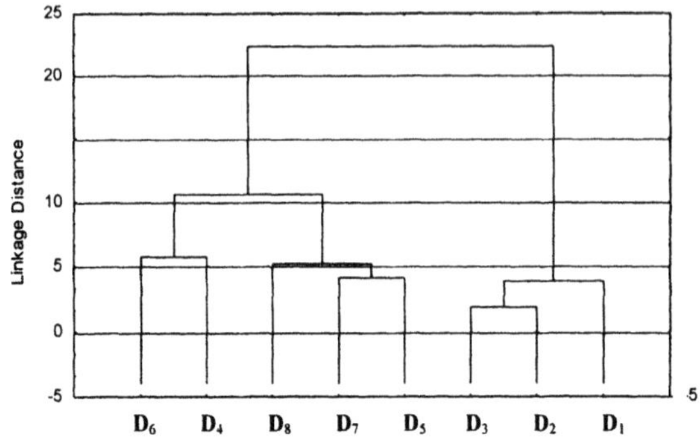

In this situation, $\mathbf{D}_8$, $\mathbf{D}_7$ and $\mathbf{D}_5$ present similar performances, as shown in Figure 1. The other homogeneous clusters are ($\mathbf{D}_6$, $\mathbf{D}_4$) and ($\mathbf{D}_3$, $\mathbf{D}_2$, $\mathbf{D}_1$).
This result may suggest that $\mathbf{D}_8$ is the preferable measure, being more parsimonious than $\mathbf{D}_7$ and $\mathbf{D}_5$. In fact, none of the criteria, necessary to specify $\mathbf{D}_7$ e $\mathbf{D}_5$, is required to calculate $\mathbf{D}_8$.

4. Conclusions

In this paper, several dissimilarity measures relevant to the *geometric* approach to the comparison of time trajectories have been illustrated. These dissimilarities can be used in dynamic cluster analysis (Carlier 1986, 1991; D'Urso, Vichi, 1998; Kosmelj, Batagelj, 1990) and in other multiway procedures (Coppi, 1998).

References

Carlier, A. (1986) Factor Analysis of Evolution and Cluster Methods on Trajectories, *COMPSTAT*, Physica-Verlag, 140-145.

Carlier, A. (1991) About Distances for Clustering Longitudinal Multivariate Data, in *Third Conference of the IFCS*, Edinburgh, Scotland.

Coppi, R. (1998) *Lectures in Multivariate Analysis*, Dipartimento di Statistica, Probabilità e St. Appl., Università di Roma "*La Sapienza*", Serie C, in Italian.

D'Urso, P., Gastaldi, T. (1999) Classification of Time Trajectories Based on A "Polygonal" Dissimilarity Measure, *submitted.*

D'Urso, P., Vichi, M. (1998) Dissimilarities Between Trajectories of a Three-Way Longitudinal Data Set, in: *Advances in Data Science and Classification*, Rizzi A., Vichi M., Bock, H. H. (Eds.), Springer-Verlag, 585-592.

Gordon, A.D. (1990) Constructing Dissimilarity Measures, *J. of Class.*,7,257-269.

Joly, S., Le Calvé, G. (1995) Three-Way Distances, *J. of Class.*, 12, 191-205.

Kosmelj, K., Batagelj, V. (1990) Cross-Sectional Approach for Clustering Time Varying Data, *Journal of Classification*, 7, 99-109.

Saporta, G., Lavallard, F. (1996) *L'Analyse des Données Évolutives. Méthodes et Applications*. Éditions Technip, Paris.

Ultramine Spaces in Classification[1]

Donatella Vicari
Dipartimento di Statistica, Probabilità e Statistiche Applicate
Università degli Studi di Roma "La Sapienza"
e-mail: vicari@pow2.sta.uniroma1.it

Abstract: Statistics is the first field of science where the notion of ultrametricity appeared outside mathematics. In fact, the particular geometric configuration of the ultrametric space finds optimal application in hierarchical cluster analysis methods. Moreover, it is interesting to consider the dual space of ultrametrics, which is induced by the "dual" ultrametric inequality, known as *ultramine* inequality. In this paper some properties of ultramine functions are analyzed and some algorithms are proposed to derive different ultramine approximation matrices of the dissimilarities between elements.

Keywords: Hierarchical clustering, Ultrametric, Ultramine.

1. Ultrametric and ultramine spaces

Ultrametrics play an important role in hierarchical cluster analysis algorithms because of the bijection between ultrametrics and dendrograms, due to the singular structure they induce in the metric space.
The aim of this paper is to study the "*dual*" space of ultrametrics, which is known as *ultramine* space (Diday, 1986) and analyze some of its properties, which can give different information on the dissimilarities between elements in a classification context.
Formally, an *ultrametric* space (E,u) is a set E endowed with a function $u: E \times E \rightarrow \Re^+$ such that:

i) $u(e_i,e_j) = 0 \Leftrightarrow e_i \equiv e_j$, $\forall (e_i,e_j) \in E \times E$;

ii) $u(e_i,e_j) = u(e_j,e_i)$, $\forall (e_i,e_j) \in E \times E$;

iii) $u(e_i,e_j) \leq \max[u(e_i,e_k),u(e_k,e_j)]$, $\forall (e_i,e_j,e_k) \in E \times E \times E$.

Note that because of the inequality iii), the set of metrics includes the set of ultrametrics.
In the study of ultrametrics a main theorem on triangles is given: a metric space is ultrametric if and only if all triangles are isosceles with narrow bases.
As a consequence, several theorems are given, which imply a singular geometric structure: each ball (i.e., hypershere) defined in E contains smaller balls with

[1] The paper is financially supported by MURST.

reduced radius; vice-versa, every ball is included in balls with longer radius; balls are either disjoint or included one within the other.
This structure induces nested partitions of E and, consequently, a hierarchical configuration which justifies why ultrametrics find their statistical application in hierarchical cluster analysis.
Let us consider a set E endowed with a symmetric function m: $E \times E \rightarrow \Re^+$. The ultrametric inequality iii) can be substituted by its dual expression:

$$m(e_i,e_j) \geq \min[m(e_i,e_k),m(e_k,e_j)], \qquad \forall(e_i,e_j,e_k) \in E \times E \times E \qquad (1)$$

called *ultramine inequality*.
An ultramine is a *similarity* measure, because by fixing $j=i$ in (1), it holds:

$$m(e_i,e_i) \geq m(e_i,e_k), \qquad \forall(e_i,e_k) \in E \times E,$$

which implies the definition of a similarity measure, together with non-negativity and simmetry of m.
Thus, the duality between ultrametrics and ultramines implies the transformation of a dissimilarity measure (metric) into a measure of similarity.
The geometric structure of an ultramine space (E,m) is quite different from the ultrametric one. Two main properties highlight the differences (Diday, 1986):

Property 1: A space is ultramine if and only if all triangles defined in E are isosceles with bases longer or equal than sides.
Property 2: The set of inverse ultrametrics coincides with the set of ultramines.

Property 2 allows to determine an ultramine approximation matrix **M** of a dissimilarity matrix **D**, simply by taking the reciprocal of all non-null entries:

$$m(e_i,e_j) = [u(e_i,e_j)]^{-1}, \qquad \forall u(e_i,e_j) \neq 0,$$

where **U** is any ultrametric approximation matrix of **D**.

2. Ultramine and ultrametric approximations of D

The application of property 2 allows to determine an ultramine approximation **M** of a dissimilarity **D**:

$$d(e_i,e_j) \xrightarrow{\text{By any hierarchical algorithm}} u(e_i,e_j) \rightarrow [u(e_i,e_j)]^{-1} = m(e_i,e_j),$$
$$\forall\, (e_i,e_j) \in E \times E : d(e_i,e_j) \neq 0 \ ,$$

which is "far" from **D** with respect to any measure of closeness, because **M** has a reciprocal scale with respect to **D**.
A better fit can be obtained by taking, at first, the reciprocal of every dissimilarity value of **D** and then, the reciprocal of every ultrametric value of **U**, which is provided, for example, by any hierarchical algorithm:

$$d(e_i,e_j) \rightarrow \left[d(e_i,e_j)\right]^{-1} \xrightarrow{\text{By any hierarchical algorithm}} u(e_i,e_j) \rightarrow \tag{2}$$
$$\rightarrow \left[u(e_i,e_j)\right]^{-1} = m(e_i,e_j), \qquad \forall\ (e_i,e_j) \in E \times E : d(e_i,e_j) \neq 0.$$

The strategy (2) better preserves the greatest initial dissimilarities, since it works with the inverse elements of **D**, and it provides an ultramine **M** "closer" to **D**, according to any norm of the difference between **D** and **M**.
We now discuss on new algorithms which provide hierarchical partitions of E with ultramine structure.
Starting from a dissimilarity matrix **D** and by considering the well-known formula by Lance and Williams (1967), we may derive an equivalent expression which determines hierarchical partitions of E in an ultramine space:

$$d(C_iC_j,C_k) = \Big\{\alpha[d(C_i,C_k)]^{-1} + \beta[d(C_j,C_k)]^{-1} + \gamma[d(C_i,C_j)]^{-1}$$
$$+ \delta\left|[d(C_i,C_k)]^{-1} - [d(C_j,C_k)]^{-1}\right|\Big\}^{-1}, \tag{3}$$

where C_i and C_j are the clusters involved in the fusion at a single step, C_k is any other group and α, β, γ, δ are the parameters.
The expression in (3) allows to determine different ultramine approximation matrices **M** of **D**. In particular, by setting the values of parameters, the "dual" restatement of the well-known clustering algorithms can be defined.
An ultramine **M** can be also derived by considering any monotone non-increasing function f of elements of **U**, such as, for example:

$$m(e_i,e_j) = a^{-u(e_i,e_j)},\ a > 0, \quad \text{or} \quad m(e_i,e_j) = \max_{k,l} u(e_k,e_l) - u(e_i,e_j)$$
$$\forall (e_i,e_j) \in E \times E.$$

For such a function f, the expressions in (2) and (3) can be generalized to define new algorithms generating ultramines matrices:

$$d(e_i,e_j) \rightarrow f\left[d(e_i,e_j)\right] \xrightarrow{\text{By any hierarchical algorithm}} u(e_i,e_j) \rightarrow f^{-1}\left[u(e_i,e_j)\right] = m(e_i,e_j),$$
$$\forall\ (e_i,e_j) \in A \subseteq E \times E$$

$$d(C_iC_j,C_k)=f^{-1}\Big(\alpha\, f(d(C_i,C_k))+\beta\, f(d(C_j,C_k))+\gamma\, f(d(C_i,C_j))$$
$$+\delta\big|f(d(C_i,C_k))-f(d(C_j,C_k))\big|\Big)$$

where f^{-1} is the inverse function of f and A is the subset of $E\times E$ where f is defined. Because of the link between ultrametrics (ultramines) and dissimilarities (similarities), most functions which map dissimilarities into similarities (Gordon, 1981) can be applied to any ultrametric **U** to provide an ultramine **M**. In fact, from a geometrical point of view, such functions transform every isosceles triangle with a narrow base, in a triangle which is still isosceles, but with the base longer than the sides.

3. A subset of ultramines

It can be interesting to consider the extended set of *proximity* measures P. Formally, we define *proximity* space (E,p), a set E endowed with a function $p{:}A\rightarrow\Re^+$, such that:

$$p(e_i,e_j)=p(e_j,e_i),\ \forall(e_i,e_j)\in A\subset E\times E,$$

where A is the subset of pairs of distinct elements.
Both *similarity* and *dissimilarity* functions, which mathematically differ just for the value they assume on pairs of identical elements, belong to P.
Figure 1 displays the links among distances, ultrametrics and ultramines belonging to the set of proximities P.

Figure 1: *Links among distances, ultrametrics and ultramines*

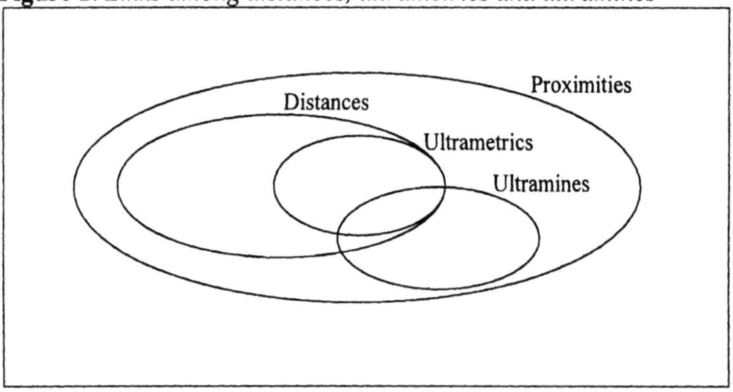

A proximity is a *distance* if it verifies the triangle inequality and, more strictly, it is an *ultrametric* if the ultrametric inequality holds.
In ultramine space (E,m) the triangular inequality does not necessarily hold: in this case, any triplet defines a "generalized" isosceles triangle.

It can be mathematically interesting to analyze ultramines belonging to the intersection with metrics, i.e. ultramines defined on pairs of different elements which satisfy the triangle inequality. The following condition is given.

Theorem
Let T_u be any isosceles triangle in an ultrametric space (E,u). In the ultramine space (E,m) induced by (2), the triangle inequality holds, if and only if T_u has the length of its side not greater than twice the length of its base.
Proof
In the ultrametric space (E,u) let T_u be the isosceles triangle of vertices (e_i, e_j, e_k), for any triplet of distinct elements and, without loss of generality, let $u(e_i, e_j)$ be its base's length. It holds:
$u(e_i, e_j) \leq u(e_i, e_k) = u(e_j, e_k)$.
In the ultramine space (E,m) induced by (2), let T_m be the isosceles triangle with the same vertices. By imposing the triangle inequality on T_m, it holds:
$m(e_i, e_j) \leq 2\, m(e_i, e_k)$, or equivalently, for T_u:
$[u(e_i, e_j)]^{-1} \leq 2\, [u(e_i, e_k)]^{-1} \Leftrightarrow u(e_i, e_k) \leq 2\, u(e_i, e_j)$.

For a monotone non-increasing function f of elements of **U** the above condition on T_u becomes, more generally: $u(e_i, e_k) \leq f^{-1}\, [2\, f[u(e_i, e_j)]]$.
An algorithm to determine an ultramine satisfying the triangle inequality is derived, by modifying the general procedure by Lance and Williams.
Starting from a dissimilarity matrix **D**, at a single step, let C_i and C_j be the nearest clusters and, consequently, the ones involved in the fusion and C_k be any other group.
Let $d(C_iC_j,C_k)$ be any dissimilarity by Lance and Williams (for example, the distance between clusters adopted in single or complete linkage). Then, let us define:

$$d^*(C_iC_j,C_k) = \min\{d(C_iC_j,C_k);\ 2d(C_iC_j)\} \tag{4}$$

where $d(C_i,C_j)$ is the shortest distance between elements belonging to C_i and C_j, respectively.
The expression (4), used recursively in a clustering algorithm, allows to get an ultrametric **U** such that the matrix **M,** derived by taking the reciprocal of any entry of **U**, satisfies the triangle inequality for any triplet of distinct elements.

4. An illustrative example

Ultramine and ultrametric approximations of the same matrix **D** give different and additional statistical information. It is interesting to point out that **U** preserves the smallest dissimilarities of **D** and it generally "deforms" the greatest ones, while in **M** small dissimilarities are deformed and the greatest ones are preserved.

Let **D** in Table 1 be the dissimilarity matrix (subjectively ranked from 1 to 5) between pairs of 6 portraits of members belonging to the same family, where even if parents do not resemble each other in face, they may be connected and classified through their relatives.
Let **U** in Table 1 be the ultrametric approximation provided by the average linkage and **M** the ultramine approximation of **D** provided by the strategy (2), but where the reciprocal function has been more properly substituted by the ranking in the reverse order.

Table 1: *Dissimilarity, Ultrametric and Ultramine matrices: data from Tamura (1971)*

D	**Ch1**	**Gm2**	**Ch3**	**P4**	**Ch5**	**P6**
Ch1	0	3	5	4	4	3
Gm2	3	0	1	4	1	1
Ch3	5	1	0	4	3	4
P4	4	4	4	0	1	5
Ch5	4	1	3	1	0	4
P6	3	1	4	5	4	0

U	**Ch1**	**Gm2**	**Ch3**	**P4**	**Ch5**	**P6**
Ch1	0	3.9	3.9	3.9	3.9	3.9
Gm2	3.9	0	1	3.7	3.7	2.9
Ch3	3.9	1	0	3.7	3.7	2.9
P4	3.9	3.7	3.7	0	1	3.7
Ch5	3.9	3.7	3.7	1	0	3.7
P6	3.9	2.9	2.9	3.7	3.7	0

M	**Ch1**	**Gm2**	**Ch3**	**P4**	**Ch5**	**P6**
Ch1	-	1.8	5	3.7	2.8	3.7
Gm2	1.8	-	1.8	1.8	1.8	1.8
Ch3	5	1.8	-	3.7	2.8	3.7
P4	3.7	1.8	3.7	-	2.8	5
Ch5	2.8	1.8	2.8	2.8	-	2.8
P6	3.7	1.8	3.7	5	2.8	-

Legenda: *Ch=Child; Gm=Grandmother; P=Parent.*

It is interesting to consider how the information from **U** and **M** can reveal different aspects on the data. We deduce from **U** that Grandmother (Gm2) looks like Child 3 (Ch3), just as Parent 4 is similar to Child 5 (Ch5); moreover, the third Child (Ch1) is the most dissimilar from the rest of the family (ultrametric distance equals to 3.9 from any other member).
On the other hand, **M** also reveals that Ch1 looks very dissimilar (ultramine value equals to 5) in particular from Child 3, just as the two parents do, as we expect.
Moreover, **M** is better than **U** as an approximation of **D**. In fact, the corresponding measures of fit, given by the squared Euclidean norm of the differences, are 30.24 and 34.82, respectively.

Figure 2: *Indexed hierarchy from* **U** *and anti-indexed hierarchy from* **M**

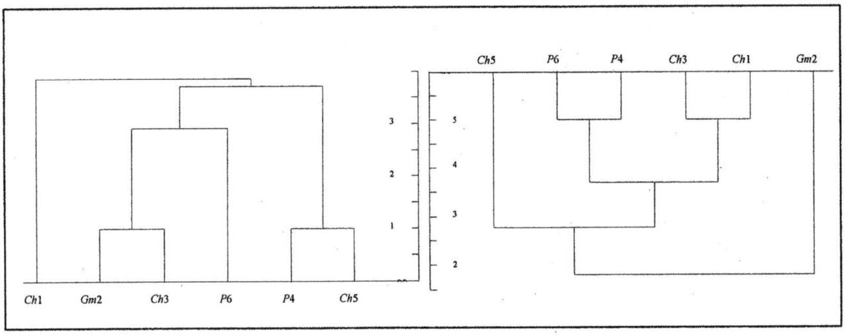

Figure 2 displays the hierarchy and the anti-hierarchy associated to matrices **U** and **M**, respectively. By cutting the dendrogram at level 3, the partition in 3 groups is derived:

$P_{\mathbf{U}} = \{(Ch1); (Gm2, Ch3, P6); (P4, Ch5)\}$.

Analogously, by cutting the other tree at level m^*=3, the following partition yields:

$P_{\mathbf{M}} = \{(Ch5); (Ch1, Ch3, P4, P6); (Gm2)\}$.

Note that, since any ultramine highlights differences more than resemblances, the second partition $P_{\mathbf{M}}$ identifies 3 "classes" of very different elements, which give a complementary information to the partition $P_{\mathbf{U}}$.
A matrix $\mathbf{M}^*$ (Table 2) with off-main diagonal entries assuming only 2 different values can be associated to $P_{\mathbf{M}}$, with reference to the value m^* which induces $P_{\mathbf{M}}$ itself, by setting:

$$m^*(e_i,e_j) = \begin{cases} 1 & \text{if } m(e_i,e_j) \le m^* \\ 2 & \text{if } m(e_i,e_j) > m^* \end{cases} \qquad \forall (e_i,e_j) \in E \times E .$$

Table 2: *Two-level Ultramine matrix corresponding to* m^*=3.

$\mathbf{M}^*$	*Ch*1	*Gm*2	*Ch*3	*P*4	*Ch*5	*P*6
*Ch*1	-	1	2	2	1	2
*Gm*2	1	-	1	1	1	1
*Ch*3	2	1	-	2	1	2
*P*4	2	1	2	-	1	2
*Ch*5	1	1	1	1	-	1
*P*6	2	1	2	2	1	-

Legenda: *Ch=Child; Gm=Grandmother; P=Parent.*

$\mathbf{M}^*$ is still an ultramine matrix that approximates **M** and **D**.

Since entries set to 1 indicate that the corresponding pairs of objects are similar, a covering (i.e. a collection of possibly overlapping subsets of the set E) is detected by the 4 groups of alike members of the family:

$$C = \{(Ch1, Gm2, Ch5); (Gm2, Ch3, Ch5); (Gm2, P4, Ch5); (Gm2, Ch5, P6)\},$$

where the intersection set ($Gm2$, $Ch5$) identifies the two members of the family, constituting the nucleus of the family, in terms of resemblance in face.

5. Conclusions

Ultramines, defined as the dual ultrametrics, have been considered, not only as complementary structure, but as independent approximations on given proximities.

Generally speaking, while ultrametrics better point out linkages between classes of similar elements, ultramines better explain linkages between classes of dissimilar elements, bringing out differences more than resemblances.

Hence, they can be properly used either to represent data characterized by great contrasts or, for example, to better highlight the largest deviations and interpret data and their representations (i.e. factorial analyses).

Moreover, when preliminarly used on large data sets, ultramine approximations of the original dissimilarities can help to better detect the initial seeds to be used in the successive applications of non-hierarchical algorithms of classification.

Further developments may regard, on one hand the characterization of the geometric structure induced by ultramines and, on the other hand, the search of optimal approximations of given dissimilarities to get a better fit to the data.

References

Diday, E. (1986) *Nouvelles représentations graphiques en classification automatique*, Rapport de recerche INRIA n. 78150, Rocquencourt, France.

Gordon, A. (1981) *Classification*, Monographs on Applied Probability and Statistics, Chapman and Hall.

Lance, G. N., Williams W. T. (1967) *A general theory of classification sorting strategies: I: Hierarchical systems II: Clustering systems*, The Computer Journal, 9-10, 373-380, 271-277.

Tamura, S., Higuchi S., Tanaka, K. (1971) *Pattern Classification Based on Fuzzy Relations*, IEEE Trans. on Syst., Man and Cybern., vol. SMC-1, 1, 61-66.

Vicari, D. (2000) *Ultramine*, Encyclopedia of Statistical Sciences, S. Kotz et al. (Eds.), 14, (to appear).

Genetic Algorithms and Clustering: an Application to Fisher's Iris Data[1]

Roberto Baragona, Claudio Calzini
Dipartimento di Sociologia, Università La Sapienza, Via Salaria 113, I-00198 Roma, Italy, e-mail: baragona@uniroma1.it

Francesco Battaglia
Dipartimento di Statistica, Probabil. e Stat. Appl., Università La Sapienza, Piazzale Aldo Moro 5, I-00185 Roma, Italy, e-mail: battag@pow2.sta.uniroma1.it

Abstract: Fisher's iris data constitute a hard benchmark for clustering procedures, and attracted much work based on statistical methods and new approaches related to evolutionary algorithms and neural networks. We suggest two genetic algorithms effective for simultaneously determining both the optimal number of groups and the assignment of items to groups. The grouping genetic algorithm proposed by Falkenauer (1998) forms the basis of our method, where the variance ratio criterion and the Marriott's method provide two fitness functions that both allow for fast computation and include the number of groups explicitly as a parameter. Specialized crossover operators, specific for each of the two fitness functions, are designed to accelerate convergence and minimize the number of iterations. Some simple implementations of our genetic algorithms are presented, that allow to classify correctly as many iris plants as the best alternative procedures proposed for this data set. Therefore genetic algorithms seem to constitute a good alternative choice for handling clustering problems.

Keywords: Classification, Crossover, Encoding, Evolutionary Computation, Fitness Function, Mutation

1. Introduction

In a classical paper Fisher (1936) considered a data set consisting of 150 items of iris specimen belonging to different varieties: 50 *setosa*, 50 *versicolor*, and 50 *virginica*. Four measurements were taken on each of them. Fisher noted that whilst the *setosa* group is neatly discriminated, the other two groups contain items which are similar and difficult to distinguish. Further interest arose about this data set. Mezzich and Solomon (1980) list several contributions and rank them according to several optimality criteria. Friedman and Rubin (1967) reached a correct classification for all but three items (two *versicolor* and one *virginica*), the

(1)The present paper is financially supported by Ministero della Ricerca Scientifica e Tecnologica, Italy best results to date. Also, iris data have been analyzed by means of evolutionary methods (the Beagle algorithm of Forsyth, 1986) and neural networks (see Adorf and Murtagh, 1988).

Genetic algorithms have been introduced by Holland (1975), though a pioneering contribution of Box (1957) is worth to mention, and have been successfully adopted to handle problems with high computational complexity. Some fields of applications are listed in Mitchell (1996, pp. 15-16). Pattern recognition, machine vision, image reconstruction, identification of objects in images, seem to be the fields where genetic algorithms have been used most frequently, as far as clustering is concerned. Clustering, partitioning and grouping problems are characterized by high degree of complexity, especially if the number of groups is unknown, and genetic algorithms have been proposed as useful tools for searching for solutions (see Jones and Beltramo, 1991). However, few contributions are found in the statistical literature (see Chatterjee, Laudato and Lynch, 1996), where no examples are reported of applications of genetic algorithms to cluster analysis. A stochastic searching algorithm was examined by Fortier and Solomon (1966) which seems to envisage the use of mutation in genetic algorithms. As a matter of fact, unless the set of items and the number of partitions are smallest, an exhaustive approach is simply not feasible. Genetic algorithms are likely to perform more efficiently than other discrete search methods, in that they do not process a single path to solution, but handle simultaneously a population of candidate solutions. These individuals are evolved in a parallel fashion, though controlled interaction may take place, which is aimed at improving the average fitness. So, the search for solutions is performed in a region larger than other methods may explore. In addition, the stochastic nature of the genetic approach is likely to act in such a way that the risk for the algorithm to be trapped in local optima is greatly reduced.

In most applications, however, the encoding does not look to be well tailored for the specific clustering problem, as primary interest is focused on items instead of on groups. Furthermore, the number of groups is assumed to be known, or a preliminary analysis is required to be performed to find the best number of clusters. The grouping genetic algorithm (*GGA*) proposed by Falkenauer (1998) seems to constitute the best procedure to capture the very structure of the clustering problem. Falkenauer showed how to implement his procedure for some particular problems, but suggested resorting to local problem-dependent heuristics to handle different problems. As a matter of fact, to be effective for classifying the iris data, the fitness function and the genetic operators have to be given a peculiar formulation that we provide in the present paper. We discuss how genetic algorithms may be employed to partition the iris data into optimal number of non overlapping groups and report how our applications produced similar results in comparison with the best published ones. We found that genetic algorithms perform efficiently and constitute a simple and effective alternative for handling clustering problems. The plan of the paper is as follows. In Section 2 a brief account of applications of genetic algorithms to clustering problems is given. The *GGA* is outlined in Section 3, and our proposal of two fitness functions and heuristic crossovers is introduced in Section 4. Application to the iris data is presented in Section 5. In Section 6 conclusions are drawn.

2. Clustering by genetic algorithms

A genetic algorithm evolves a finite discrete population of individuals, each of which has to encode a single solution to the clustering problem. We assume that every item has to belong to one and only one group of the partition and the assignment of items to groups is made by attempting to maximize the fitness (objective) function. Encoding is a crucial point which may drive the choice of what genetic operators to be used in conjunction with, and several methods have been proposed. Some of them, that we may mention here, include the linear code (often in its group-numbers form) and the Boolean matching code, permutation encoding, binary string, and ordinal and ordered representations. Redundancy and context insensitivity are envisaged by Falkenauer (1998, Chapter 4) as serious drawbacks of such encoding techniques. In addition, we may notice that, when these encoding methods are adopted, either the number of groups has to be held fixed, or, even if specific devices are employed to allow the number of groups to vary, this event is unlike to occur in practice, because the number of groups is never taken explicitly into account. So the algorithm is almost always confined to search the optimum for some given groups' number. As a matter of fact, there are as many optima as the numbers of groups ranging from one to the maximum allowed, but only one is the global optimum we are seeking for.

3. The grouping genetic algorithm

We shall outline the basic steps of the *GGA*. The encoding, crossover and inversion are reported as introduced in Falkenauer (1998, Chapter 5). The procedure of reproduction and the mutation operator are chosen by us, among the alternatives therein proposed, as the best ones for our specific problem. Let n denote the number of items to classify, p the number of measurements taken on each item (variables), and g the maximum allowed number of clusters.
Let s be the size of the population that is being processed by the genetic algorithm. Each individual h, $h=1,\ldots,s$, represents a candidate solution, which is characterized by a number of groups k, $1\leq k\leq g$, and a partition (mutually exclusive) of the n items into the k groups. The chromosome for encoding such a candidate solution is a string of length $n+k$ of integer values. Genetic operators, but mutation, are applied only to the genes at the loci $n+1$, $n+2$, ..., $n+k$. So, for the ease of exposition, we will consider two distinct chromosomes, the first one of fixed length n, and the second one of variable length k. The first one follows the group-numbers straightforward encoding: each gene relates to a specific item, and may take a value between 1 and g, indicating the cluster to which the item itself belongs. The second one encodes the clusters' labels as genes. Therefore its length

k is variable, and ranges from 1 through g. The allelic values of its genes still are integers within the range $[1,g]$.
The so-called roulette wheel is the most common way of modeling the reproduction of the individuals in a given population. Copies of an existing individual are generated with probability proportional to its fitness function. In a population of s individuals, each of which has fitness function f_h, h=1,...,s, the expected number of copies of the h-th individual is $sf_h/\Sigma f_h$, or f_h/f^*, where f^* is the average fitness $\Sigma f_h/s$. So individuals characterized by above average fitness have a probability of reproduction higher than the remaining ones. In our implementation, the new population entirely replaces the old one, except the individual, which, in the past population, had the highest fitness function. If this individual exited the population, it is recovered to replace the individual that, within the current population, has the worst fitness function. This is our choice for the implementation of the elitist strategy. Rudolph (1994) advocated using the elitist strategy as necessary for the genetic algorithm to converge.
A kind of two-point crossover is performed on the second chromosomes of the parent's individuals. Two cutting points are randomly selected for each parent. Then two children are formed, the one by inserting the genes in between the crossing sites of parent two just before the first crossing point of the parent one, the other by reversing roles of the parents individuals. If necessary, the resulting groups have to be adapted according to the hard constraints and the fitness function. At this stage, local problem-dependent heuristics may be applied. The two children replace both parents in the new population. Typically, crossover applies only to a portion of the population. The percentage of pairs to be selected for crossover is the crossover's rate p_c. The candidate pairs of individuals are $s/2$. The number of crossovers, which are performed at a given step, is a binomial random variable with parameters $s/2$ and p_c. So, it turns out that the expected number of crossovers is $p_c s/2$.
The mutation operator acts so that a small number of genes are allowed to change at random their allelic values. The purpose of mutation is to maintain diversity among individuals in the population. Furthermore, it is the only way to recover some solutions that were lost all along the evolutionary path of the population, or to explore new regions of the solutions' space. Also, mutation is a useful operator as far as convergence of the genetic algorithm is concerned (Rizzi, 1997). In general, mutation is often defined as the smallest possible modification of a chromosome. For the grouping problem, it seems appropriate to consider that a mutation occurs when an item is moving from a group to another. In practice, with a small probability p_m, any gene of the first chromosome of each individual may change its allelic value to another one chosen randomly among the allelic values within the second chromosome.
Inversion is sharing features from both mutation and crossover. In fact, like mutation, it applies to a single individual. Then, it requires two crossing sites be chosen at random, so as to look alike two-point crossover. The genes between the crossing sites are reversed in order. Each individual is assigned a usually small probability for inversion p_i to occur. There seems a general agreement to exist that

it should only be applied when the interpretation of each gene is not locus dependent, just like happens for the second chromosome, which lists the groups' labels. Each gene, therein, refers to a group whose meaning depends only on what items are belonging to it. Inversion does not change the composition of the groups. The reason for using it resides in that, the promising genes (well performing groups), if close together, are more likely to be transmitted to individuals in the next generation.

4. Two fitness functions

Let B, W and T denote, as usual, the between-groups, within-groups and total sum of squares $p \times p$ matrices. We shall introduce two fitness functions that we tried as criteria for separating the iris data. The first one is aimed at minimizing trace(W), the second one det(W). Note that, under Gaussian multivariate model, the classification maximum likelihood approach reduces essentially to minimize det(W), if the observations within all groups are assumed to have common covariance matrix, and to minimize trace(W), if the common covariance matrix is diagonal (see Banfield and Raftery, 1993). When the covariance matrices may differ from group to group, then the evaluation of det(W_j), $j=1,\ldots,k$, where W_j is the sum of squares matrix of the items that belong to the j-th group, is too time-consuming with respect to the (at present) available computing resources. In addition, Scott and Symons (1971) noticed that, in practice, the maximum likelihood methods would always partition the data into the maximum number g of partitions allowed. Methods for determining the optimal number of groups that are including as well a criterion for optimal assignment of items to groups seem to be only the variance ratio criterion (*VRC*) (Calinski and Harabasz, 1974), and the Marriott's method (*MM*) (Marriott, 1971). Required computations may be performed quickly enough to make their use in conjunction with genetic algorithm viable in practice.

We designed two different versions of *heuristic crossover*, according to what fitness function is concerned.

The variance ratio criterion $VRC=\{\text{trace}(B)/(k-1)\}/\{\text{trace}(W)/(n-k)\}$ offers directly a suitable fitness function, because it takes only positive values and has to be maximized for optimal partition of items into groups. The crossover may be adapted to the fitness function, in order to try to accelerate the convergence towards the solution. Following Falkenauer (1998, pp. 100-101), the items occurring twice are put aside, and each is re-assigned to the group with nearest centroid in terms of squared Euclidean distance. This is a well-known device in non-hierarchical clustering algorithms, which leads to increase trace(B). As $B+W=T$, and k is fixed, this implies that the *VRC* increases too.

Marriott's method consists of finding that partition that minimizes $k^2\text{det}(W)$. The latter is a positive function, but a fitness function must be a non-decreasing one. So we may define $MM=\text{det}(T)/\{k^2\text{det}(W)\}$ as a suitable fitness function, where det(T) is constant over partitions and independent of k (Marriott, 1971, p. 503).

The crossover is performed as follows. The items, put aside because are being assigned to two different groups, are each re-assigned to the valid group whose mean minimize the Mahalanobis' generalized distance from the item itself. We used the formula reported in Marriott (1971, p. 508), where the within-groups dispersion matrix is computed by taking into account only the items which are not involved in the re-assignment step. Moreover, the co-ordinates of the groups' centroids are not updated at each re-assignment, for computations would turn out to be too much cumbersome, in the presence of little improvements.

5. Application to iris data

Since the work of Fisher (1936) many authors noticed that the species *setosa* does not overlap with the other two species whilst these latter are overlapping somewhat. Evidence of this circumstance is given by plots of the data and by exploratory methods (see Cerioli and Zani, 1999). A variety of clustering methods has been tentatively entertained in order to produce a partition of the iris data, which may exactly match the three species. We shall label the plants by taking consecutively down the rows of the originally published table by Fisher. So, labels 1-50 denote iris *setosa*, 51-100 *versicolor* and 101-150 *virginica*. Friedman and Rubin (1967, Section 6.2) applied the min trace(W) and the max det(T)/det(W) criteria to this data. The first one gave, for fixed k=3, 10 misclassified items, partly *versicolor* and partly *virginica*. For k=4, still *setosa* resulted as a separate group, then a group was obtained including only *virginica*, a group only of *versicolor* except for plant 107 of *virginica*, and a group containing a mixture of *versicolor* and *virginica*. The second criterion yielded a partition into 3 groups, which recovered the three species except for 3 plants: 71 and 84 went with *virginica*, and 134 went with *versicolor*. Another criterion they applied, consisting of maximizing trace($W^{-1}B$), produced as well, for fixed k=3, a partition with 3 misclassified items (but not the same as before). Duran and Odell (1974, p. 103), by using Mahalanobis' distance and k=9, obtained 22 misclassifications, and, by using Euclidean distance and k=17, 8 misclassified plants. A clustering method based on neural networks was proposed by Adorf and Murtagh (1988). Example on the iris data with k=3 was reported, where their method led to correctly classify 48 iris *setosa* in the first group, 35 *versicolor* in the second one, and 33 *virginica* in the third group. However, 34 plants were misplaced. Everitt (1993, p. 116) reported application of normal mixtures for the analysis of the iris data, with only 5 misclassified of *versicolor* added to *virginica*. 5 misclassifications were reported as well by Fraley and Raftery (1998) as a result of *EM* algorithm.

Of course, genetic algorithms cannot perform better than the fitness function they are being used with. Their usefulness resides in that they are able to find solutions that have not been possibly reached by other methods. As far as problems of this kind are concerned, where the space of solutions cannot be explored exhaustively, we shall never be confident that the global optimum is actually achieved, but some confirmation or some improvement may as well be obtained. For implementation

of a genetic algorithm, quite a few parameters are needed to specify. We made our choice mainly on the guidelines provided by Mitchell (1996, Section 5.6). We assumed s=200 as population size, the rates of crossover, mutation and inversion p_c=0.8, p_m=0.005, and p_i=0.1 respectively. The number of items is n=150 and that of measurements is p=4. For the maximum number of groups, we considered both g=3 and g=5. The initial population of s individuals was generated as follows: for each individual, the number of groups k was chosen as a random integer uniformly ranging from 1 through g; then, each item was uniformly randomly assigned to one of the k available groups. Renumbering was applied for the groups' labels to coincide with the integers 1,2,...,k. Number of iterations has been 50,000 for both fitness functions, though a much smaller number proved to be really needed.
Let us consider the *VRC*-based genetic algorithm and let g=5 are the maximum allowed number of clusters. In Table 1 the assignment of species to groups is displayed which maximized the fitness function.

Table1: *Assignment of species to groups on VRC-based fitness function*

group	*setosa*	*versicolor*	*virginica*	
1	50			50
2		48	14	62
3		2	36	38
	50	50	50	150

Figure 1: *Increasing of VRC over iterations*

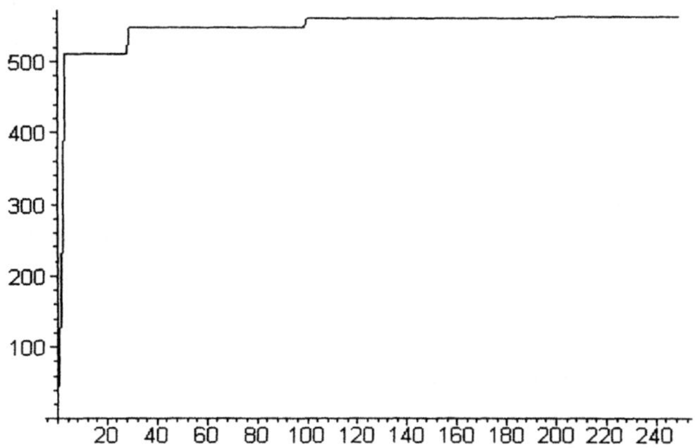

The first group is composed only of *setosa*, the second one is a mixture of 48 *versicolor* and 14 *virginica*, the third group is composed of 36 *virginica* and the *versicolor* 53 and 78. Summing up, we have 16 misclassified items. Nevertheless, the maximum of the fitness function turns out to equal 561.63, whilst the figure reported by Friedman and Rubin for their min trace(W) criterion, given k=3, translated to *VRC*, corresponds to 560.43, in spite of the fact that, in the latter

case, the misclassified items are only 10. This circumstance is not surprising, however, as we may compute 487.33 the *VRC* for the partition which assigns exactly the three species to each of three groups. In Figure 1 the fitness is plotted as a function of the iterations. Its behavior seems to support that genetic algorithms may quickly find promising regions of the search space, with sudden increasing of the fitness function, followed by slower heuristic problem-dependent search for the true optimum (see Mitchell, 1996, p. 124). The maximum is achieved in correspondence of the 200-th step. Note that the genetic algorithm include simultaneously in its search all candidate solutions with k ranging from 1 through 5. This means that the *VRC* points at k=3 as the optimal number of group.
Let us now take the *MM*-based genetic algorithm with g=3 into account. After 200 iterations, exactly the same result, obtained by Friedman and Rubin by using their max det(T)/det(W) criterion, is achieved. The best fitness figure turns out to be 5.04. Note that the fitness computed when assigning each species to a separate group equals 4.74.
For g=5, results yielded by the *MM* criterion are displayed in Table 2. The optimal number of groups turns out to equal k=4, and the corresponding fitness function is equal to 6.8, a figure that encompasses the one which we can compute from the value 74.13 reported by Friedman and Rubin for max det(T)/det(W), that they obtained by assuming 4 groups. This seems to indicate that genetic algorithm may perform a better search towards the true optimum. The fitness function increases as displayed in Figure 2. The maximum value is reached after 109 iterations. By looking at Table 2, we may observe that *setosa* are well separated again, and form a group on their own, whilst *virginica* are split in two groups, the first one adding *versicolor* 84 and the second one *versicolor* 71. The remaining group is solely composed of *versicolor*. So, we have only 2 misclassified items. These findings are in agreement with Scott and Symons (1971, p. 394), who supported the use of max det(T)/det(W) as a criterion appropriate as far as iris data are concerned.

6. Concluding remarks

Fisher's iris data cannot be partitioned neatly into species as far as three groups are assumed. If we allow the number of clusters, and the assignment of items to them, both vary, the misclassified items may reduce to only 2, at the expense of splitting iris *virginica* into two groups. We found that genetic algorithms can perform efficiently the search moving around the space of solutions in a stochastic way driven by the genetic evolutionary mechanism. In order to avoid being trapped in local optima, variable length chromosomes, all along with grouping encoding and heuristic crossover, seem most efficient, since the number of clusters is explicitly modeled and the algorithm may take advantage from knowledge of the distance, properly assessed according to the fitness function, of items from groups' means.
So, genetic algorithms seem to constitute a good alternative choice for handling clustering problems. In fact, most statistical methods are able to explore only a

small region of the solutions' space, so that they are likely to find out local optima, and not the global one. On the other hand, some new proposals, such as neural networks and Gibbs sampler (Bensmail, Celeux, Raftery and Robert, 1997) need much heavier both design and computation. Genetic algorithms may explore the whole space of solutions and their implementation with respect to a given problem seems to be more straightforward. Computations are less cumbersome than for other methods, since the basic steps turn out to be very simple. In fact, most time is spent in the evaluation of the fitness function, a task that any algorithm must accomplish anyway.

Table 2: *Assignment of species to groups on MM-based fitness function*

group	*setosa*	*versicolor*	*virginica*	
1	50			50
2		48		48
3		1	17	18
4		1	33	34
	50	50	50	150

Figure 2: *Increasing of MM criterion over iterations*

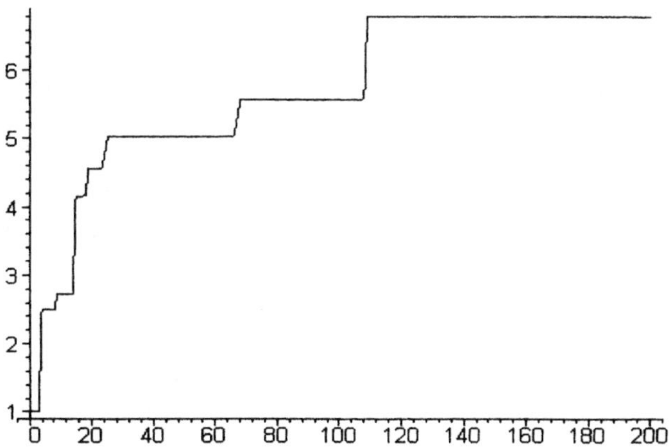

References

Adorf, H.-M., Murtagh, F. (1988) Clustering based on neural network processing, in: *Compstat 1988*, Physica-Verlag Heidelberg for IASC (International Association for Statistical Computing), 239-243.

Banfield, J. D., Raftery, A. E. (1993) Model-based Gaussian and non-Gaussian clustering, *Biometrics* 49, 803-821.

Bensmail, H., Celeux, G., Raftery, A. E., Robert, C. P. (1997) Inference in model-based cluster analysis, *Statistics and Computing* 7, 1-10.

Box, G. E. P. (1957) Evolutionary operation: a method for increasing industrial productivity, *Applied Statistics* 6, 81-101.

Calinski, T., Harabasz, J. (1974) A dendrite method for cluster analysis, *Communications in Statistics* 3(1), 1-27.

Cerioli, A., Zani, S. (1999) Exploratory methods for detecting high density regions in cluster analysis, in: *Book of Short Papers CLADAG99*, Meeting held in Rome, Italy, 5-6 July 1999.

Chatterjee, S., Laudato, M., Lynch, L. A. (1996) Genetic algorithms and their statistical applications: an introduction, *Computational Statistics & Data Analysis* 22, 633-651.

Duran, B. S., Odell, P. L. (1974) *Cluster Analysis: A Survey*, Springer, New York.

Everitt, B. S., (1993) *Cluster Analysis* (Third edition), Edward Arnold, London.

Falkenauer, E. (1998) *Genetic Algorithms and Grouping Problems*, Wiley, New York.

Fisher, R. A. (1936) The use of multiple measurements in taxonomic problems, *Annals of Eugenics* 7, 179-188.

Forsyth, R. (1986) Evolutionary strategies, in: *Machine Learning: Applications in Experts Systems and Information Retrieval*, Forsyth & Rada, (Eds.), Ellis Horwood, London.

Fortier, J. J., Solomon, H. (1966) Clustering procedures, in: *Multivariate Analysis*, Paruchuri & Krishnaiah (Eds.), 493-506.

Fraley, C., Raftery, A. E. (1998) *MCLUST: Software for Model-Based Cluster Analysis*, Technical Report no. 342, University of Washington, The Statistics Department, Seattle, WA.

Friedman, H. P., Rubin, J. (1967) On some invariant criterion for grouping data, *Journal of the American Statistical Association* 63, 1159-1178.

Jones, D. R., Beltramo, M. A. (1991) Solving partitioning problems with genetic algorithms, in: *Proceedings of the Fourth International Conference on Genetic Algorithms*, Belew & Booker (Eds.), Morgan Kaufmann Publishers, San Mateo, California, 442-449.

Marriott, F. H. C. (1971) Practical problems in a method of cluster analysis, *Biometrics* 27, 501-514.

Mezzich, J. E., Solomon, H. (1980) *Taxonomy and Behavioral Science: Comparative Performance of Grouping Methods*, Academic Press, London.

Mitchell, M. (1996) *An Introduction to Genetic Algorithms*, The MIT Press, Cambridge, Massachusetts.

Rizzi, A. (1997) A convergence theorem for genetic algorithms, *Metron* 55, 69-83.

Rudolph, G. (1994) Convergence analysis of canonical genetic algorithms, *IEEE Transactions on Neural Networks, Special Issue on Evolutionary Programming*, 1-11.

Scott, A. J., Symons, M. J. (1971) Clustering methods based on likelihood ratio criteria, *Biometrics* 27, 387-397.

Using Radial Basis Function Networks for Classification Problems

Isabella Morlini
Istituto di Statistica, Università di Parma, Via J.F. Kennedy 6, I-43100 Italy.
E-mail: isabella.morlini@unipr.it

Abstract: multi-layer perceptron is now widely used in classification problems, whereas radial basis function networks (RBFNs) appear to be rather less well known. Purpose of this work is to briefly recall RBFNs and to allow a synthesis of theirs best features. The relationships between these networks and other well-developed methodological tools for classification, both in neural computing and in statistics, are shown. The application of these networks to the forensic glass data set, which is not new in literature (Ripley, 1994; 1996), try to lay out what is common and what is distinctive in these networks and other competitive methods and to show, through empirical validation, the networks performance.

Keywords: Classification, Learning Vector Quantization, Mixture Analysis, Neural Networks, Radial Basis Functions.

1. Introduction

This work is an attempt to demonstrate the suitability of radial basis function networks (RBFNs) for classification problems. It provides an overview of these networks and shows how they work and differ from other well-developed methodological tools. The paper is structured as follows. This section briefly recalls the architecture of a RBFN. Section 2 shows the differences between the widely used multi-layer perceptron (MLP). Section 3 describes how RBFNs can be used for classification purposes and demonstrates an interesting geometrical property. This section also points out the relationships with other statistical methods and discusses the argument for or against the use of normalisation. Some new insights are given, regarding the comparison between RBFNs and mixture analysis. Section 4 proposes an application to the forensic glass data set and, showing numerical results, gives empirical validation of the networks performance. Further challenges are given as concluding remarks.

RBFNs (Broomhead and Lowe, 1988; Moody and Darken, 1989) are a special class of networks with a single hidden layer, in which the activation functions for the output units are taken to be linear and the outputs are represented as linear combinations of a set of m predefined non-linear basis functions. Unlike the MLP, that is based on units which compute the scalar product of the input vector and a weight vector, the activation of a RBFN hidden neuron is determined by the distance between the input vector and a prototype vector, also called centre. The RBFN mapping computed by the *k-th* output unit when input vector $\mathbf{x}$ is presented

(k=1, …, c), can be formalised as follow:

$$y_k(\mathbf{x}) = \sum_{j=1}^{m} w_{kj} \phi_j (\| \mathbf{x} - \mu_j \|) + w_{k0} \tag{1}$$

where μ_j is the vector determining the centre of basis function ϕ_j. Usually the distance is taken to be Euclidean and several form of basis functions can be considered, the most common being the Gaussian:

$$\phi_j (\| \mathbf{x} - \mu_j \|) = \exp\left(- \frac{\| \mathbf{x} - \mu_j \|^2}{\tau_j^2} \right) \tag{2}$$

where the smoothing parameter τ_j determines the width of the hidden unit.
A necessary requirement for the network to be able to globally approximate systems as complex as the basis function's local models, is to form a partition of unity for the input space, so that every point is covered to the same degree. This partition of unity ensures that every point in the input space is due only to the vector parameter **w**, weighting the basis functions activation, and is reached through the basis function normalisation. A normalised RBFN (Moody and Darken, 1989) has lateral connections between different hidden units in the network diagram, so that the output becomes:

$$y_k(\mathbf{x}) = \sum_{j=1}^{m} w_{kj} \frac{\phi_j (\| \mathbf{x} - \mu_j \|)}{\sum_{j=1}^{m} \phi_j (\| \mathbf{x} - \mu_j \|)} \tag{3}$$

In these networks the output of the hidden units are forced to sum to unity. Thus, since each output unit computes a weighted sum in which the weights are the input received from the hidden units and the inputs are the output's node parameters, the output values must lie within the range of the output layer weights. The outputs of a normalised Gaussian RBFN can be written as:

$$y_k(\mathbf{x}) = \sum_{j=1}^{m} w_{kj} \frac{\exp\left(- \| \mathbf{x} - \mu_j \|^2 / \tau_j^2 \right)}{\sum_{j=1}^{m} \exp\left(- \| \mathbf{x} - \mu_j \|^2 / \tau_j^2 \right)} \tag{4}$$

in which the hidden unit's outputs are multiple logistic functions, that reduce to the simple logistic when there are only two hidden nodes.
Training an RBFN usually consists of two phases. The position of the centres in input space is first determined by using unsupervised procedures (like k-means cluster analysis and self-organising-maps) or by selecting a subset of the input vectors (subset selection or orthogonal least squares are often used for this purpose). The widths are then set by using a number of heuristics. In the second stage the basis functions are kept fixed and the estimation of the coefficients w_{kj} which minimise the sum-of-squares error becomes a linear optimisation task and

can be performed by using matrix pseudo-inverse techniques or the conventional least squares algorithm. To avoid overfitting, regularised subset selection (Orr, 1993) can be used in the first training phase when a weight decay penalty term is added to the error function. Some alternative methods, based on the gradient descent technique, are discussed in Bishop (1995) and Berthold and Hand (1999).

A major problem related to RBFNs, as like as to weighted piecewise constant models in general, is termed the *curse of dimensionality*. Since the network is forced to use exponentially increasing number of basis functions to approximate a given system as the input dimension increase, if one must work with a limited number of training data, then increasing the dimensionality of the space rapidly leads to the point where the data is very sparse, in which case the network provides a very poor representation of the system (Bishop, 1995). In many applications, therefore, reducing the number of input variables in a pre processing stage (and with feature extraction techniques) may lead to improved network performances for a given data set (for an example, see Morlini, 1999).

2. Comparisons Between the Multi-Layer Perceptron

RBFNs offer several advantages over the MLP, which regard the training, the activations of the hidden units and the mapping function implemented.

Since the hidden layer parameters control the bell-shaped output functions of the hidden units, the goal of training is to arrange these units so that they cover the portions of input space where example actually occur, with their density proportional to the density of examples in each area, and so that their widths provide an appropriate degree of overlap. The hidden layer is thus self-organising and its parameters depend only on the distribution of the inputs, not on the mapping from inputs to outputs. These features, which justify the two-phase approach, do not hold for the MLP, where supervised learning is used to set the values of all parameters and the minimum error in the mapping function must be found by iterative techniques, like the gradient descent or second-order methods. Using these techniques requires a significantly higher level of mathematical expertise, as well as more computational time and a solution to the problem of local minima.

The difference between the hidden unit's activation functions can be interpreted geometrically: while in a MLP the isoactivation contours of the hidden unit are parallel hyperplanes, in a RBFN with Gaussian basis functions based on Euclidean distance they are concentric hyperspheres (hyperellipsoids if the RBFN is based on Mahalanobis distance). If the basis functions are Gaussians, the hidden units assume a localised nature: the network forms a representation in the space of hidden units which is local with respect to the input space, because, for a given input vector, only few hidden units will have significant activations.

RBFNs are attractive because, if the basis functions are placed close to the data, they often give more sensible extrapolation then sigmoidal networks, which create quite unjustified divisions of the input space in regions where no data are provided and, unlike the MLP, they can be easily adapted to estimate data density

(Specht, 1990; Hwang *et al.*, 1994). In a normalised RBFN, if the second-layer parameters are set within a reasonable range (such as the range of the target values) the outputs will be within the same range for all possible inputs, even when the net is extrapolating. No comparably useful bounds exist for the MLP.

3. Radial Basis Function Networks for Classification

While a MLP can separate classes by using hidden units which form hyperplanes in the input space, and for this reason should be related to discriminant analysis, a RBFN is able to model each class distribution by local kernel functions, and should be rather compared with kernel discriminant analysis or mixture analysis. Normalised RBFNs which exactly perform kernel discriminant analysis are called *probabilistic neural networks* (PNNs) and are set up as follows. A localised hidden unit is created for each input vector $\mathbf{x}_n$ of the training set (n=1, ..., N), the centres are set equal to the data point $\mathbf{x}_n$, the second layer weights w_{kj} are set equal to 1 if the point $\mathbf{x}_n$ belongs to class k, 0 otherwise (alternatively, these weights can be adjusted for the prior probabilities of each class). The only parameters to be learned, namely the widths τ_j of the basis functions, are heuristically chosen. If the optimal widths are recognised to be different for basis functions in different regions of input space, than the smoothing parameter τ_j of unit j can be set to a small, positive fraction α of the distance δ_j to the nearest other unit ($\tau_j = \alpha\delta_j$). In this case, the network presents an interesting geometrical property that we demonstrate as follows. Let N be the number of Gaussian basis functions (corresponding to the number of data point in the training set), then the network hidden layer converts every d-dimensional input $\mathbf{x}$ into a N-dimensional vector

$$\mathbf{y}' = (\phi_1(\|\mathbf{x} - \mu_1\|), \ldots, \phi_N(\|\mathbf{x} - \mu_N\|)), \tag{5}$$

which forms the input of the output layer. Due to normalisation, for every vector $\mathbf{x}$ the activation of unit j is inside the unit interval: $0 \leq \phi(\|\mathbf{x} - \mu_j\|) \leq 1$. Consequently, the vector $\mathbf{y}'$ always lies inside the N-dimensional unit cube. For each data point $\mathbf{x}_n$ exactly one localised hidden unit is maximally activated (the unit with its centre at μ_n=$\mathbf{x}_n$) whereas for all other units the following holds:

$$\phi_j(\|\mathbf{x}_n - \mu_j\|) \leq \frac{\exp(-1/\alpha)}{\sum_{j=1}^{N} \phi_j(\|\mathbf{x} - \mu_j\|)} \quad j \neq n \tag{6}$$

So, for small values of α, all but one components of the activation vector are very close to zero. In the limiting case, when α tend to 0, this vector lies on those corners of the N-dimensional unit cube which have a distance 1 to the origin. Let a particular partition into two groups be considered and let the second layer weights w_{kj} be equal to 1 if the point $\mathbf{x}_j$ belongs to class k, 0 otherwise. If a point is classified following the optimal classification rule, namely as belonging to group k if the value of the corresponding output unit is bigger than 0.5, then no misclassification of the training data can occur if, for α, the condition

$$0.5 < (N\text{-}1)\ \frac{\exp(-1/\alpha)}{\sum_{j=1}^{N}\phi_j(\|\mathbf{x}-\mu_j\|)} \tag{7}$$

is satisfied. If (7) holds then, for each training data, the classification derived from its corresponding hidden unit can not be changed by the combined activity of all the (N-1) other units. Therefore, a PNN have the useful ability to tell whether a test case is similar (that means, has a high density) to any of the training data. If not, the network is extrapolating and the output classification should be viewed with scepticism, due to the poor generalisation capability of the system.

A normalised Gaussian RBFN with equal widths in each hidden unit can be also thought as a smoothed variant of the learning vector quantization (LVQ) (Kohonen, 1988; Ripley, 1996). In LVQ, the hidden units are also called *codebook* (or *reference*) *vectors* and the degree of fit of a particular solution is assessed by the sum of squares differences between examples and the corresponding reference vector. So, LVQ amounts to the nearest-neighbour classification on the reference vectors and normalised Gaussian RBFNs can be viewed as a smoothed variant that uses not just the single nearest neighbour, but a weighted average of near neighbours. As the width of the hidden units approaches zero, the weights approach one for the nearest neighbour and zero for all other codebook vectors.

The hidden layer of a RBFN can thought as a Gaussian mixture model constituted by a common pool of basis functions equal for every density and representing the class conditional distributions. Bishop (1995, pp.179-182) gives important insights to this view and demonstrates how the outputs of a normalised RBFN can be interpreted as the Bayesian posterior probabilities of class membership. However, Bishop's approach has two drawbacks. Although, from a theoretical point of view, the output values of a RBFN can be considered as posterior probabilities of class membership, when least squares are used to set the second layer parameters there is no guarantee that these values lie in the range [0,1]. They can be forced to sum to one when the 1-of-c coding scheme is used for target data, but not to lie in this range. The most important drawback is perhaps due to normalisation. Normalisation is desirable for classification problems, since at every point in the input space the sum of the basis function is forced to sum to unity so that, in mixture underlying model, the activation of each basis function can be interpreted as the posterior probability of the presence of corresponding feature in input space. However, normalisation leads to a number of side effects which are described in Shorten and Murray-Smith (1994). The first one is that when the basis functions are Gaussians, the normalisation results the whole of the input space being covered and not just the region of the input space defined by the training data. The second one is that basis functions with different widths can become multi-modal, meaning that theirs activations increase as the distance function between the input vector and the centre decreases (this phenomenon is called 'reactivation' of the basis functions). A final side effect, which also concerns basis functions with different widths, is that the maxima may no longer be at theirs centres. These three side effects make the argument for or against the use of normalisation far more complicated than it is often assumed. Comparison between mixture analysis and RBFNs should be more detailed, since the purposes of these two methods are different. A Gaussian mixture model is more suited for estimating the parameters of a population which is known to be composed of a fixed number of homogeneous sub-populations. It directly models the class distributions by Gaussian mixtures in the sampling paradigm. The hidden layer of a RBFN can be viewed as a Gaussian mixture model. However, unsupervised procedures which are often used to set the basis function parameters, like k-means, partition a data set deterministically into subgroups and the number of these sub-populations is heuristically determined. The hidden layer of a RBFN should be rather used to learn about the class distributions and to estimate the number of sub-clusters in the training data, when this number is unknown. Procedure k-means can be seen as a particular limit of the expectation-maximisation (EM) algorithm often used in mixture analysis. It can be shown that in case of Gaussian basis functions with a common width parameter τ and in the limit $\tau \to 0$, the EM update formula for a basis function centre reduces to the k-means update formula. However, means and variances of the k-clusters are not in general considered as estimators of the parameters of the component densities. Similarly, the mixing coefficient w_{kj}, which are determined by the EM algorithm in mixture analysis, are given by least squares in the RBFN and should be motivated from a geometrical point of view rather than from the principle of maximum likelihood. From a technical point of view, the EM algorithm is a re-estimation procedure which may involve some problems in the solution of the equation system, especially for very large data sets: the computational time and costs, the eventual convergence to a singular estimate of the covariance matrix (that will cause the algorithm to fail), the absence of convergence or the convergence to a local maximum. Some of these problems can be overcome with a constrained maximum solution and the availability of good computer programs. These matters do not hold in a RBFN in which the position of the centres in input space is determined by selecting a subset of the input vectors and the second-layer weights are given by least squares. If the basis functions are selected by k-means, and the Euclidean distance is used, convergence is guarantee but the problem of local minima still persists. A final observation relates to the assumption of multivariate normality of the density function in each group. In Gaussian mixture models this density function appears explicitly in the update formula. On the contrary, calculus performed by a RBFN can be justified without assuming normality or any other particular distribution. The example of Section 4 demonstrates that the performances of a RBFN can be really better, given the non-normal nature of the data set.

4. An Application to a Real Data Set

The example comes from forensic testing of 214 fragments of glass and the data set is taken from Murphy and Aha (1995). Each observation has a measured refractive index and composition (weight percent of oxides of sodium, magnesium, aluminium, silicon, potassium, calcium, barium and iron). The fragments were originally classed as seven types, one of which was absent in this data set. Categories which occur are window float glass (Wf), window non-float glass (Wnf), vehicle window glass (Veh), containers (Con), tableware (Tabl) and vehicle headlamps (Head). Observations belonging to each category are, in order, 70, 76, 17, 13, 9 and 29. In Ripley (1994) several analysis over this data set are performed, omitting headlamp glass and grouping together containers and tableware. The analysis reported here overlaps that in Ripley (1996) which considers all original groups and asses different methods by 10-fold cross validation. We use the same random partition to compute error rates of a (non-normalised) RBFN. Particularly attention must be paid to define precisely what procedure is to be cross-validated. We could use the same basis functions for each cross-validation experiment and fit the second layer weights at each run. Alternatively, both basis functions and second layer weights could be optimised in each run. The first option means that the fitting is done slightly more quickly but the results can be rather biased. For this reason, we chose to draw the number of basis functions and theirs centres from the input data of the training set in each cross-validation experiment. Therefore, models are made up of Gaussian basis functions drawn from different sets of candidates. Forward selection is used to determine the best subset of input vectors. The procedure starts with an empty subset to which is added one basis function at a time, the one which most reduces the sum-of-squares error between output and target values in the training set. To decide when stopping adding further basis functions, the error in the test set is also monitored. Although the training error never increases as extra functions are added, the test error eventually stops decreasing and starts to increase as over-fit sets in. That is the point at which to cease adding basis functions to the network. Forward selection is, of course, a non linear algorithm but has the following advantages: there is no need to determine the number of hidden units in advance; the number of hidden units and their centres are optimised at the same time and the computational requirements are relatively low. This method only set the basis function centres, and the width parameters must be chosen using some other heuristics. In order to give a relatively smooth representation of the distribution of the training data, the width parameters are chosen to be equal and to be given by the average distance between input cases. This distance is used in all cross-validation runs. Variables are standardised to zero mean and unit variance before any process. With a 22.4% cross validated percentage of misclassified observations, the RBFN outperforms most other well-known classification methods whose results, reported in Ripley (1996), are summarised in Table 1.

The cross validated confusion matrix is reported in Table 2. Fits for discriminant analysis and mixture analysis are rather poor. This is not surprising, given the very non-normal nature of the data set and the presence of rare classes which are too small to fit even a single normal density. MLP and non-parametric methods like LVQ and classification tree show useful gains over parametric methods, but still worse performances than the RBFN.

As a final remark, it must be noted that the application may not be fully representative, since a deeper comparison between performances should be carried out by means of more sophisticated techniques, such are, for example, those based on simulations or bootstrapping.

Table 1: *Estimated error rates*

Method	Error
Classification and regression tree	32.2%
Learning vector quantization	29.9%
Linear discriminant analysis	38.0%
Logistic discriminant analysis	36.0%
Multivariate adaptive regression spline	32.2%
Mixture analysis	30.8%
MLP (2 hidden units)	30.4%
MLP (4 hidden units)	23.8%
MLP (8 hidden units)	26.2%
Projection pursuit regression	42.0%
RBFN	22.4%
1 nearest neighbour	23.4%

Table 2: *Confusion matrix*

	Wf	Wnf	Ve	Ta	He	Co
Wf	62	6	2	0	0	0
Wnf	13	59	1	0	0	3
Ve	4	7	6	0	0	0
Ta	0	2	0	7	0	0
He	2	2	0	0	24	1
Co	1	3	1	0	0	8

References

Berthold M., Hand, D.J. (1999) *Intelligent Data Analysis: An Introduction,* Springer, Berlin.

Bishop C.M. (1995) *Neural Networks for Pattern Recognition*, Clarendon Press, Oxford.

Broomhead D.S., Lowe, D. (1988) Multivariable functional interpolation and adaptive networks, *Complex System* 2, 321-335.

Hwang J.N., Lay, S.R. and Lippman, R.P. (1994) Nonparametric Multivariate Density Estimation: A Comparative Study, *IEEE Transactions on Signal Processing*, 2795-2810.

Kohonen T. (1988) Learning vector quantization, *Neural Networks*, 1, 3-13.

Murphy P.M., Aha, D.W. (1995) *UCI Repository of Machine Learning Databases,* Irvine, CA: University of California, Dept. of Information and Computer Science. Available from ics.uci.edu in directory pub/machine-learning-databases.

Moody J., Darken, C.J. (1989) Fast learning in networks of locally-tuned processing units, *Neural Computation* 1(2), 281-294.

Morlini I. (1999) Radial basis function networks with partially classified data, *Ecological Modelling Journal*, V. 120, Elsevier Science, Oxford, 109-118.

Orr M.J.L. (1993) Regularised centre recruitment in radial basis function networks. Research Paper, 59, Centre for Cognitive Science, Edinburgh University.

Ripley B.D. (1994) Neural networks and related methods for classification, *Journal of the Royal Statistical Society, B* 56(3), 409-456.

Ripley B.D. (1996) *Pattern Recognition and Neural Networks*, Cambridge University Press, Cambridge.

Shorten R., Murray-Smith, R. (1994) On normalising basis functions networks, *4th Irish Neural Networks Conference*, University College, Dublin.

Specht D.F. (1990) Probabilistic neural networks, *Neural Networks* 3, 109-118.

Unconditional Latent Budget Analysis: a Neural Network Approach*

Roberta Siciliano
Dep. of Mathematics and Statistics, University of Naples Federico II
Monte S. Angelo, via Cintia, 80126 Napoli, Italy
roberta@unina.it

Ab Mooijaart
Dep. of Psychology, Leiden University, The Netherlands
Wassenaarseweg 52, 2333 AK Leiden, PO 9555, 2300 RB Leiden
mooijaart@fsw.LeidenUniv.nl

Abstract: The latent budget model is a reduced rank model for the analysis of compositional data. The model can be also understood as a *supervised neural network model* with weights interpreted as conditional probabilities. Main advantage of this approach is that a classification rule for budget data can be defined for new observed cases. In this paper, a *constrained (weighted) least-squares algorithm* – which is alternative to the one already introduced in literature for standard latent budget model – is proposed for the estimation of the parameters. A distinction is made between *conditional latent budget analysis* (the standard approach) and *unconditional latent budget analysis* (the neural network approach).

Keywords: mixture model, supervised learning, least-squares estimation.

1. A supervised neural network for latent budget analysis

The latent budget model is a reduced-rank probability model to decompose a table with constant row-sum data, namely compositional data such as for example time budgets and conditional proportions derived from a contingency table (de Leeuw and van der Heijden, 1991). The decomposition is a mixture of K nonnegative factorial terms satisfying restrictions likewise conditional probabilities. The latent budget model can be written as

$$\Pi = \mathbf{A}\mathbf{B}^{T},\ \mathbf{A}\mathbf{1}_K = \mathbf{1}_I,\ \mathbf{1}_J{}^{T}\mathbf{B} = \mathbf{1}_K{}^{T}, \tag{1}$$

* *This research project is supported by MURST funds.*

where Π is the I x J probability matrix of *theoretical budgets*, **A** is the I x K matrix of *mixing parameters* $a_{ik} \geq 0$ and **B** is the J x K matrix of *latent budgets* $b_{jk} \geq 0$. The model is fitted to a table of I observed budgets over J response categories. In this respect, the latent budget model can be used for dependence analysis to describe the relations between the two variables. A suitable supervised neural network can be defined for latent budget analysis, namely a *double-layer perceptron* in figure 1 where one hidden layer connects the input layer with the output layer (Bishop, 1995).

Figure 1: *A double-layer perceptron for latent budget analysis.*

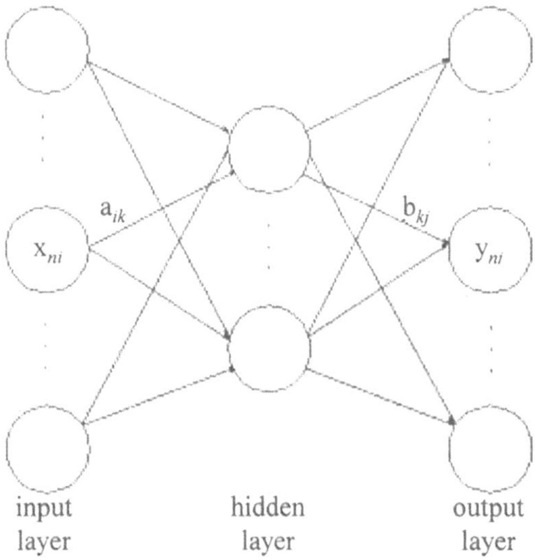

Let **X** be the N × I matrix whose N rows contain the N observations of an input I-vector which includes the indicator variables associated to the explanatory categories; let **Y** be the N × J matrix whose N rows contain the N observations of an output J-vector which includes the indicator variables associated to the response categories. The matrix **A** contains the weights a_{ik} before the hidden layer and the matrix $\mathbf{B}^T$ contains the weights b_{kj} following the hidden layer.
For a *double-layer perceptron* the output can be defined in general as follows

$$y_{nj} = \psi_1 (\Sigma_k b_{kj} \psi_2 (\Sigma_i x_{ni} a_{ik}) + d_j) + e_{nj},$$

for n = 1, ..., N observations, where ψ_1 and ψ_2 are the transfer functions associated to the output layer and to the hidden layer respectively.
Although different transfer functions can be tried out, in order to relate the neural network to latent budget analysis, we consider *identity transfer functions* and the null constant term d_j so that the output can be defined as follows

$$y_{nj} = \Sigma_k (\Sigma_i x_{ni} a_{ik}) b_{kj} + e_{nj}$$

In the proposed neural network the weights are required to be nonnegative and to satisfy specific restrictions. Mixture models in neural network approach are described also by Bishop (1995). In this paper we show an application and some justification in terms of a particular type of latent budget analysis.
In matrix formulation, the neural latent budget network can be defined as

$$\mathbf{Y} = \mathbf{XAB}^{\mathrm{T}} + \mathbf{E},\ \mathbf{A1}_K = \mathbf{1}_I,\ \mathbf{1}_J{}^{\mathrm{T}}\mathbf{B} = \mathbf{1}_K{}^{\mathrm{T}}, \tag{2}$$

where $\mathbf{E}$ is the error term, $\mathbf{A}$ and $\mathbf{B}$ include nonnegative parameters. Pre-multiplying the (2) by $(\mathbf{X}^{\mathrm{T}}\mathbf{X})^{-1}\mathbf{X}^{\mathrm{T}}$ allows to specify the latent budget model (1). Notice that the model parameters are not identified since $\mathbf{AB}^{\mathrm{T}} = (\mathbf{AT})(\mathbf{T}^{-1}\mathbf{B}^{\mathrm{T}})$ for any nonsingular matrix $\mathbf{T}$. This point is extensively discussed in van der Ark (1999). In the neural network approach the latent budget structure is defined in terms of the indicator variables instead of the contingency table. The idea is to predict $\mathbf{Y}$ on the basis of $\mathbf{X}$ yielding a classification rule for new observed cases of unknown response class in latent budget analysis.

2. A constrained least-squares estimation procedure

In the following we propose a *supervised learning algorithm* based on least-squares criterion to obtain estimates of the parameters of the network (2). The function to be minimized is:

$$f = \mathrm{SSQ}\,(\mathbf{Y} - \mathbf{XAB}^{\mathrm{T}}) = \mathrm{tr}\,(\mathbf{Y}^{\mathrm{T}}\mathbf{Y} - 2\mathbf{Y}^{\mathrm{T}}\mathbf{XAB}^{\mathrm{T}} + \mathbf{BA}^{\mathrm{T}}\mathbf{X}^{\mathrm{T}}\mathbf{XAB}^{\mathrm{T}}). \tag{3}$$

Obviously, $\mathbf{X}^{\mathrm{T}}\mathbf{Y}$ is a frequency table, written as $\mathbf{F}$; $\mathrm{tr}\,(\mathbf{Y}^{\mathrm{T}}\mathbf{Y}) = N$ is the sample size; $\mathbf{X}^{\mathrm{T}}\mathbf{X}$ is a diagonal matrix, written as $\mathbf{D}_X$, with diagonal equal to $\mathbf{F1}$. So the function can be written as

$$f = N - 2\mathrm{tr}(\mathbf{F}^{\mathrm{T}}\mathbf{AB}^{\mathrm{T}}) + \mathrm{tr}(\mathbf{BA}^{\mathrm{T}}\mathbf{D}_X\mathbf{AB}^{\mathrm{T}})\,. \tag{4}$$

For minimizing the function f, it suffices to use $\mathbf{F}$, instead of the matrices $\mathbf{X}$ and $\mathbf{Y}$. For the derivatives of f with respect to $\mathbf{A}$ and $\mathbf{B}$, it holds

$$\partial f / \partial \mathbf{A} = -2\mathbf{FB} + 2\mathbf{D}_X\mathbf{AB}^{\mathrm{T}}\mathbf{B}, \tag{5a}$$
$$\partial f / \partial \mathbf{B} = -2\mathbf{F}^{\mathrm{T}}\mathbf{A} + 2\mathbf{BA}^{\mathrm{T}}\mathbf{D}_X\mathbf{A}. \tag{5b}$$

We define the following transformation using temporary parameters u_{il} and z_{lk}

$$a_{il} = \exp(u_{il}) / \Sigma_m \exp(u_{im}), \tag{6a}$$
$$b_{jk} = \exp(z_{lk}) / \Sigma_m \exp(z_{mk}). \tag{6b}$$

From these definitions it follows: $a_{il} > 0$; $\Sigma_l a_{il} = 1$ and $b_{lk} > 0$; $\Sigma_l b_{lk} = 1$. In each row of **U** one element may be fixed. The same holds for each column of **Z**. For instance, we set $u_{i1} = 0$, and $z_{1k} = 0$. This type of transformation is also used in generalized linear models (McCullagh and Nelder, 1990).
For the derivatives of f with respect to the elements of **U** and **Z**, we have

$$\partial f / \partial u_{ik} = \Sigma_l (\partial f / \partial a_{il}) (\delta^{lk} a_{il} - a_{il} a_{ik}), \text{ for } k > 1, \tag{7a}$$
$$\partial f / \partial z_{jk} = \Sigma_l (\partial f / \partial b_{lk}) (\delta^{lj} b_{lk} - b_{lk} b_{jk}), \text{ for } j > 1, \tag{7b}$$

where δ is the usual Kronecher delta.
The optimization procedure requires starting values for the matrices **U** and **Z** and consists in the following steps:
1. minimize f with respect to **U** and **Z** using a quasi-Newton method which utilizes the function evaluation for given **U** and **Z** in (4) where **A** and **B** can be found by (6a) and (6b);
2. in the same way compute the gradient vectors with respect to **U** and **Z** as in (7a) and (7b) where the derivatives with respect to **A** and **B** are defined by (5a) and (5b);
3. after having solved **U** and **Z, A** and **B** can be found by (6a) and (6b).
Mooijaart et al. (1999) discuss alternative estimation procedures to deal with the above constraints. In particular, they used the *active constraint method.* Instead, we propose a *direct method (*which is in this respect more efficient) to take account of the parameter constraints via a suitable parameter transformation (eq. 6). Main advantage of this approach is that we can use any standard optimization procedure such as for example the quasi-Newton method to estimate the model parameters under the given constraints.

3. Unconditional versus conditional latent budget analysis

The neural network model (2) for latent budget analysis is alternative to the standard model (1). This can be shown upon comparison of the optimization criteria underlying the two models.
For the least-squares estimation of the latent budget model (1) the function to be minimized is as follows:

$$f = \text{SSQ}\,(\mathbf{P} - \mathbf{AB}^T),$$

where **P** is the row conditional matrix of proportions. A disadvantage of this loss function is that the row frequencies are eliminated. So each row is equally important, even if the row frequencies may differ a lot. To overcome this disadvantage Mooijaart et al. (1999) have proposed to use a weighted least squares function. This can be written as

$$f = SSQ\,(\mathbf{W}(\mathbf{P} - \mathbf{A}\mathbf{B}^T)), \tag{8}$$

where **W** is a diagonal weighting matrix. This function can be rewritten as

$$\begin{aligned} f &= tr\,(\mathbf{P} - \mathbf{A}\mathbf{B}^T)^T\,\mathbf{W}^2\,(\mathbf{P} - \mathbf{A}\mathbf{B}^T) \\ &= tr\,(\mathbf{P}^T\mathbf{W}^2\mathbf{P} - 2\mathbf{P}^T\mathbf{W}^2\mathbf{A}\mathbf{B}^T + \mathbf{B}\mathbf{A}^T\mathbf{W}^2\mathbf{A}\mathbf{B}^T). \end{aligned} \tag{9}$$

Notice that $\mathbf{D_X P} = \mathbf{F}$ (where **F** is the matrix with unconditional frequencies, as defined above). If we choose $\mathbf{W}^2 = \mathbf{D_X}$, then the function becomes

$$f = tr\,(\mathbf{P}^T\mathbf{D_X}\,\mathbf{P} - 2\,\mathbf{P}^T\mathbf{D_X}\mathbf{A}\mathbf{B}^T + \mathbf{B}\mathbf{A}^T\mathbf{D_X}\mathbf{A}\mathbf{B}^T)$$

$$= tr\,(\mathbf{P}^T\mathbf{F} - 2\mathbf{F}^T\mathbf{A}\mathbf{B}^T + \mathbf{B}\mathbf{A}^T\mathbf{D_X}\mathbf{A}\mathbf{B}^T). \tag{10}$$

This function is equal to the function (4), up to an irrelevant constant. So the derivatives are equal in both situations, and both methods will yield the same estimates of **A** and **B**. As a result, we distinguish two cases:

I: *Conditional Latent Budget Analysis* (LBA_C): $\mathbf{W} = \mathbf{I}$. Estimation of **P** is provided by the standard model $\mathbf{AB}^T$, and the residuals are $e_{ij} = (\mathbf{P} - \mathbf{AB}^T)_{ij}$.

II: *Unconditional Latent Budget Analysis* (LBA_U) : $\mathbf{W} = \mathbf{D_X}^{1/2}$. Estimation of **F** is given by the neural network model $\mathbf{D_X AB}^T$, with residuals $e_{ij} = (\mathbf{D_X}^{-1/2}(\mathbf{F} - \mathbf{D_X AB}^T))_{ij}$.

As a result, we have shown that one special case of the weighted version of the mixture model presented by Mooijaart et al. (1999) can be understood in terms of unconditional analysis with a neural network interpretation. In this way, the choice between the latent budget model for conditional and unconditional analysis depends on the prior information about the population. In *retrospective studies*, such as in epidemiological studies, a partition of the population into J groups is assumed to be known a priori, and in addition an independent product multinomial sampling scheme is performed on the J sub-populations. In *prospective studies*, such as in pattern recognition, no knowledge about the structure of the J groups is assumed and the population is stratified on the basis of the predictor categories.

4. Examples

In order to show some applications of the neural network model we consider a data set extracted by a survey of the Bank of Italy on Family Budgets at the year 1994, that is a sample of 2000 families.

The analysis focalises on the preferred type of payment chosen by the head of the family distinguishing five profiles of consumers that have been defined as *archaic payers* (who use exclusively cash), *classic payers* (who use cheque), *evolving payers* (who use cheque and chip cards), *modern payers* (who use also credit

cards), *mixed payers* (unclassifiable group). This output can be explained by the following inputs: 1. the net income in million per year *(≤ 30, 30 -| 45, > 45)*, 2. the age *(≤ 35, 35 –| 45, 45 –| 55, 55 –| 65, > 65)*, *3.* the geographic area of residence *(North-West, North-East, Center, South, Isles)*, 4. the title of study *(none, primary school diploma, middle school diploma, high school diploma, degree)*.
Furthermore, a compound input can be built up by cross-classifying two single inputs, i.e., we consider the geographic area of residence and the net income.
Table 1a shows the estimates of different neural network models, one for each different input, whereas *table 1b* shows the estimates of the neural network model using the compound input. In both cases the proposed algorithm has been used and the number of latent budgets has been selected on the basis of the usual goodness of fit measure based on the residuals.
For the interpretation of the results, the *weights A,* which correspond to the mixing parameters, identify the groups characterizing each latent budget (looking at the highest percentages for each row), whereas the *weights B,* which correspond to the latent budgets, identify to which response the latent budget can be assigned (looking at those percentages in each column which are higher than the corresponding ones under the independence model).
When using input 1 (i.e., the net income) the first latent budget describes the archaic payers having low income whereas the second latent budget includes the other types of payers having higher income. When using input 2 (i.e., the age) a discrimination is made between payers with age higher than 55 belonging to the archaic payers group and payers with age lower than 55 belonging to the remaining groups. When using input 3 (i.e., the geographic area) it can be pointed out that the head of a family in the North-West is less archaic than others. When using input 4 (i.e., the title of study) we need three latent budgets in order to fit well the data set: the first latent budget characterized by the families where the head has got a middle school diploma describes both the classic and the evolving payers; the second latent budget characterized by the families where the head has got either an high school diploma or a degree describes well the modern payers but also the evolving and the classic payers; the third latent budget characterized by families where the head has not a good title describes the archaic payers. Finally, when using the compound input (i.e., the geographic area and the income) we find that the first latent budget describes the archaic payers, the second the classic and the evolving payers, the third the modern payers. It is interesting to notice a discrimination across the areas such for example in the South there are mainly archaic payers, whereas in the North they correspond to those having a low income.

5. Concluding remarks

This paper has provided a supervised learning algorithm for the estimation of a *double-layer perceptron* associated to the latent budget model where the weights are nonnegative and sum up to one. A direct method is proposed to deal with such constraints. The neural network approach allows to define a model for *unconditional latent budget analysis* which is alternative to the standard model for conditional latent budget analysis. Unconditional analysis should be preferred to conditional analysis when the aim is to use the latent budgets for classifying new cases of unknown response under prospective sampling scheme.

Table 1a: *Neural network estimates of Unconditional Latent Budget Models for prediction of the type of payment (output) preferred by the head of the family due to different inputs (Survey of the Bank of Italy).*

	Weights A				***Weights B***		
Input 1: income	**k=1**	**k=2**		**Output**	**k=1**	**k=2**	***Indep.***
≤30	0.00	1.00	***1.00***	*Arch.*	0.14	0.69	*0.45*
30 -| 45	0.41	0.59	***1.00***	*Class.*	0.22	0.18	*0.20*
> 45	0.89	0.11	***1.00***	*Evol.*	0.36	0.09	*0.22*
				Mod.	0.21	0.00	*0.08*
				Mixed.	0.07	0.04	*0.05*
					1.00	***1.00***	***1.00***
Input 2: age	**k=1**	**k=2**		**Output**	**k=1**	**k=2**	***Indep.***
≤35	0.59	0.41	***1.00***	*Arch.*	0.10	0.72	*0.43*
30 -| 45	0.64	0.36	***1.00***	*Class.*	0.21	0.18	*0.19*
45 -| 55	0.63	0.37	***1.00***	*Evol.*	0.40	0.06	*0.23*
55 –| 65	0.40	0.60	***1.00***	*Mod.*	0.17	0.03	*0.09*
> 65	0.07	0.93	***1.00***	*Mixed.*	0.12	0.01	*0.06*
					1.00	***1.00***	***1.00***
Input 3: area	**k=1**	**k=2**		**Output**	**k=1**	**k=2**	***Indep.***
NW	0.74	0.26	***1.00***	*Arch.*	0.13	0.70	*0.40*
NE	0.60	0.40	***1.00***	*Class.*	0.22	0.18	*0.20*
Center	0.30	0.70	***1.00***	*Evol.*	0.44	0.04	*0.20*
South	0.13	0.87	***1.00***	*Mod.*	0.14	0.05	*0.13*
Isles	0.15	0.85	***1.00***	*Mixed.*	0.07	0.03	*0.07*
					1.00	***1.00***	***1.00***

Input 4: title	**k=1**	**k=2**	**k=3**		**Output**	**k=1**	**k=2**	**k=3**	***Ind.***
none	0.00	0.00	1.00	***1.00***	*Arch.*	0.21	0.13	0.82	*0.40*
primary	0.22	0.07	0.71	***1.00***	*Class.*	0.23	0.21	0.16	*0.20*
middle	0.48	0.18	0.34	***1.00***	*Evol.*	0.47	0.27	0.01	*0.20*
high	0.30	0.61	0.09	***1.00***	*Mod.*	0.01	0.30	0.00	*0.11*
degree	0.00	1.00	0.00	***1.00***	*Mixed.*	0.08	0.09	0.01	*0.09*
						1.00	***1.00***	***1.00***	***1.00***

Table 1b: *Neural network estimates of the Unconditional Latent Budget Model for prediction of the type of payment (output) preferred by the head of the family due to a compound input (Survey of the Bank of Italy).*

	Weights A			
Input 5: area × income	**k=1**	**k=2**	**k=3**	
NW ≤30	0.49	0.48	0.03	***1.00***
NW 30-|45	0.09	0.74	0.17	***1.00***
NW >45	0.01	0.39	0.60	***1.00***
NE < 30	0.78	0.19	0.03	***1.00***
NE 30-|45	0.24	0.70	0.06	***1.00***
NE >45	0.00	0.53	0.47	***1.00***
Center ≤30	0.83	0.17	0.00	***1.00***
Center 30-|45	0.67	0.25	0.08	***1.00***
Center >45	0.27	0.39	0.34	***1.00***
South ≤30	0.99	0.00	0.01	***1.00***
South 30-|45	0.68	0.30	0.02	***1.00***
South >45	0.41	0.35	0.24	***1.00***
Isles ≤30	0.97	0.03	0.00	***1.00***
Isles 30-|45	0.71	0.08	0.21	***1.00***
Isles >45	0.30	0.00	0.70	***1.00***
	Weights B			
Output	**k=1**	**k=2**	**k=3**	***Indep.***
Arch.	0.81	0.21	0.00	*0.40*
Class.	0.15	0.32	0.11	*0.20*
Evol.	0.00	0.39	0.42	*0.20*
Mod.	0.01	0.00	0.42	*0.10*
Mixed.	0.03	0.09	0.04	*0.10*
	1.00	***1.00***	***1.00***	***1.00***

References

Bishop, C.M. (1995). Neural Networks for Pattern Recognition, Clarendon Press, Oxford.

McCullagh, P., Nelder, J.A. (1990). *Generalized Linear Models*, Chapman and Hall, Cambridge.

Mooijaart, A., van der Heijden, P.G.M., van der Ark, L.A. (1999). A least-squares algorithm for a mixture model for compositional data, *Computational Statistics and Data Analysis,* 30, 359-379.

van der Ark, A. (1999). *Contributions to Latent Budget Analysis. A Tool for the Analysis of Compositional Data*, DSWO Press, Leiden University.

PART II

MULTIVARIATE DATA ANALYSIS

Generalized Constrained Principal Component Analysis[(1)]

Pietro Amenta
Department of Econ. Sciences, Mathematics and Statistics, University of Lecce, Via per Monteroni, 73100 Lecce, Italy. E-mail: amenta@economia.unile.it

Luigi D'Ambra
Department of Mathematics and Statistics, University "Federico II", Via Cinthia, 80126 Naples, Italy. E-mail: dambra@unina.it

Abstract: This paper deals with a non-symmetrical analysis of two multiple data sets in order to study the structure of dependence among sets of variables which play different role in the analysis. This approach represents a generalization of the Constrained Principal Component Analysis (CPCA) (D'Ambra and Lauro, 1982).

Keywords: Multiple Data Sets, PCA, Non-Symmetrical Analysis.

1. Introduction

In experimental data analysis a subset of variables can play a non symmetrical role towards the others. An example of this dependence framework can be found in sensory evaluations. The traditional way of evaluating sensory influences on overall liking (Fig. 1.a) consists in studying statistical links between explanatory sensory descriptive variables (sensory attributes evaluated for each product by trained judges), and criterion hedonic variables (scores given by consumers to the same products). The commonly used method is the External Preference Mapping (MacFie and Thomson, 1988; Schlich, 1995) which consists of two separate steps: a principal component analysis (or a generalized procrustes analysis) of sensory data gives some new synthetic sensory variables (sensory components) which summarize the main sensory differences among products; afterwards, the hedonic response is regressed on these main sensory components using a quadratic model. Other approaches (Kvalheim, 1988; Huon de Kermadec F. *et al.*, 1996), are based on the Partial Least Squares (PLS) and its non-linear extensions (Huon de Kermadec F. *et al.*, 1997).

We assume the explanatory variables as well as the criterion ones are collected in multiple data sets (Kiers, 1991) ("two blocks"). More general "two blocks" dependence framework is commonly analyzed by PLS based methods. For example, the PLS analysis of the two blocks unfolded in columns-linked two-way tables (D'Ambra *et al.*, 1988). Another approach (Wold *et al.*, 1987) is based on the Lohmoller-Wold decomposition of a multiway data arrays combined with the non-linear PLS algorithms.

Figure 1: *Data frames. Scheme A* (1.a) – *Scheme B* (1.b).

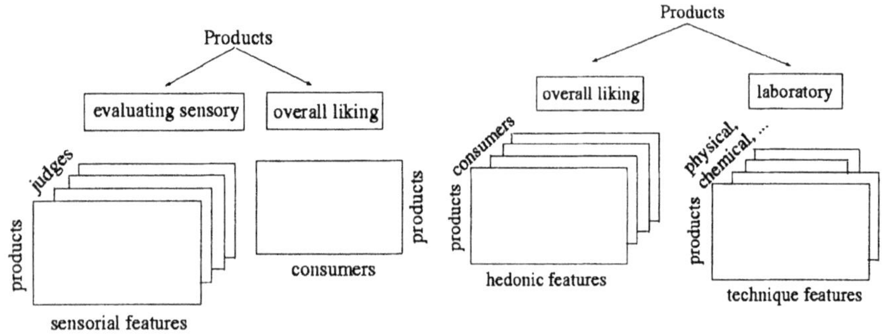

1.a 1.b

In this paper we propose a generalization of CPCA to this "two blocks" dependence framework (Fig. 1.b). The followed approach has been developed in two articulate moments: in a first moment, moving from a covariance criterium with more sets of criterion variables, we seek simultaneously different linear combinations of each explanatory variables set. We show how this approach is equivalent to consider a unique linear combination of all explanatory variables, as this one can be written as linear combination of the formers. A greater interpretative role to each explanatory set of variables is developed by looking to the different graphical positions of the statistical units and criterion variables with respect to the explanatory sets, as well as taken all together. In a second moment we highlight, in the same framework, how it is possible to evaluate the contribution of each explanatory set. All that is obtained in a single mathematical framework and not as collection of single and independent classical techniques.
An example of application concerning the evaluation of technique features influences on overall pasta liking by consumers, will be showed.

2. Generalized constrained principal component analysis

Suppose that data on K sets of explanatory variables and S criterion variables of n statistical units are represented in terms of matrices X_k $(k=1,\ldots,K)$ and Y_s $(s=1,\ldots,S)$ of orders $(n\times p_1),\ldots,(n\times p_K)$ and $(n\times q_1),\ldots,(n\times q_S)$, respectively. We suppose, without loss of generality, identity matrices for the metrics of the spaces of variables of X_k and Y_s with $D_n = diag(1/n)$, weight matrix of statistical units. We assume that X_k's and Y_s's are centered as to the weights D_n.
Let $X=[X_1|\ldots|X_K]$ and $Y=[Y_1|\ldots|Y_S]$, respectively, be K and S matrices column linked of orders $(n\times\sum_k p_k)$ and $(n\times\sum_s q_s)$. Let be, also, $W_Y = YY'$.
We denote v_k the coefficients vector $(p_k\times 1)$ of the linear combination for each

X_k such that $z_k = X_k v_k$.

Generalized CPCA (GCPCA) consists in seeking for K linear combinations z_k (or, in same way, K coefficients vectors v_k) simultaneously, such that:

$$\begin{cases} \max \sum_{i=1}^{K} \sum_{j=1}^{K} \langle Y' X_i v_i, Y' X_j v_j \rangle \\ \text{with the constraint} \sum_{k=1}^{K} \| X_k v_k \|^2 = 1 \end{cases} \quad \text{or} \quad \begin{cases} \max v'(A'A)v \\ \text{with } v'Bv = 1 \end{cases} \tag{1}$$

where $A = Y'X$, $v' = (v_1' | \ldots | v_k')$ and metric B, with

$$A'A = \begin{bmatrix} X_1'YY'X_1 & \ldots & X_1'YY'X_K \\ \vdots & \ddots & \vdots \\ X_K'YY'X_1 & \ldots & X_k'YY'X_k \end{bmatrix} \quad \text{and} \quad B = \begin{bmatrix} X_1'X_1 & \ldots & 0 \\ \vdots & \ddots & \vdots \\ 0 & \ldots & X_k'X_k \end{bmatrix}.$$

The constrained maximum problem (1) turns out to be an extension of criterium $\sup_{\sum_k \|X_k v_k\|^2 = 1} \sum_i \sum_k \langle X_i v_i, X_k v_k \rangle$ (Sabatier, 1993) with more sets of criterion variables. This criterion takes in account a unitary weight for each matrix X_k.

The solution of this constrained maximum problem leads to solve the eigen-equation

$$\sum_{k=1}^{K} P_{X_k} W_Y g = \lambda g \tag{2}$$

where $g = Xv$ and $P_{X_k} = X_k (X_k' X_k)^{-1} X_k'$. Starting from the relation

$$P_{X_k} W_Y X v = \lambda X_k v_k \tag{3}$$

(which is obtained from the expression $X' W_Y g = \lambda B v$) the coefficients vectors v_k and the linear combinations $z_k = X_k v_k$ maximizing (1) can be given by the relations $v_k = \frac{1}{\lambda}(X_k' X_k)^{-1} X_k' W_Y X v$ and $z_k = \frac{1}{\lambda} P_{X_k} W_Y X v$, respectively.

Solution eigenvectors g_l $(l = 1, \ldots, rank(\sum_{k=1}^{K} P_{X_k} W_y))$ can be written, according to (2) and (3), as sum of the linear combinations z_k: $g_l = \sum_k X_k v_k$

The role played by CPCA, which is born within a geometrical context, in eigen-equation (2) results evident, leading us to consider GCPCA as a CPCA generalization. As matter of a fact, the CPCA eigen equations in presence of one set of explanatory variables $(K = 1)$ and one set of criterion variables $(S = 1)$, as

with more than one set of the latter $(S > 1)$, are the same.
GCPCA is an extension of CPCA to the case of more than one set of explanatory variables $(S > 1; K > 1)$. We remark that GCPCA seeks, simultaneously, K constrained linear combinations (one for each X_k) taking in account also the criterion variables. In this way, each explanatory set of variables develops a greater role. A GCPCA property is to be invariant for simultaneously non-singular different transformations of matrices X_k. Furthermore, according to the CPCA properties, different classical techniques proposed in literature can be considered as particular cases of GCPCA depending on the nature of the variables (quantitative and/or qualitative coded according to the disjunctive form) and on the number of sets of variables.

3. Technique features influences on overall pasta liking by consumers

Frequently firm have to define which mix of technical parameters mostly influence the liking of a product. The need of marketing teams is to know precisely each parameter's influence on acceptability and to obtain in this way directives to correct the defaults of their products. The selection of the "optimal" mix of parameters may be not a simple question.
In order to evaluate technique features influences on overall pasta liking by consumers (Fig. 1.b) we have analyzed the statistical links between some chemical and physical descriptive variables and some hedonic variables evaluated on 36 different brands of pasta.
These 36 brands were rated for technical descriptive variables (*physical*: **color, temperature**, **humidity** and **ashes**; *chemical*: **gluten, protein** and **acidity**) by a laboratory and these data were collected in two matrices, X_1 and X_2, of order (36×4) and (36×3) respectively. Besides the same brands were rated for some hedonic variables: pile (**pile**), consistency (**cons**) and stickiness (**stic**), by 11 trained consumers with scores given on a discrete scale from 0 to 6.
These hedonic data were all collected in a three-way matrix Y of order $(36 \times 3 \times 11)$. It is evident that pasta liking by consumers is affected by the technical structure and we assumed the set of variables, gathered all together, describing the latter as reference space. In this framework, our aim was to point out the role of each explanatory set of variables.
The results of GCPCA referred to all technical variables and hedonic evaluations are reported in Fig. 2 which relate to the factorial plane 1-2 (65.94 % of explained inertia). Following the Daudin's approach (1988), a stability analysis has been performed to evaluate, by bootstrap, the goodness of the results: the orientation of the first factorial plane seems not considerably modified over 300 replications.
The reference (physical/chemical characteristics) and the criterion variables (hedonic variables for each consumer) which have shown the greatest correlations with the factorial axes, are reported in Fig. 2.a and Fig. 2.b respectively.

Figure 2.a: *(P)hysical–(C)hemical Variables. 1° factorial plane.*

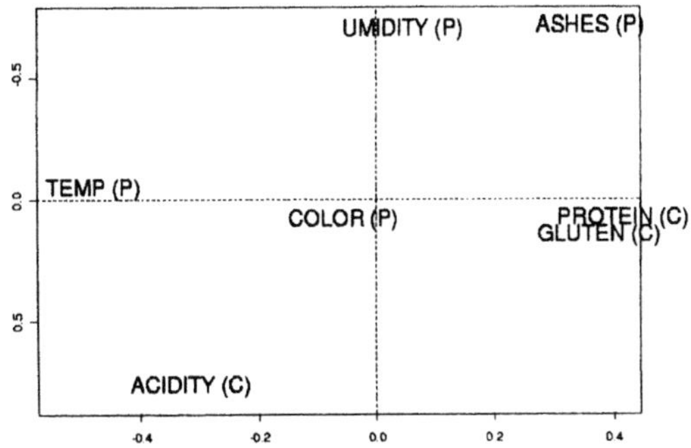

Figure 2.b: *Hedonic evaluations / Physical-Chemical Variables. 1° factorial plane.*

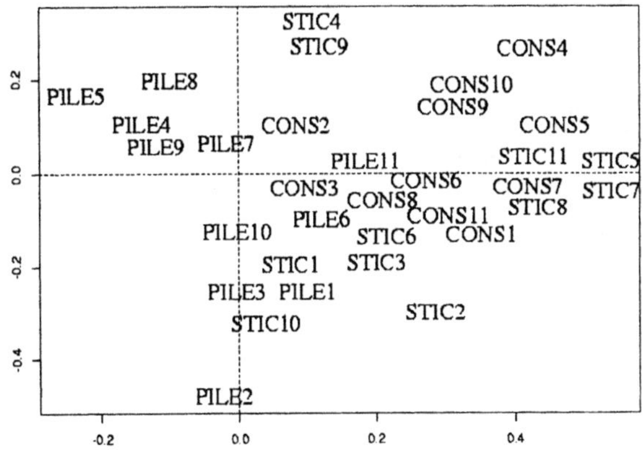

The interpretation, by the logical overlap of the pictures, is quite immediate and it allows to point out the consumers peculiarities of technical descriptive variables. The first axis of Fig. 2.a can be explained by the gluten/protein-temperature opposition (due to the denature of the formers respect to high values of the latter) while the second one by the contrast between humidity/ashes-acidity with a neutral position of color on both axes.

From the position of the reference variables, it comes out how some consumers have had a similar behavior for the pile (Fig. 2.b) respect to the temperature (taking into account all explanatory variables) while gluten/protein and ashes would be at the base of many consistency and stickiness evaluations.

Physical set seems be more influential in hedonic evaluations by consumers on both axes. This aspect can be seen in a more stable position of hedonic evaluations by consumers in Fig 3.a (physical variables) respect to the same point in Fig 3.b (chemical variables) taking into account their original positions in Fig 2.b.

Figure 3.a: *Hedonic evaluations / Physical Variables. 1° factorial plane.*

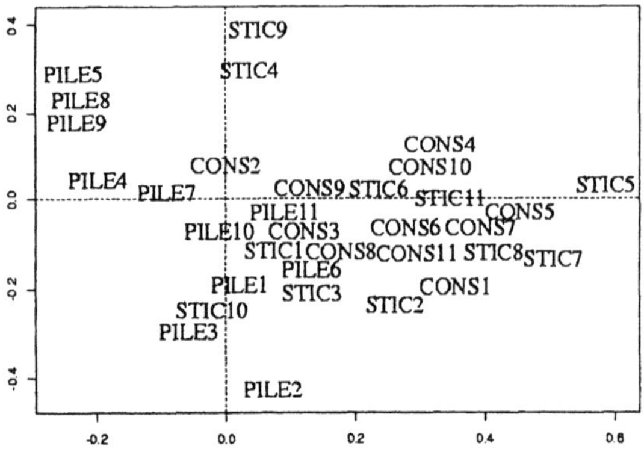

Figure 3.b: *Hedonic evaluations / Chemical Variables. 1° factorial plane.*

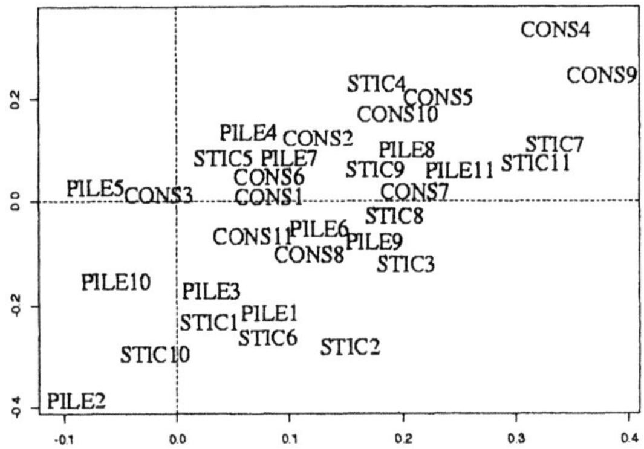

These considerations have to be taken into account on overall pasta liking by consumers. From Fig. 4 we see how some consumers have had a similar behavior, respect to the brands 25 and 8, for the pile (Fig. 2.b) according to high values of temperature (Fig. 2.a). In the same way, high values of gluten/protein (and ashes) would be at the base of many consistency and stickiness evaluations for the brands 36, 24, 27, 18, 17 and 23. In the meantime, Acidity and Umidity (second axis) have an influence on similar consistency and stickiness evaluations for the brands 16, 22, 10, 4 and 33, 32, 35, 30 and 26, respectively.
Figures 4.b, 5.a and 5.b could be read in the same way as Fig 3 considering the different contributions of physical and chemical variables. The result of GCPCA referred to all brands (Fig. 4) highlights three groups. Fig. 5.a is not able to show this discrimination while Fig. 5.b highlights two groups with a clearer specification for the brands: group, with brand 25, with high-medium values of Gluten and Protein and low values of Acidity, and the group, with brand 8, with high values of Acidity and low values of Gluten and Protein and a more evident contribution of all chemical variables for the brand 19. Moreover, Fig. 5.a and

Fig. 5.b show the physical and chemical effects on different brands positions with respect to their original ones in Fig 4.

Figure 4: *Pasta liking / Physical-Chemical Variables. 1° factorial plane.*

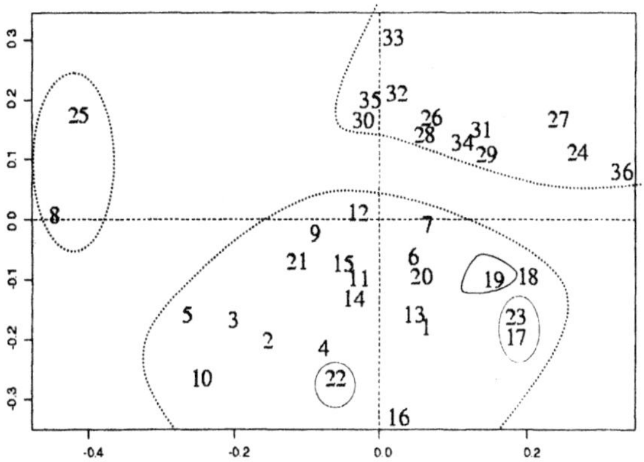

Figure 5.a: *Pasta liking / Physical Variables. 1° factorial plane.*

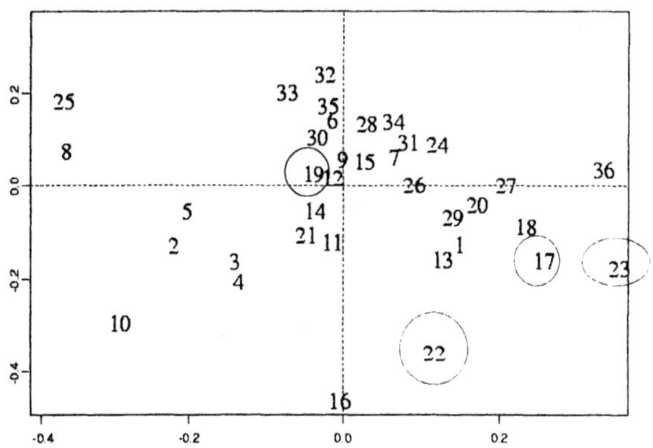

Figure 5.b: *Pasta liking / Chemical Variables. 1° factorial plane.*

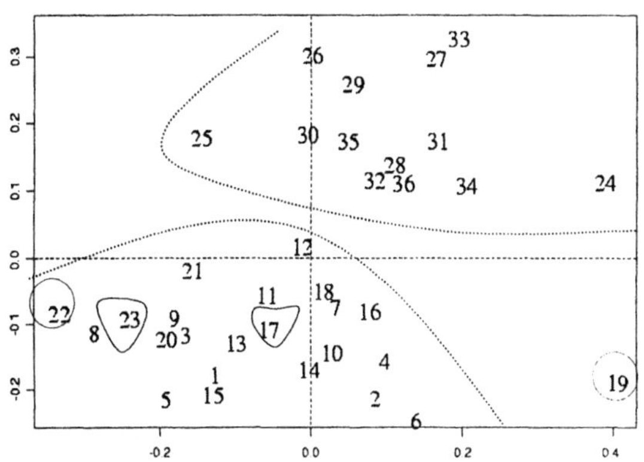

4. Conclusion

In this paper, we have proposed a generalization of CPCA to the "two blocks" dependence framework. A more flexible approach is to develop a single mathematical framework with external information on statistical units, criterion and explanatory variables. This development makes the "two blocks" multivariate analysis of structured data to fit the questions raised by the experimenter.
A different PLS based approach can be, also, developed to overcome the case with more statistical units respect to the number of variables which involves the not availability of the projection operators P_{X_k}. This approach, will be based on the criteria $\max \sum_{k=1}^{K} \mathrm{cov}(X_k v_k, Yg)^2$ (D'Ambra, Sabatier, Amenta, 1998) subject to the constraints $v_k' v_k = 1$ and $g'g = 1$.

References

D'Ambra L., Lauro N.C. (1982) Analisi in componenti principali in rapporto ad un sottospazio di riferimento, *Rivista di Statistica Applicata*, n.1, vol.15.

D'Ambra L., Sabatier R., Amenta P. (1998) Analisi fattoriale delle matrici a tre vie: sintesi e nuovi approcci, (invited lecture) *Atti XXXIX Riunione SIS.*

Daudin J.J, Duby C., Trecourt P. (1988) Stability of principal component analysis studied by bootstrap method. *Statistics*, 19, 2, 241-258.

Huon de Kermadec F., Durand J.F., Sabatier R. (1996) Comparaison de méthodes de régression pour l'étude des liens entre données hédoniques, in Third Sensometrics Meeting, E.N.T.I.A.A., Nantes.

Huon de Kermadec F., Durand J.F., Sabatier R. (1997) Comparison between linear and nonlinear PLS methods to explain overall liking from sensory characteristics, *Food Quality and Preference*, vol. 8, n. 5/6.

Kiers H.A.L. (1991) Hierarchical relations among three way methods *Psychometrika,* 56.

Kvalheim O.M. (1988) A partial least squares approach to interpretative analysis of multivariate analysis, *Chemometrics and Intelligent Laboratory System*, 3.

MacFie H.J.H, Thomson D.M.H. (1988) Preference mapping and multidimensional scaling methods, in: *Sensory Analysis of Foods*. Elsevier Applied Science, London.

Schlich P. (1995) Preference mapping: relating consumer preferences to sensory or instrumental measurements, in: *Bioflavour*, INRA, Dijon.

Wold S., Geladi P., Esbensen K., Ohman J. (1987) Multi-way principal components and PLS-analysis, *Journal of Chemometrics*, vol.1.

(1) *This research has been supported with a grant C.N.R. 1998 (Prof. L. D'Ambra).*

Interaction Terms in Homogeneity Analysis: Higher Order Non-Linear Multiple Correspondence Analysis [1]

Rosaria Lombardo
Institute of Management Research and Quantitative Methodology- S U N- Italy-
e-mail: Rosaria.Lombardo@unina2.it

Jan van Rijckevorsel
Department of Statistics -TNO PG Leiden -The Netherlands-
e-mail: JLA.vanRijckevorsel@pg.tno.nl

Abstract: This article aims to generalise Homogeneity Analysis (*HA*, van Rijckevorsel & de Leeuw 1988) by the introduction and optimal selection of interaction terms as linear manifolds of univariate and multivariately transformed variables (Costanzo & van Rijckevorsel 1994).
Optimal interaction terms are a general way to introduce higher order non-linear Multiple Correspondence Analysis.

Keywords: Homogeneity Analysis, Interaction terms, Multivariate B-spline, Chi-square, Burt tables, Multi-way tables.

1. Introduction

When the data do not come from a multinormal distribution and thus the linear correlation coefficient does not coincide with the correlation ratio *Principal Component Analysis* (*PCA*) is stochastically misleading. To make *PCA* stochastically significant one can maximise correlation coefficient (Kojak, 1987) and in order to investigate more complex interrelations among variables one can introduce interaction terms and identify interaction terms that maximise the correlation coefficient.
Historically there exist two different approaches to Correspondence analysis (Gifi 1990): the Principal Component Analysis of categorical data (*HA*) and the bivariate analysis of a contingency table.
Here we study the non linear dependence among categorical variables by Homogeneity Analysis, also known as *Multiple Correspondence Analysis*, and identify the optimally fitting interaction terms.
The linearity assumptions are relaxed by using the most common form of coding i.e. crisp coding or coding by dummy variables. Various attempts have been presented to generalise *PCA* (Gnanadesikan & Wilk, 1969) including both non-linearly transformed variables and "constructed" variables which are functions of the original variables, useful to reveal the model underlying the data.

The higher order non-linear *MCA*, which detects dependencies i.e. interactions among variables, uses univariate and multivariate transformations. The use of interaction terms in *HA* means looking for functions of two or more variables. The order number of interaction depends on the number of variables involved in the analysis. In this paper the adopted functions are linear combinations of Basis spline functions which are also known as tensor products.

2. Interactions in homogeneity analysis

Homogeneity Analysis using crisply coded variables, approximates the matrix of chi-squared distances between objects (or between categories). Martin (1980) proved that the fuzzy coding of two random variables with a known distribution, is an asymptotic approximation of the crisp coding in terms of the total chi-squared test value. Van Rijckevorsel and de Leeuw (1993) illustrate that crisp and fuzzy coding of two random variables with a normal distribution are asymptotic approximations of the product moment correlation.

In this paper we prefer crisp rather than fuzzy coding because it is extremely simple and has a straightforward relationship with the representation by bivariate contingency tables. The association between any pair of categorical variables can be expressed in terms of the classical chi-square statistic. Lancaster (1951) proposed a detailed analysis of deviations from independence to study the association among variables based on the decomposition of *chi-square* statistic.

For example in a set of m=4 variables, the total chi-square has been split up dimension-wise (six interactions of the first order, involving the bivariate marginals, four interactions of the second order among three variables and one of the third order with respect to all the variables), identifying the proportion of *chi-square* accounted for each term. Partitionings of the total chi-square test value of all bivariate contingency tables are interesting because they can discriminate the amount of homogeneity due to the size of the problem from the amount of homogeneity attributed to the model parameters. The former are called the *inertia parameters*, the latter the *goodness of fit* parameters (van Rijckevorsel 1987, p.155). In a broad sense *HA* aims to optimize the homogeneity of categorical or categorised variables (the distinction between continuous and categorical data is formally non existent after coding). The departure from homogeneity is measured by the loss function:

$$m^{-1}\sum_{j=1}^{m} SSQ(g(\mathbf{h}_j)-\mathbf{X}) \tag{1}$$

where **H** is the original data matrix (of order n,m, with m the number of variables, n the number of objects) and $\mathbf{h}_j$ its generic column; the function $g(\mathbf{h}_j)=\mathbf{G}_j\mathbf{y}_j$ represents the transformation of variable j by the univariate B-spline basis and $\mathbf{G}_j=\{\mathbf{g}_1,\ldots,\mathbf{g}_k\}$ is the (n,k_j) coding matrix, $\mathbf{X}_{(n,p)}$ the component scores (object or factor scores) matrix with p components; $\mathbf{y}_j$ the weighting coefficients (component or factor loadings). Equation (1) measures the departure of the transformed variables from an hypothetical common variable. It is minimised with respect to the $\mathbf{X}$ and $\mathbf{y}_j$ matrices by an Alternating Least Squares (*ALS*) algorithm. Introducing interaction terms, *HA* assumes the following modified loss function (Costanzo & van Rijckevorsel, 1994):

$$\sigma_2 = m^{-1}\sum_{j=1}^{m} SSQ(\mathrm{g}(\mathbf{h}_j) - \mathbf{X}) + m^{*(-1)}\sum_{j=1}^{m^*} SSQ(\mathrm{g}^*(\mathbf{h}_j) - \mathbf{X}) = \sigma_1 + \sigma_1^* \quad (2)$$

with m^* the number of interaction terms, to be determined for every specific data context, and the function $g^*(\mathbf{h}_j) = \mathbf{G}^*_j \mathbf{y}_l$ with $\mathbf{G}^*_j = \otimes_d \mathbf{G}_{j \le d}$ the tensor product of the univariate B-spline functions (Schumaker, 1981) which is equal to the multivariate interaction (with $d = 2,..,m^*$ number of univariate spline involved).

The total inertia of a data-set is proportional to the sum of all eigen-values of

$$\mathbf{G}^T\mathbf{D}^{-1}\mathbf{G} \quad (3)$$

where $\mathbf{G} = [\mathbf{G}_1 | \mathbf{G}_2 | ... | \mathbf{G}_m]$, and $\mathbf{D}$ is the diagonal partitioned matrix multivariate extension of $\mathbf{D}_j = (\mathbf{G}_j^T\mathbf{G}_j)$. The chi-square distance is the squared difference between two rows of $\mathbf{GD}^{-1/2}$ (of order n,mk).

By expression (2) we are able to identify the particular variables which enter into the model which either enter purely in an additive way or are also involved in interactions with other variables. The interpretation of the model is greatly facilitated through the ANOVA decomposition model (see also Friedman, 1991).

In order to investigate about the amount of homogeneity due to the size of the problem we check the total inertia of a data-set solving the eigensystem (4), which has the same solutions of equation (3)

$$\mathbf{D}^{-1/2}(\mathbf{C} - \mathbf{Duu'D}/n)\mathbf{D}^{-1/2} = \mathbf{W\Lambda\Lambda}^T \quad (4)$$

where the Burt matrix $\mathbf{C} = \mathbf{G}^T\mathbf{G}$ contains all pair-wise contingency tables and the univariate margins. Notice that, because of the centring of $\mathbf{C}$, the trivial solution $\mathbf{w} = \mathbf{u}\sqrt{n}$ cannot occur. The non-diagonal sub-matrix of (4) contains the scaled bivariate marginals of the variables j and l

$$\mathbf{D}_j^{-1/2}(\mathbf{C}_{jl} - \mathbf{D}_j\mathbf{uu'D}_l/n)\mathbf{D}_l^{-1/2}$$

After multiplying this matrix with $\sqrt{n}$, its elements equal the discrepancy between the observed and the expected values, divided by the square root of the expected values. So the sum of squares of a non-diagonal sub-matrix of (4) equals the chi-square statistics divided by n (Gifi 1990).

For a diagonal sub-matrix of (4):

$$\mathbf{D}_j^{-1/2}(\mathbf{D}_j - \mathbf{D}_j\mathbf{uu'D}_j/n)\mathbf{D}_j^{-1/2}$$

the sum of squares is $\mathrm{k}_j - 1$ (it is idempotent sub-matrix) and for all sub-matrices is equal to $\sum_{j=1}^{m} k_j - m$.

Then the sum of squares of (4) is equal to

$$\sum_{j=1}^{m} k_j - m + \sum_{i \neq l} \chi_{il}^2 / n = \sum_j \lambda_j^4$$

so it is shown (van Rijckevorsel & de Leeuw, 1988) that

$$\sum_{i \neq l} \chi_{il}^2 / n = \sum_j (\lambda_j^2 - 1)^2 \qquad (5)$$

In conclusion we can observe that *HA*, when studied through Burt tables, does not allow to take into account interactions among variables but only their bivariate margins. Instead, the relationship between *HA* and chi-square statistics may be more completely studied through Burt tables plus the multivariate margins. This permits the evaluation of the global measure of dependence computing interaction among variables of higher order. When we add the multivariate interactions as separate variables, we practically handle a multi-way contingency table, which takes into account higher order interaction among variables.

3. The computational procedure

The algorithm we propose for the optimal selection of the interaction terms uses the recursive partitioning algorithm proposed by Friedman (1991), in the optimal interaction detection phase.

Since the loss σ_2 can be written as the sum of the two losses σ_1 and σ_1* we can split up the problem of simultaneously considering univariate and multivariate transformations. In optimising the spline transformations we distinguish between the univariate and the multivariate case. In the univariate case, the knot sequence (number and location) has been adaptively optimised by van Rijckevorsel and Tessitore (1993), and by Lombardo *et al.* (1997). In the multivariate case the situation is more complicated, because the optimal knots found for the univariate case could lead to the presence of empty areas in the multivariate scatter of the interaction variable. The problem of empty areas is linked to the correlation between the variables and to the choice of the B-spline basis.

In this first exercise, in order not to increase the computational burden of the procedure and avoid the curse of dimensionality, we fix the degree of the spline and the sequence of the knots for both univariate and multivariate B-spline transformations.

Preliminary phase:

The procedure starts from the computation of the linear principal components **X**.

Forward phase:

- Code the original variables by crisp coding (with splines of degree zero) or fuzzy variables (with splines of degree 1).
- Measure the lack of fit function σ_1.
- Compute the optimal interaction among all the possible ones for each order number of interaction.
- Introduce the optimal interaction as new variable in the model.
- Measure the lack of fit of the new model.

- Evaluate $\| \sigma_2 - \sigma_1 \| > \varepsilon$ for a fixed small value of ε ($\varepsilon = 5\%$).
- If the decision is to include a certain interaction, then this term is added as a new variable in the model and successively an update version of σ_1 i.e. σ_1^+ will be considered. This decision is based on the percentage of increment in the explained variability or on the percentage of decrement of the lack of fit function.
- Continue until all interaction terms of different order number are evaluated.
- The final model is estimated on the basis of the new interaction variables.

Backward phase:

The backward pruning procedure takes all interaction orders into account.

- Prune basis function one variable at time.
- Refit the model after each removal
- Keep track of the goodness of fit for each model
- Choose the model in the sequence having the largest goodness of fit measure (*GCV* criterion).

This is a trade-off problem between the accuracy of the summarisation and the risk of overfitting related to an increasing number of interaction variables. The adopted criterion is a modified form of the *Generalised Cross-Validation* (*GCV*, Craven and Wahba, 1979) which is the average-squared residual of the fit to the data divided for a penalty to account for the increased variance associated with increasing model complexity (maximum number of interaction terms *r*):

$$GCV(R) = \frac{1}{n} \frac{\sum_{i=1}^{n} m^{-1} \sum_{j=1}^{m} SSQ(g_R(h_{ij}) - x_{ij})}{[1 - \hat{C}(R)/n]^2}$$

here the dependency of the function $g(.)$ on the number of interactions $R=r,...,0$, is explicitly indicated, and $\hat{C}(R) = tr\left(\mathbf{G}\left(\mathbf{G}^T\mathbf{G}\right)^{-1}\mathbf{G}^T\right) + dR$ represents a suitable increasing cost function of the dimension space. In the regression context, Friedman and Silverman (1989) reports reasons for choosing $2 \le d \le 4$.

4. Application

In this section we present an example to study the association among four variables. The data are listed in "Chip computer", an Italian magazine the issue of September 1999. The data table 1 refers to 20 printers by four variables: **Standard equipment (D)**, (1=optimum, 2=good, 3=sufficient), **Ergonomics (E)**, (1=optimum, 2=good, 3=sufficient), **Manual textbook knowledge (M)**, (1=optimum, 2=good, 3=sufficient) and **Price (P)**. The printer brands are: 1) Minolta PagePro 8L (**LMIN**), 2) Epson Epl-5700 (**LEPS**), 3) Xerox DocuPrint

P8e (**LXER**), 4) Oki Okipage 8p (**LOKI**), 5) Kyocera Fs-680 (**LKYO**), 6) Hewlett-Packard LaserJet 1100 (**LHP**), 7) Samsung Mi-85 Plus (**LSAM**), 8) Lexmark Optra E-310 (**LLEX**), 9) Oki Okipage 6W (**LOKI**), 10) Canon Lbp-660 (**LCAN**), 11) Hewlett-Packard DeskJet 895 Cxi (**CHP**), 12) Epson Stylus Color 640 (**CEPS**), 13) Hewlett-Packard DeskJet 815 C (**CHP2**), 14) Epson Stylus Color 740 (**CEPS2**), 15) Olivetti Jp883 (**COLI**), 16) Canon Bjc-6000 (**CCAN**), 17) Canon Bjc 2000 (**CCAN2**), 18) Xerox DocuPrint Xj6C (**CXER**), 19) Lexmark Jetprinter 3200 (CLEX), 20) Xerox DocuPrint Xj8C (**CXER2**).

Table 1: *The Printers*

	D	E	M	Price
LMIN	2	2	1	852
LEPS	2	3	1	1079
LXER	3	2	2	956
LOKI	2	3	3	956
LKYO	3	2	3	1074
LHP	2	3	1	1076
LSAM	1	2	1	1050
LLEX	2	2	2	1020
LOKI	3	2	3	514
LCAN	2	2	1	704
CHP	1	1	2	876
CEPS	2	1	2	502
CHP2	1	2	2	499
CEPS2	2	2	2	646
COLI	2	3	2	494
CCAN	2	2	2	708
CCAN2	3	2	2	299
CXER	3	3	2	431
CLEX	2	2	2	420
CXER2	3	3	2	538

We adopt the crisp coding with three knots in fixed location. We fit the model in two sequential steps, with and without interaction terms.

Table 2: *Interactions of the 1st order*

First order Interactions	Lof
D*E	1231025
D*M	979505,4
D*Price	250275,9
E*M	991282,4
E*Price	241570,7
M*Price	189816,7

At first we choose the optimal interaction of the first order. We evaluate the lack of fit of the model for each of the six interaction terms: 1) **D*E**, 2) **D*M**, 3) **D*Price**, 4) **E*M**, 5) **E*Price**, 6) **M*Price** (see table 2). The term which has the least lack of fit (*LOF*) is **M*Price** (see table 2).

The interactions of the second order are four: 1) **D* E*M,** 2) **D*E*Price** 3) **D*M*Price**; 4) **E* M*Price**. We show that the optimal interaction of the second order is the third interaction (**D*M*Price)** (see table 3).

Table 3: *Interactions of the 2nd order*

Second order Interactions	Lof
D*E*M	881970,3
D*E*Price	235122,5
D*M*Price	184083,7
E*M*Price	185668,0

In table 4 we can read the different values of model fit in case of: 1) no interactions, 2) model (1) plus one interaction of the first order, 3) model (2) plus one interaction of the first and one of the second order, 4) model (3) with all interactions of the first, second and third order. We observe the naturally increased value of model fit to data after adding interactions (r=3).

Table 4: *The Fit measure of the models*

	Fit of Model
Model without interaction	0,94
(1) with the 1st order interaction	1,76
(2) with the 1st and 2th order interaction	3,16
(3) with all the1st , 2nd, and 3th order interaction	4,74

Keeping in mind equation 5, we compute the phi-square statistics at first on the model of the original variables and at second on the model which include interactions terms (table 5).

Table 5: *The Phi-square statistics of the models*

	Phi-square
Model without interaction	1,63
(1) with the 1st order interaction	26,58
(2) with the 1st and 2nd order interaction	152,18
(3) with all the 1st , 2nd and 3th order interaction	501,92

At end in order to avoid overfitting problem we run the backward procedure which allow to choose the optimal model. We see in table 6 that models with the largest order number of interactions have the minimum *GCV* value, while the model with the first order interaction (**M*Price**) has in the sequence the largest value of *GCV* statistics.

Table 6: *The GCV statistics of the models*

	GCV statistics with d=3
Model without interaction	0.0032
(1) with the 1st order interaction	0.0057
(2) with the 1st and 2nd order interaction	1.2e-024
(3) with all the 1st, 2nd and 3th order interaction	1.1e-028

5. Conclusion

With non-linear zero degree B-spline transformations of the data and optimal selection of the most important interactions, we are able to identify the most important interaction terms and thus visualise the underlying model in non parametric-way. The change to a first degree B-spline basis will increase the computational burden but this study proves feasibility of this non-parametric approach. By the strategy of Friedman's algorithm (Friedman 1991) we select the optimal interactions. This procedure allows to evaluate exactly the global measure of dependence and can be seen as particularly useful for categorised multi-way data tables which have higher order interaction terms.

References

Craven P., & Wahba G. (1979). Smoothing noisy data with spline functions. Estimating the correct degree of smoothing by the method of generalised cross-validation. *Numer.Math.* 31, 317-403.

Costanzo G.D. & Rijckevorsel J.L.A. van (1994). *Interaction in Non-Linear Component Analysis Contributions to Statistics.* Statistical Theory and Computational Aspects of Smoothing W. Hardle, M.G. Schimek, eds. Springer.

Friedman J.H. (1991). Multivariate Adaptive Regression Splines. *The Annals of Statistics*, 19,1:1-141.

Friedman J.H. & Silverman B.W. (1989). Flexible parsimonious smoothing and additive modeling. *Technometrics*, 31,3-39.

Gifi A. (1990). *Non Linear multivariate analysis*. Chichester: Wiley.

Gnanadesikan R. & Wilk M.B. (1969). Data analytic methods in multivariate statistical analysis. *Multivariate analysis* II. Krishnaiah PR, NewYork.

Lancaster H.O. (1951). Complex contingency tables treated by the partition of the chi-square. *Journal of Royal Statistical Society, Series B*, 13, 242-249.

Lombardo R., D'Ambra L., and Tessitore G. (1997). An algorithm for detecting the number of knots in Constrained PCA. *Contributed papers*, ISI 1997.

Martin J.F. (1980). Le Codage Flou et ses Applications en Statistique..*Thèse de Doctorat de3ème cycle*, Universitè de Pau et des Pays de l'Ardour.

Rijckevorsel J.L.A. van & Tessitore G. (1993). An algorithm for MACCA. *Contributed papers*, vol. 2, ISI 49th session.

Rijckevorsel J.L.A. van & de Leeuw J. (1988). *Component and Correspondence Analysis*. Chicester: Wiley.

Rijckevorsel J.L.A. van (1987). *The application of horseshoes and fuzzy coding in multiple correspondence analysis*. Leiden: DSWO Press.

Schumaker, L. (1981). *Spline Functions: Basic Theory*. New York: Wiley.

(1) The present paper has been financially supported by a grant from SUN "Young Reseacher Project" (responsible R. Lombardo).

Perturbation Models for Principal Component Analysis of Rainwater Pollution Data

Pietro Mantovan - Andrea Pastore
Dipartimento di Statistica - Università Ca' Foscari di Venezia
S. Polo, 2375 - I-30125 Venezia (Italy)

Abstract: The study of the correlation matrix between ion concentration and the principal component analysis on the related variance matrix are widely used to explore the presence of contamination patterns in rainwater. The paper shows that the covariance between ion concentrations is a perturbed measure, and that the total conductivity can be interpreted as the perturbation factor. Then, the paper describes some strategies for measuring and removing the perturbation and how, by removing this effect, correct contamination patterns can be identified. A summary of the results of an application on data measured by the monitoring network of the Veneto region is proposed.

Keywords: principal component analysis, compositional data, rainwater pollution.

1. Introduction

Many programs have been developed recently in order to monitor the quality of precipitation. One of the principal aims in this context is to identify, for a certain territorial area, some contamination patterns and the related causes. Possible causes are: proximity of a see or of a lake, of industrial settlements, of towns, of roads, and so on. In order to identify contamination patterns, using data collected from monitoring network, the variance matrix of the ion concentration measures of the pollutants is considered. Then, usually, principal component analyses are performed on the basis of that variance matrix. But it is possible to prove that the variability of all the ion concentration measures is influenced, sometimes greatly, by a common factor and that the total conductivity can represent this *perturbation* factor. Because of this perturbation, the analysis of variance (or correlation) matrix between ion concentration measures cannot correctly highlight the contamination patterns.

The paper, after a discussion of these aspects, describes some strategies for measuring and removing the perturbation and how, by removing this effect, correct contamination patterns can be identified. Finally, some results on data measured by the monitoring network of the Veneto region, in Italy, are reported.

2. A perturbation model for the rainwater pollution data: strategies for perturbation removal

Many papers reported examples of correlation matrix between ion concentration measures from monitoring network of different areas (for instance: Zeng and Hopke, 1989; Saylor *et al.*, 1992; Mantovan *et al.*, 1995). The common context is a set of variable all positively correlated. This fact suggests the presence of a common factor that influences the variability of all the variables. An exploratory analysis of the Veneto region data (Mantovan *et al.*, 1995) highlighted that in all the considered stations most of the total variability could be explained by a principal component very closed (in terms of linear correlation) with the total conductivity. This fact happens also for stations which are located in quite different areas and probably related to different pollution sources. A reasonable suspicion is that the total conductivity could be taken either as the common factor (which explains most of the total variability), or at least as a good approximation of it.

We propose to model these quantities as follows. Let $\mathbf{w} \in \mathrm{R}^k$ be the random vector of the ion concentrations (expressed as $\mu\mathrm{eq} \cdot \mathrm{l}^{-1}$) and c the total conductivity (expressed as $\mu\mathrm{S} \cdot \mathrm{cm}^{-1}$), measured at a rain event. The ion concentrations can be suitably expressed as measures of specific conductivity in a random vector $\mathbf{x}$, obtained with the transformation: $\mathbf{x} = \Lambda\mathbf{w}$ where Λ is a diagonal matrix whose entries are *coefficients of equivalent ionic conductivity at infinite dilution*. So, the vector $\mathbf{y}$:

$$\mathbf{y} = c^{-1}\mathbf{x} , \tag{1}$$

represents the composition of the total conductivity with respect to the different pollutants. The model (1) highlights as the composition $\mathbf{y}$ could be interpreted as a unperturbed measure, where the total conductivity c is the perturbation factor. It is reasonable to assume c to be "partially" independent from the composition $\mathbf{y}$. The total contamination c could depend on many exogenous variables (for instance: the volume of the precipitation), while the composition is mainly due to specific characterization of the location.

A discussion of the problem of principal component analysis with compositional data may be found in Aitchison (1983, 1986) along with a number of suggestions for alternative solutions. Data presenting in form of composition are common, for example, in chemistry and in geology. An example of application on government budget data can also be found in Jackson (1981, pp. 382). In this paper, we consider some aspects of the problem explicitly related with the variability of the the "total", that is the sum of the variables considered for the analysis.

Let us suppose to be interested in the factorial solution of the unperturbed vector $\mathbf{y}$, when only $\mathbf{x}$ is observable and only the first moments of c are known (or they can be estimated). Then, we need a strategy to approximate the unperturbed variance matrix $\Sigma = \mathrm{VAR}(\mathbf{y})$ from the perturbed one $\mathbf{S} = \mathrm{VAR}(\mathbf{x})$. A solution, based on some issues of matrix perturbation theory, has been proposed and discussed by Mantovan and Pastore (1996a, 1996b). From the model (1), under the assumptions:

$$\mathrm{COV}(c,\mathbf{y}) = \mathbf{0},\ \mathrm{COV}(c^2,\mathbf{yy}') = \mathbf{0}, \tag{2}$$

a decomposition of the perturbed variance matrices $\mathbf{S} = \mathrm{VAR}(\mathbf{x})$ can be derived, such that the matrix Σ can be computed as follows:

$$\Sigma = (\sigma^2 + \mu^2)^{-1}\mathbf{S} - z\mathbf{mm}' \tag{3}$$

where, $\mu = \mathrm{E}[c]$, $\sigma^2 = \mathrm{var}[c]$, $z = \sigma^2/(\sigma^2 + \mu^2)$ and $\mathbf{m} = \mathrm{E}[\mathbf{y}] = (1/\mu)\mathrm{E}[\mathbf{x}]$. The scalar z, $0 \le z \le 1$, can be considered as a measure of the effect of the perturbation factor c. Expliciting $\mathbf{S}$ in equation (3), it is possible to observe that $\mathbf{S}$ tends to become a matrix with all positive elements, and this is coherent with what real data suggest.
More detailed results about the relation between the correlation matrices $\mathrm{CORR}(\mathbf{x})$ and $\mathrm{CORR}(\mathbf{y})$ have been obtained (Mantovan and Pastore, 1996b, Theorem 1.). Conditions of invariance between the Jordan spectral forms of $\mathbf{S}$ and Σ have been also proposed.
Another strategy for perturbation removal is based on a log-normality assumption of the vector ($\mathbf{x}'$, c). The log-normal model bas been widely used for approximating the distribution of pollutant concentration measures (Crow, 1988). With this assumption, (Pastore, 1998), it is possible to become to the following expression for the VAR(LN($\mathbf{y}$)):

$$\mathrm{VAR}(\mathrm{LN}(\mathbf{y})) = \mathrm{VAR}(\mathrm{LN}(\mathbf{x})) - \mathbf{T}_z - \mathbf{T}_\alpha \mathbf{T}'_\alpha \tag{4}$$

where: $\mathbf{T}_z = -\ln(1-z)\mathbf{1}_{k,k}$ (being $\mathbf{1}_{k,k}$ a matrix of order k,k with all elements equal to one), $\mathbf{T}_\alpha = \mathbf{1}_{k,1}\gamma'$, with $\gamma = (\gamma_1,\ldots,\gamma_k)'$, and $\gamma_j = \mathrm{cov}[\ln(c),\ln(y_j)]$, $j = 1,\ldots,k$. It should be noted that $\mathbf{T}_z$ is a matrix which depends only on z (therefore it depends only the perturbation factor c). $\mathbf{T}_\alpha$ is matrix of rank ≤ 2, and is a function of parameters which essentially specify the dependence between c and $\mathbf{y}$. The variance matrix VAR($\mathbf{y}$) can be obtained from VAR(LN($\mathbf{y}$)) following standard properties of the multivariate log-normal distribution (see, for example, Johnson and Kotz, 1972).
These two strategies are different. The second one requires the assumption, sometimes not suitable, of log-normality, while it does not depend on assumptions on the dependence between the unperturbed random vector and the perturbation factor, required instead by the first strategy.
The assumption of log-normality allows to specify a dependence between the perturbation factor c and the unperturbed vector $\mathbf{y}$, via the parameters γ_j. With the

first strategy, the assumption (2), not so far from the independence between c and $\mathbf{y}$, is required. In the applications, the assumption (2) could not be exactly satisfied and, in this case, the expression (3) does not ensure Σ to be a positive semi-definite matrix.

3. Some results of an application to Veneto region data

An example of application has been carried out on data sampled from the Veneto region monitoring network. The network, consisting of eight stations, was established in 1987, in line with EMEP (European Monitoring Environmental Program) declarations. More details about the network and the collected data can be found in the papers by Zilio Grandi and Szpyrkovicz (1991) and Mantovan *et al.* (1995). In this last paper, for each of the eight stations of the network, the first principal component of the perturbed vector results to be correlated with the total conductivity. So, for example, a station located in a mountain zone appears to be very similar other station located in a rural plain zone or near the Adriatic sea.
For these data, the values of the unperturbed vector $\mathbf{y}$ are available, and then it has been possible to compare the principal component analysis of the "true" VAR($\mathbf{y}$) matrix with those of the unperturbed variance matrices obtained with the proposed perturbation removal strategies. In this paper some results of a complete analysis are reported in order to illustrate how the perturbation factor can *obscure* the contamination patterns. As an example, table 1 and 2 give the results respectively for station 1 (located in a mountain zone near Belluno) and 4 (located in Caorle, very closed to the Adriatic sea); the correlation between "observed" variable and the first two principal components are reported.

Table 1: *results of perturbed and unperturbed analyses (first and second principal components) for the station 1 of the Veneto region network.*

analysis		Perturbed analysis		true unperturbed analysis		unperturbed anal. A (matrix pert. theory)		unperturbed anal. B: (log-normal distrib.)	
principal component		I p.c.	II p.c.	I p.c.	II p.c.	I p.c.	II p.c.	I p.c.	II p.c.
% of explained variance		0.71	0.10	0.45	0.23	0.42	0.23	0.49	0.14
correlation between observed variable and principal components	HCO_3	0.45	0.48	0.50	0.54	0.11	0.53	0.98	0.00
	H	-0.25	-0.23	-0.99	0.06	-0.89	0.13	-0.80	0.33
	SO_4	0.95	-0.01	0.11	-0.36	0.77	-0.12	-0.21	-0.30
	NO_3	0.89	-0.22	0.17	-0.51	0.54	-0.47	-0.18	-0.57
	NH_4	0.84	-0.46	0.28	-0.86	0.27	-0.86	0.00	-0.87
	Cl	0.57	0.26	-0.03	0.25	-0.04	0.45	0.04	0.49
	Mg	0.79	0.29	0.37	0.38	0.35	0.42	0.42	0.15
	Na	0.45	0.30	0.06	0.24	-0.22	0.46	0.08	0.46
	Ca	0.89	0.43	0.59	0.67	0.82	0.50	0.68	0.19
	K	0.62	0.26	0.24	0.24	0.18	0.33	0.35	0.19
correlation between total conductivity and principal components		0.97	0.05	0.29	-0.06	0	0	0	0

In the analysis of station 1, it is quite evident that the first "true" unperturbed principal component is not heavily correlated with the total conductivity (that is the perturbation factor c) and also the differences between the results of the perturbed analysis and those of the "true" unperturbed one are clear. Both the approximate unperturbed analyses highlight the role of H in the first principal component and that of NH_4 in the second one. The second component of the unperturbed analysis based on the log-normal distribution seems to be different from the "true" one for the weights of HCO_3 and Ca. The graphs of the factorial plane of the three unperturbed analyses are shown in Figure 1. The factorial plane provided by the analysis based on the log-normal distribution assumption seems to be more similar to the true one than the plane provided by the analysis based on the matrix perturbation theory.

Figure 1: *graphical representation of unperturbed analyses for stations 1 of the Veneto region network (correlations between observed variables and principal components)*

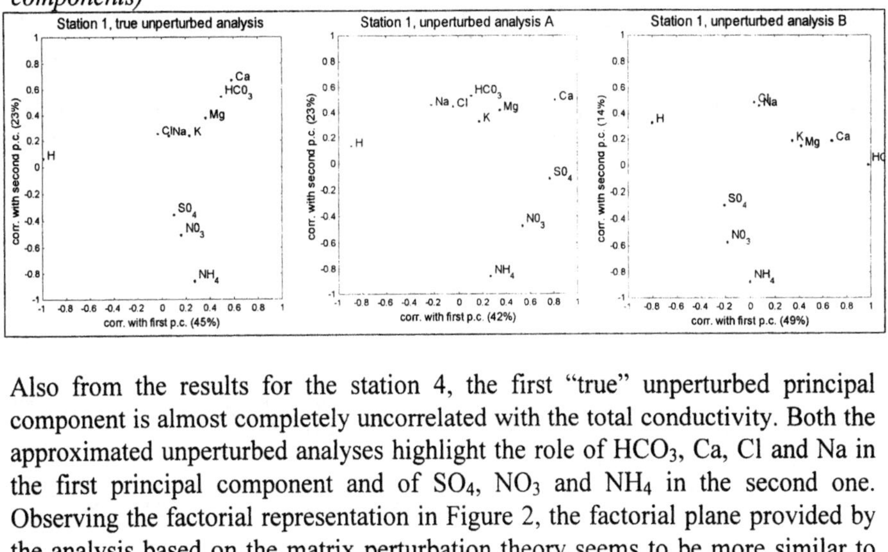

Also from the results for the station 4, the first "true" unperturbed principal component is almost completely uncorrelated with the total conductivity. Both the approximated unperturbed analyses highlight the role of HCO_3, Ca, Cl and Na in the first principal component and of SO_4, NO_3 and NH_4 in the second one. Observing the factorial representation in Figure 2, the factorial plane provided by the analysis based on the matrix perturbation theory seems to be more similar to the true one than the plane provided by the analysis based on the log-normal distribution assumption.

In the analysis of both these stations, with the strategy based on the log-normal distribution, the specification of simple hypothesis of dependence between the perturbation factor c and the unperturbed variables $\mathbf{y}$ does not seem to behave better.

For the other stations of the Veneto region network, the results are quite similar. The unperturbed analysis based on the matrix perturbation theory gives a better reconstruction of the true factorial planes when the assumptions (2) are verified. Otherwise, the analysis based on the log-normal distribution can be preferred.

Table 2: *results of perturbed and unperturbed analyses (first and second principal components) for the station 4 of the Veneto region network.*

analysis		Perturbed analysis		true unperturbed analysis		unperturbed anal. A (matrix pert. theory)		Unperturbed anal. B: (log-normal distrib.)	
principal component		I p.c.	II p.c.	I p.c.	II p.c.	I p.c.	II p.c.	I p.c.	II p.c.
% of explained variance		0.59	0.22	0.50	0.28	0.48	0.31	0.46	0.28
correlation between observed variable and principal components	HCO_3	0.72	-0.41	0.66	-0.66	0.74	-0.46	0.87	-0.43
	H	-0.20	0.10	-0.20	0.17	-0.23	0.24	-0.32	0.23
	SO_4	0.79	-0.07	0.18	0.86	0.11	0.63	-0.12	0.77
	NO_3	0.72	-0.06	0.31	0.67	0.22	0.72	0.10	0.76
	NH_4	0.53	-0.07	0.19	0.74	0.01	0.87	-0.04	0.74
	Cl	0.55	0.81	-0.94	-0.26	-0.71	-0.63	-0.77	-0.51
	Mg	0.80	0.17	0.16	-0.08	-0.17	0.41	0.18	0.03
	Na	0.46	0.80	-0.85	-0.26	-0.78	-0.54	-0.68	-0.61
	Ca	0.94	-0.27	0.78	-0.44	0.84	-0.41	0.79	0.01
	K	0.67	0.21	-0.07	0.02	-0.15	0.02	-0.07	-0.04
correlation between total conductivity and principal components		0.97	0.17	-0.03	0.09	0	0	0	0

Figure 2: *graphical representation of unperturbed analyses for stations 4 of the Veneto region network (correlations between observed variables and principal components)*

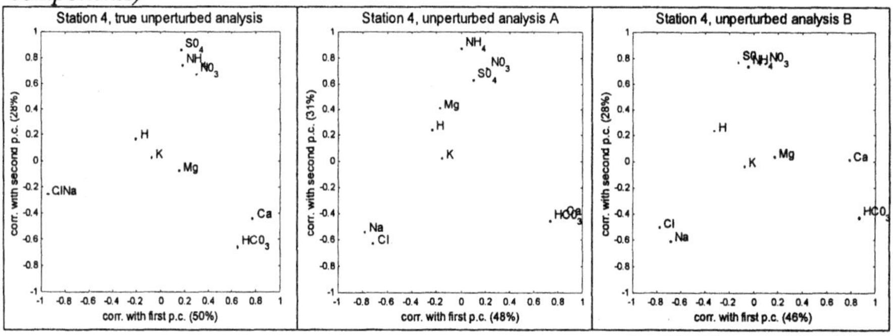

More detailed considerations can be proposed on the basis of the comparison of the results of different stations. Figure 3 contains graphical factorial representations of perturbed and true unperturbed analyses for stations 1 and 4. These stations are located in quite different areas and one might expect to find different patterns of contamination.

Graphs in the first row of the picture are those related to perturbed analyses, and they show similar patterns, where H is alone and not well represented in the factorial plane, while the other variables are well represented and have high values of correlation with the first principal component. That is, the patterns of the two stations seem to be similar, despite the different geographical location of the same stations. The unperturbed analyses, (graphs in the second row of Figure 2) on the contrary, show different patterns that can be connected to the different location of the stations.

Figure 3: *graphical representations of perturbed and true unperturbed analyses for stations 1 and 4 of the Veneto region network (correlations between observed variables and principal components)*

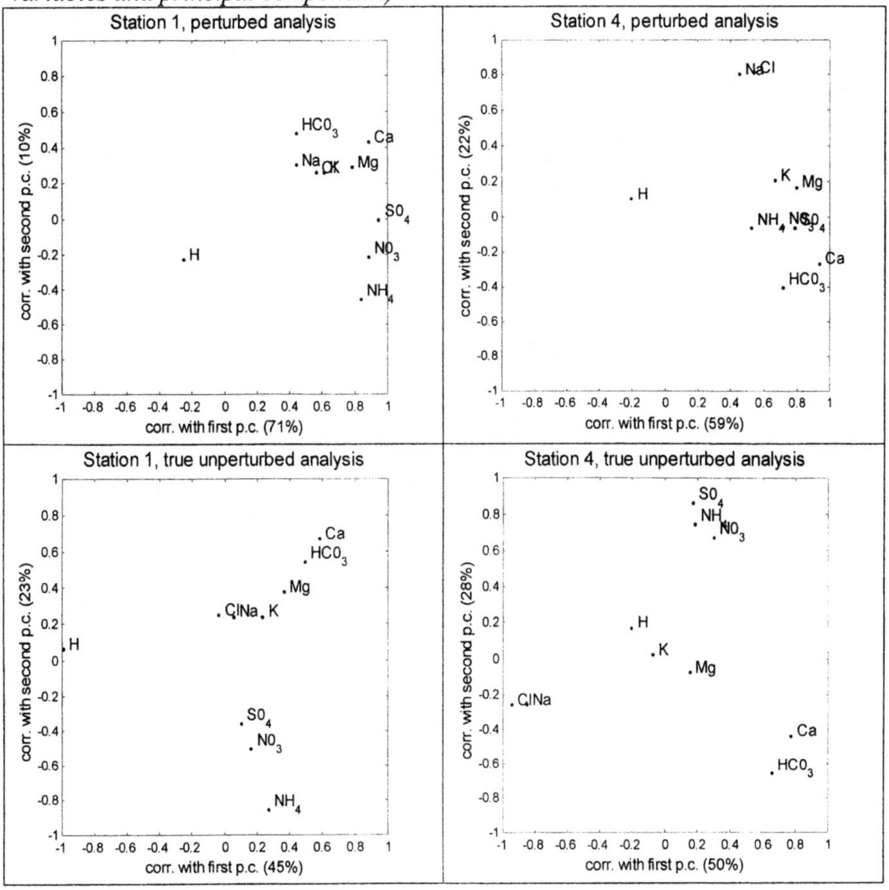

4. Conclusions

The proposed methods are based on some considerations on experimental data and on a particular model for the contamination process of rainwater. These methods appear to be appropriate to detect and remove the perturbation factor, in order to find significant and unperturbed patterns of contamination.

The solution based on the matrix perturbation theory gives better results when an assumption of uncorrelation between unperturbed variables and perturbation factor is verified. The solution based on the log-normal distribution assumption for the considered random vectors allows to model this dependence, but in practice the specification of these values of correlation is not easy to do.

References

Aitchison, J. (1983) Principal component analysis of compositional data, *Biometrika*, 70, 57-65.

Aitchison, J. (1986) *The Statistical Analysis of Compositional Data*, Chapman and Hall, New York.

Crow E.L. (1988) Applications in Atmospheric Sciences, in: *Lognormal distributions, theory and applications*, E.L. Crow and K. Shimizu (Eds.), Dekker, New York, 331-356.

Jackson J.E. (1991) *A User's Guide To Principal Components*, Wiley, New York.

Johnson N.L.,Kotz S. (1972) *Distributions in Statistics: Continuous Multivariate Distributions*, Wiley, New York.

Mantovan P, Pastore A. (1996a) Alcuni problemi di identificazione di soluzioni fattoriali perturbate, *Società Italiana di Statistica, Atti della XXXVIII Riunione Scientifica*, Maggioli editore, 61-68

Mantovan P., Pastore A. (1996b) A Perturbation Model for Factorial Solutions with Application *Metron*, LIV(3-4), 39-63.

Mantovan P., Pastore A., Szpyrkovicz L., Zilio Grandi F. (1995) Characterization of rainwater quality from the Veneto region network data using Multiway Data Analysis, *The Science of the Total Environment*, 164, 27-43.

Pastore A. (1998) A Perturbation Removal Strategy for Factorial Solution based on the Log-Normal Distribution, *VI Conference of IFCS, Book of Short Papers*, 229-232.

Saylor R.D., Butt K.M., Leonard K.P. (1992) Chemical Characterization of Precipitation from a Monitoring Network in the Lower Ohio Valley, *Atmospheric Environment*, 1147-1156.

Zeng Y., Hopke P.H. (1989), A study of the sources of acid precipitation in Ontario, Canada, *Atmospheric Environment*, 1499-1509.

Zilio Grandi F., Szpyrkovicz L. (1991), Air Pollution Monitoring Network for the Venice Region: Preliminary Results for the Rain Quality, *Toxicological and Environmental Chemistry*, 29, 281-296.

Core Matrix Rotation to Natural Zeros in Three-Mode Factor Analysis

Roberto Rocci
Dipartimento di Scienze Economiche, Gestionali e Sociali
University of Molise

Abstract: This paper presents a new rotation method to simplify the interpretation of the core matrix in three-mode factor analysis. The rotated solution is compared, theoretically and empirically, with the TUCKALS solution (Kroonenberg, 1994).

Keywords: Three-mode data, Factor analysis, Rotation.

1. Introduction

Three-mode factor analysis is a useful tool to analyse and summarise a three-mode data matrix. It was first proposed by Tucker (1966) and further developed by Kroonenberg & De Leeuw (1980) as an extension of the usual two-mode principal component analysis based on the Eckart-Young decomposition approach (Eckart and Young, 1936). A different approach has been taken by Bloxom (1968, 1984) who reformulated the model in a covariance structure form as a third order factor analysis model. Bentler, Poon, and Lee (1988) have shown how to compute the estimates of the parameters using the widely available software for structural equations models.

Basically, three-mode factor analysis decomposes the data into different sets of factors or components, one for each mode, linked by a three-mode matrix named *core*. In both approaches, the parameters involved in the model are not uniquely determined. This feature, like in the two mode case, can be used to select a convenient solution to make the interpretation easier. This paper focuses on this point and proposes a new method to rotate the factors or components which yield an easier interpretation of the core matrix by setting some of its elements to zero.

In what follows we consider the Eckart-Young approach to three-mode factor analysis adopted by Kroonenberg and De Leeuw under the name three-mode principal component analysis. We limit our attention to this case but the rotation method that we are proposing could be also applied to the core matrix of any model in covariance structure form, for which the core array is determined up to non-singular transformations in three directions.

2. Three-mode principal component analysis

Three-mode Principal Component Analysis (TPCA) was developed by Kroonenberg and De Leeuw (1980) starting from the three-mode factor analysis model proposed by Tucker (1966). The model can be considered a generalisation of the standard two-mode principal component analysis (PCA) to the analysis of three-mode data matrices regarding the observation of J quantitative variables on I units in K different occasions. It provides three different sets of components, one for each mode of classification of the data, and is based on the following approximate factorisation of the data matrix

$$x_{ijk} = \sum_{p=1}^{P}\sum_{q=1}^{Q}\sum_{r=1}^{R} a_{ip} b_{jq} c_{kr} g_{pqr} + e_{ijk} \tag{1}$$

where x_{ijk} is the value of variable j on unit i at occasion k, e_{ijk} denotes a residual term, $\mathbf{A}=[a_{ip}]$, $\mathbf{B}=[b_{jq}]$ and $\mathbf{C}=[c_{kr}]$ are matrices of component weights for units, variables and occasions, respectively, while $\mathbf{G}=[g_{pqr}]$ is the so-called *core matrix*. The role of the core can be explained rewriting formula (1) in matrix notation as

$$\text{vec}(\mathbf{X}) = (\mathbf{A} \otimes \mathbf{B} \otimes \mathbf{C})\text{vec}(\mathbf{G}) + \text{vec}(\mathbf{E}) \tag{2}$$

where $\mathbf{X}=[x_{ijk}]$ and $\mathbf{E}=[e_{ijk}]$, the symbol $\otimes$ refers to the (right) Kronecker product, with, for example, $\mathbf{A} \otimes \mathbf{B} = [a_{ip}\mathbf{B}]$, and vec($\mathbf{Y}$) denotes a vector having the elements of the matrix $\mathbf{Y}$ arranged in such a way that the first indices run slowly and the last run fast. From (2) it is clear that the element g_{pqr} of the core is the regression coefficient of the triad of components $\mathbf{a}_p \otimes \mathbf{b}_q \otimes \mathbf{c}_r$. In other words, the core resembles the importance of each combination of components in the reconstruction of the data. The parameters are estimated minimising the sum of squared residuals. To this end Kroonenberg and de Leeuw (1980) proposed an alternating least squares algorithm to minimise

$$f(\mathbf{A},\mathbf{B},\mathbf{C},\mathbf{G}) = \left\|\text{vec}(\mathbf{X}) - (\mathbf{A} \otimes \mathbf{B} \otimes \mathbf{C})\text{vec}(\mathbf{G})\right\|^2$$

for fixed numbers of components in each mode.

It is easy to show that the TPCA parameters are not uniquely determined. In fact, for non-singular matrices $\mathbf{D}$, $\mathbf{E}$ and $\mathbf{F}$ we can write

$$\begin{aligned}
\text{vec}(\hat{\mathbf{X}}) &= (\mathbf{A} \otimes \mathbf{B} \otimes \mathbf{C})\text{vec}(\mathbf{G}) \\
&= (\mathbf{A} \otimes \mathbf{B} \otimes \mathbf{C})(\mathbf{D} \otimes \mathbf{E} \otimes \mathbf{F})^{-1}(\mathbf{D} \otimes \mathbf{E} \otimes \mathbf{F})\text{vec}(\mathbf{G}) \\
&= (\mathbf{A}\mathbf{D}^{-1} \otimes \mathbf{B}\mathbf{E}^{-1} \otimes \mathbf{C}\mathbf{F}^{-1})[(\mathbf{D} \otimes \mathbf{E} \otimes \mathbf{F})\text{vec}(\mathbf{G})] \\
&= (\tilde{\mathbf{A}} \otimes \tilde{\mathbf{B}} \otimes \tilde{\mathbf{C}})\text{vec}(\tilde{\mathbf{G}})
\end{aligned}$$

where $\hat{\mathbf{X}} = [\hat{x}_{ijk}]$ denotes the fitted part of the data. The TUCKALS algorithm (Kroonenberg, 1994) computes the least squares solution iteratively. It is interesting to note that after convergence, the columns of $\mathbf{A}$, $\mathbf{B}$ and $\mathbf{C}$ correspond to the unit-length eigenvectors of the following symmetric matrices of scalar products

$$\mathbf{S}_{\mathrm{A}} = \left[\sum_{jk} \hat{x}_{ijk} \hat{x}_{ljk} \right], \mathbf{S}_{\mathrm{B}} = \left[\sum_{ik} \hat{x}_{ijk} \hat{x}_{ilk} \right] \text{ and } \mathbf{S}_{\mathrm{C}} = \left[\sum_{ij} \hat{x}_{ijk} \hat{x}_{ijl} \right],$$

while, given **A**, **B** and **C,** the core elements can be simply determined by linear regression, see formula (2). It should be noted that **A**, **B** and **C** are column-wise orthonormal and hence **A**⊗**B**⊗**C** is column-wise orthonormal as well. It implies

$$g_{pqr} = (\mathbf{a}_p \otimes \mathbf{b}_q \otimes \mathbf{c}_r)' \operatorname{vec}(\mathbf{X}) \tag{3a}$$

and

$$\sum_{ijk} \hat{x}_{ijk}^2 = \sum_{pqr} g_{pqr}^2 \, . \tag{3b}$$

That is, each squared element of the core represents the contribution of each triad of components to the explained sum of squares. It is also important to note that, when K=1, the three-mode matrix degenerate in a two-mode matrix. In this case the factorization coincides with the singular value decomposition and the model with ordinary two-mode PCA.

TPCA model is very general and can be applied in a large variety of situations. However, this generality represents also a drawback because the core is a three-mode matrix and can be easily interpreted only if it shows a "simple" structure. For this reason, several authors proposed suitable procedures to achieve structures in the core that would be easy to analyse. They followed mainly two different approaches: constrain some elements to zero (e.g. Kiers, 1992; Kiers, Ten Berge & Rocci, 1997; Harshman & Lundy, 1996; Rocci, 1992); rotate the solution (e.g. Kiers, 1997, 1998). In this paper we introduce a new method to simplify the interpretation of the core following both approaches. In fact, our proposal is to constrain some core elements to zero but without any loss of fit. In other words, we propose to rotate the solution to obtain some elements of the core equal to zero but maintaining the columnwise orthonormality of the component matrices. The method will be discussed in detail in subsequent sections.

3. Natural zeros

Our rotation method is based on a particular full factorisation of the core matrix. In the next section we will show how to rotate the TUCKALS solution using this factorisation, while in this section we discuss its main properties. In details, we show how it is possible to construct a full factorisation of a three-way matrix **X** into a model of the form (2) where **A**, **B** and **C** are orthonormal matrices and **G** has some elements equal to zero. We use the symbol **X** for three-way matrices to be factorised because these results were not specifically derived for core matrices.

Proposition. Let $\mathbf{X}=[x_{ijk}]$ be a $I\times J\times K$ matrix, where $I\leq J\leq K$, and $\mathbf{A}=[a_{ip}]$, $\mathbf{B}=[b_{jq}]$ and $\mathbf{C}=[c_{kr}]$ be three orthonormal matrices, of orders $I\times I$, $J\times J$ and $K\times K$, respectively, such that

$$[(\mathbf{a}_h \otimes \mathbf{b}_h \otimes \mathbf{c}_h)' \operatorname{vec}(\mathbf{X})]^2 = \max_{\mathbf{a},\mathbf{b},\mathbf{c}}[(\mathbf{a} \otimes \mathbf{b} \otimes \mathbf{c})' \operatorname{vec}(\mathbf{X})]^2 \text{ for } h=1,2,\ldots,I, \quad (4a)$$

$$[(\mathbf{a}_I \otimes \mathbf{b}_h \otimes \mathbf{c}_h)' \operatorname{vec}(\mathbf{X})]^2 = \max_{\mathbf{b},\mathbf{c}}[(\mathbf{a}_I \otimes \mathbf{b} \otimes \mathbf{c})' \operatorname{vec}(\mathbf{X})]^2 \text{ for } h=I+1,\ldots,J, \quad (4b)$$

under the constraints

$$\|\mathbf{a}\| = \|\mathbf{b}\| = \|\mathbf{c}\| = 1,$$
$$\mathbf{A}_h'\mathbf{a} = \mathbf{B}_h'\mathbf{b} = \mathbf{C}_h'\mathbf{c} = \mathbf{0},$$

where $\mathbf{A}_h$ denotes the matrix obtained by $\mathbf{A}$ keeping only the first h-1 columns with the convention $\mathbf{A}_0 = \mathbf{0}$. Then we have

$$g_{hhl} = g_{hlh} = g_{lhh} = 0 \text{ if } l > h, \quad (5a)$$
$$g_{lhl} = g_{llh} = 0 \text{ if } l > h, \quad (5b)$$

where $g_{pqr} = (\mathbf{a}_p \otimes \mathbf{b}_q \otimes \mathbf{c}_r)' \operatorname{vec}(\mathbf{X})$.

Proof. For sake of brevity, we show only that $g_{hhl} = 0$ if $l > h$. The other cases can be shown in the same way. The statement follows noting that if $g_{hhl} \neq 0$ then $\mathbf{c}_h$ does not satisfy (4a). In fact, indicating with $\dot{\mathbf{c}} = \lambda^{-1}(g_{hhh}\mathbf{c}_h + g_{hhl}\mathbf{c}_l)$, where $\lambda = \sqrt{g_{hhh}^2 + g_{hhl}^2}$, we have $\dot{\mathbf{c}}'\dot{\mathbf{c}} = 1$, $\mathbf{C}_h'\dot{\mathbf{c}} = \mathbf{0}$ and

$$\begin{aligned}
&[(\mathbf{a}_h \otimes \mathbf{b}_h \otimes \dot{\mathbf{c}})' vec(\mathbf{X})]^2 = \\
&= [\lambda^{-1} g_{hhh}(\mathbf{a}_h \otimes \mathbf{b}_h \otimes \mathbf{c}_h)' vec(\mathbf{X}) + \lambda^{-1} g_{hhl}(\mathbf{a}_h \otimes \mathbf{b}_h \otimes \mathbf{c}_l)' vec(\mathbf{X})]^2 \\
&= (\lambda^{-1} g_{hhh}^2 + \lambda^{-1} g_{hhl}^2)^2 \\
&= g_{hhh}^2 + g_{hhl}^2 \\
&\geq g_{hhh}^2 = [(\mathbf{a}_h \otimes \mathbf{b}_h \otimes \mathbf{c}_h)' vec(\mathbf{X})]^2
\end{aligned}$$

where the equality holds if and only if $g_{hhl} = 0$. ♦

The previous proposition can be used to fully decompose a three-way matrix. In fact, **A**⊗**B**⊗**C** is square orthonormal and then there exists a unique vector vec(**G**) satisfying (2) with **E**=**0**. The elements of **G** are given by (3a) and some of them are zero according to the rules (5). For instance, a three-way 3×3×3 matrix can be exactly decomposed with a core of the form

$$\begin{bmatrix} \times & 0 & 0 \\ 0 & \times & \times \\ 0 & \times & \times \end{bmatrix} \begin{bmatrix} 0 & \times & \times \\ \times & \times & 0 \\ \times & 0 & \times \end{bmatrix} \begin{bmatrix} 0 & \times & \times \\ \times & 0 & \times \\ \times & \times & \times \end{bmatrix},$$

while for a 2×4×6 matrix we have

$$\begin{bmatrix} \times & 0 & 0 & 0 & 0 & 0 \\ 0 & \times & \times & \times & \times & \times \\ 0 & \times & \times & \times & \times & \times \\ 0 & \times & \times & \times & \times & \times \end{bmatrix} \begin{bmatrix} 0 & \times & \times & \times & \times & \times \\ \times & \times & 0 & 0 & 0 & 0 \\ \times & 0 & \times & 0 & 0 & 0 \\ \times & 0 & 0 & \times & 0 & 0 \end{bmatrix}.$$

In practical applications we can easily solve the optimisation problem (4) by means of the standard algorithms already proposed for the PARAFAC (Harshman, 1972) or CANDECOMP (Carroll & Chang, 1970) models. In fact those models, when we require only one factor or component, are fitted minimising the loss function

$$f(\lambda, \mathbf{a}, \mathbf{b}, \mathbf{c}) = \|\mathrm{vec}(\mathbf{X}) - \lambda \mathbf{a} \otimes \mathbf{b} \otimes \mathbf{c}\|^2 \tag{6}$$

under the constraints $\|\mathbf{a}\| = \|\mathbf{b}\| = \|\mathbf{c}\| = 1$. Minimising (6) respect to λ, it is easy to show that f attains a minimum when the function

$$\tilde{f}(\mathbf{a}, \mathbf{b}, \mathbf{c}) = [(\mathbf{a} \otimes \mathbf{b} \otimes \mathbf{c})' \,\mathrm{vec}(\mathbf{X})]^2$$

attains a maximum. The general optimisation problem (4) for each h can be solved by means of one of the above cited algorithms with linear constraints (see for example Carroll, Pruzansky S. & Kruskal, 1980). In the following section we will see how to use the factorisation here shown to rotate the TUCKALS solution.

4. Rotation to natural zeros

The results shown in the previous section can be used to simplify the interpretation of the core in the following way. First of all we compute, by means of the TUCKALS algorithm, a solution obtaining

$$\mathrm{vec}(\hat{\mathbf{X}}) = (\mathbf{A} \otimes \mathbf{B} \otimes \mathbf{C})\,\mathrm{vec}(\mathbf{G}). \tag{7}$$

Afterwards, we compute a full factorisation of the core **G** following the previous proposition and obtaining

$$\text{vec}(\mathbf{G}) = (\mathbf{D} \otimes \mathbf{E} \otimes \mathbf{F})\text{vec}(\widetilde{\mathbf{G}}). \tag{8}$$

Combining (7) and (8) we have

$$\text{vec}(\hat{\mathbf{X}}) = (\mathbf{A} \otimes \mathbf{B} \otimes \mathbf{C})(\mathbf{D} \otimes \mathbf{E} \otimes \mathbf{F})\text{vec}(\widetilde{\mathbf{G}}) = (\widetilde{\mathbf{A}} \otimes \widetilde{\mathbf{B}} \otimes \widetilde{\mathbf{C}})\text{vec}(\widetilde{\mathbf{G}})$$

It is important to note that the rotated component matrices are columnwise orthonormal as in the original TUCKALS solution and then the interpretation of the core elements, based on (3) remains unaltered. We note also that when K=1 the rotated solution again coincides with the SVD. From these considerations, we conclude that the rotation yields an easier interpretation of the core maintaining the main properties of the model.

5. Example

In this section we consider two examples from a paper by Kroonenberg (1984, pag.88). For the same three-mode data matrix there as been computed two different representations of the TPCA model, one is obtained choosing two components for each mode (sum of squares explained 59%) and the other with three (sum of squares explained 68%). The core matrix of the 2×2×2 solution is give in table 1.

Table 1: 2×2×2 solution

	TUCKALS				Rotated			
Core	19.9	5.8	-6.7	3.0	-21.7	0.0	0.0	-2.7
Matrix	-5.8	13.5	-2.1	7.7	0.0	-11.6	2.1	12.1
Explained	30.6	2.6	3.5	0.7	36.4	0.0	0.0	0.6
Variation	2.6	14.1	0.3	4.6	0.0	10.4	0.3	11.3

It should be noted that the "simplicity" of the core is not destroyed after the rotation. In fact, if we classify the core's elements of the original solution into three subsets according to their values (disregarding sign), we have a class containing the elements with the strongest influence, they are g_{111}, g_{221} and g_{222}, a class formed by the elements having a medium influence, g_{121}, g_{211}, g_{112} and g_{222}, and, finally, the class of remaining elements having low influence. Adopting the same classification for the rotated core computed by means of our method, we see that the "middle" class disappear because three elements of it are zeroed while the remaining one increases its influence from 4.6% to 11.3% of explained variation and then goes to the first class. The same comments can be also done in the 3×3×3 case (see table 2).

Table 2: 3×3×3 solution.

TUCKALS									
Core Matrix	20.1	4.5	0.6	-6.5	2.8	2.6	1.1	-6.7	-2.7
	-4.8	13.7	-2.2	-2.0	6.8	0.1	-1.7	-0.8	0.0
	-2.0	-2.3	-0.5	-7.1	-0.3	0.6	-4.9	-1.3	0.0
Explained Variation	30.9	1.5	0.0	3.2	0.6	0.5	0.1	3.4	0.6
	1.8	14.3	0.4	0.3	3.5	0.0	0.2	0.0	0.0
	0.3	0.4	0.0	3.9	0.0	0.0	1.8	0.1	0.0
Rotated									
Core Matrix	-21.6	0.0	0.0	0.0	-3.6	-3.2	0.0	6.0	2.1
	0.0	-11.7	2.1	2.1	11.2	0.0	2.0	0.0	-0.3
	0.0	2.0	0.7	6.3	0.0	0.5	6.4	-1.0	0.1
Explained Variation	35.6	0.0	0.0	0.0	1.0	0.8	0.0	2.7	0.3
	0.0	10.5	0.3	0.3	9.5	0.0	0.3	0.0	0.0
	0.0	0.3	0.0	3.0	0.0	0.0	3.1	0.1	0.0

In this two examples we have seen how the proposed method of rotation adds clarity to the interpretation of the core's elements giving a strongest evidence of the pattern of relations between components.

6. Final remarks

In this paper we introduced a new method to simplify the interpretation of three-mode factor analysis. The idea is to rotate the components in order to obtain some specified core elements equal to zero. Very recently, and independently, Henrion (2000) has shown how to rotate the components to obtain $P + Q + R - 3$ elements equal to zero in a $P \times Q \times R$ core matrix maintaining the columnwise orthonormality of the component matrices. Our work can be considered a generalisation of this result. Finally, we note that we can obtain more zeros by rotating the components without imposing any orthonormality constraints (see for example Murakami et al.). However, a larger number of zeros does not necessarily guarantee a simpler interpretation of the core because property (3) does not hold and the component matrices can be nearly singular. More research is needed to clarify the problems of oblique transformations of core arrays..

References

Bentler P.M., Poon W., Lee S. (1988). Generalized multimode latent variable models: implementation by standard programs. *Computational Statistics & Data Analysis,* 6, 107-118.

Bloxom B. (1968). A note on invariance in three-mode factor analysis. *Psychometrika*, vol. 33, n.3, 347-350.

Bloxom B. (1984), "Tucker's three-mode factor analysis", in Law H.G. et al. (eds.), *Research Methods for Multimode Data Analysis*. Praeger, New York, pp. 104-121.

Carroll, J.D., Chang, J.J. (1970). Analysis of individual differences in multidimensional scaling via an N-way generalization of Eckart-Young decomposition, *Psychometrika*, 35, 283-319.

Carroll J.D., Pruzansky S., Kruskal J.B. (1980). CANDELINC: A general approach to multidimensional analysis of many-way arrays with linear constraints on parameters. *Psychometrika*, 45, 3-23.

Eckart C., Young C. (1936). The approximation of one matrix by another of lower rank. *Psychometrika*, 1, 211-218.

Harshman R.A. (1972). PARAFAC2: Mathematical and technical notes. *U.C.L.A. Working paper in Phonetics*, 22.

Harshman R.A., Lundy, M.E. (1996). Uniqueness proof for a family of models sharing features of Tucker's three-mode factor analysis and PARAFAC/CANDECOMP. *Psychometrika*, 61, 133-154.

Henrion R. (2000). On global, local and stationary solutions in three-way data analysis. *Journal of Chemometrics*, 14, 261-274.

Kiers H. A. L. (1992), TUCKALS Core Rotations and Constrained TUCKALS modelling *Statistica Applicata*, 4, 659-667.

Kiers H.A.L., ten Berge J.M.F., Rocci R. (1997), Uniqueness of three-mode factor models with sparse cores: the 3×3×3 case, *Psychometrika*, 62, 349-374.

Kiers, H.A.L. (1997). Three-mode orthomax rotation. *Psychometrika*, 62, 579-598.

Kiers, H.A.L. (1998). Three-way SIMPLIMAX for oblique rotation of the three-mode factor analysis core to simple structure. *Computational Statistics & Data Analysis*, 28, 307-324.

Kroonenberg P.M., de Leeuw J. (1980). Principal components analysis of three-mode data by means of alternating least squares algorithms. *Psychometrika*, 45, 69-97.

Kroonenberg P.M. (1984). Three-mode principal components analysis: Illustrated with an example from attachment theory, in: *Research methods for multimode data analysis*, Law H.G. et al. (eds.), Praeger, New York.

Kroonenberg P.M. (1994). The TUCKALS line: A suite of programs for three-way data analysis. *Computational Statistics & Data Analysis*, 18, 73-96.

Murakami, T., Ten Berge, J.M.F. & Kiers, H.A.L. (1998). A case of extreme simplicity of the core matrix in three-mode principal components analysis. *Psychometrika, 63,* 255-261.

Rocci R. (1992), Three-mode factor analysis with binary core and orthonormality Constraints, *Journal of the Italian Statistical Society*, 1, pp. 413-422.

Tucker L. R., (1966). Some mathematical notes on three-mode factor analysis. *Psychometrika*, vol. 31, n.3.

A Factorial Technique for Analysing Textual Data with External Information

Simona Balbi
Dipartimento di Matematica e Statistica
Università di Napoli "Federico II", sb@unina.it

Giuseppe Giordano
Dipartimento di Istituzioni Economiche e Finanziarie
Università di Macerata, pino@dms.unina.it

Abstract: The paper aims at proposing a method allowing to take into account information on context, when we are analysing lexical tables by means of factorial techniques, as Correspondence Analysis. Such a kind of information, external to the main data structure, can concern where and how words are used, but, moreover, can concern their (syntactical, grammatical, etc.) role inside the *corpus*. Here a methodological tool has been proposed: a technique based on projections on subspaces spanned by two sets of variables related to *fragments* and *words*. The matrix to be analysed, called *inter-reference matrix*, measures the importance of the association between the external information on words and fragments. The final outputs are graphical representations that enrich the results of textual data analysis.

Key-words: Singular Value Decomposition; Biplot; Lexical Tables.

1. Introduction

The aim of the paper is to propose a peculiar technique, which enriches results obtained by classic textual data analysis methods. In such a context, exploratory techniques are widely used in order to identify and describe association, or dependence, between structural variables and the vocabulary adopted (e.g. with survey data, that means cross-classifying answers to one, or more, closed questions and answers to an open-ended one).

It is well known, indeed, that one of the critical point in performing textual data analysis concerns with identifying proper textual units. In other words, that means to build the reference vocabulary, which is strictly connected to the field of application and the peculiarity of the study.

Usually, *corpora* are severely pre-processed, before the vocabulary can be considered ready and, consequently, the matrix to be analysed can be prepared. The subjectivity that characterises pre-processing texts makes necessary an effort for saving at least something of the lost information.

In the following, after considering the common transformation adopted in building the *vocabulary*, and the usual principal axes techniques for analysing textual data (par.2), we propose a strategy for introducing information concerning with the context in which the text has been produced (par. 3). The proposal will be finally presented on an Italian wines advertisement campaign (par. 4).

2. Textual units and textual data analyses

In textual data analysis, one of the most interesting methodological tool is correspondence analysis of lexical tables (Lebart, *et al.*, 1998), i.e. tables cross-classifying *fragments* of the analysed *corpus*, and the corresponding *textual units*. The first question to be solved in order to build such a table consists in choosing the textual units. Lebart (1981) suggests the use of *graphical forms* (i.e. series of non-delimiting characters, whose boundaries at both ends are delimiters): they are easy to be stored and not influenced by recoding. Note that one graphical form can represent more than one word (e.g. in English, *sole*, as a noun, *flatfish*, or as an adjective, *not married* or *the only one*). Furthermore, Bolasco (1993) criticises the different weight of the parts of speech, due to their nature: adverbs, or adjectives, have just one form, while nouns two (singular, or plural) and verbs more, according to tense, and person.
Classically, in the lexicometric approach, text are pre-processed, by the so-called lemmatisation, which consists in making all words homogeneous in gender, number, and verbal form. Homogeneity means loss of part of the information present in the original *corpus*, diminishing the *variability* of expressions, but, in the meantime, lemmatisation has good results in terms of *standardising* the observed units. Moreover, words normally exist in context, in syntactical relationship with other words within a phrase or a sentence. Thus, other textual units have been proposed (*textual forms*, Bolasco, 1993; *repeated segments*, Salem, 1984; *quasi-segments*, Becue, Peiro, 1993). The choice of the proper textual unit has to be acceptable at the same time to linguists and to statisticians. However, as noticed by Muller (1968, in Lebart, *et al.,* 1998), their needs are often contradictory. Nowadays there is no a definite solution for building the proper vocabulary for a textual data analysis.
Once the lexical table has been built, usually, Correspondence Analysis is performed in order to obtain a synthesis of the association structure in the contingency table which crosses items of one (or more) categorical variable(s) and textual units, by decomposing Pearson's ϕ^2 index. Graphical representations adopt a weighted Euclidean matrix, known as χ^2 distance, which assign heavy weights to unusual words. Balbi (1995) proposes the use of non-symmetrical correspondence analysis (Lauro, D'Ambra, 1984) in order to emphasise the different roles played by the two ways of the table. As a consequence of the chosen method, distances on a factorial plane are measured in a usual Euclidean metric: the identified structure is mainly influenced by common words. In both the analyses, the attention is paid to the association structure in the lexical table.

3. The proposed method

Here we propose a strategy consisting in dealing with external information structures, in order to enrich the information in the main data set, i.e. the usual lexical table. The results achieved by means of the proposed technique allow to enhance associations between the categories of the criterion variables, introduced both on rows and columns of the lexical table. Furthermore, they allow representing association onto factorial subspaces.

3.1 The data structure

Let $\mathbf{T}_{(p\times n)}$ be the matrix to be analysed, whose general term is $t_{i,j}$, number of times the i-th *word* has been used in the j-th *fragment* (open answer, document, advertisement). Here, with *word* we extensively intend the textual units and with *fragment* each component in which consists the *corpus*. For our aims, it is possible to consider two additional informative structures, inherent to rows and columns of matrix $\mathbf{T}$. Let $\mathbf{X}_{(k\times n)}$ be the indicator matrix defining a partition of the n fragments with respect to the k categories of one or more criteria related to the different context in which the *fragment* has been used.
The product $\mathbf{L} = \mathbf{X}\mathbf{T}'$ defines the aggregated contingency table usually adopted in textual data analysis. The matrix $\mathbf{L}$ cross-classifies lexical profiles and homogeneous strata of fragments (Lebart, Salem 1994; when we want to analyse vocabulary with respect to stratification, looking for dependence, Balbi, 1995).
Moreover, let $\mathbf{Z}_{(p\times q)}$ be the matrix, inherent to the auxiliary information about p *words* (e.g. knowledge about pre-processing, or syntactical roles, or rhetorical effects). Its general element is the number of times that a word appears in a specific context. Thus, the cross-product matrix $\mathbf{M}$:

$$\mathbf{M} = \mathbf{X}\bar{\bar{\mathbf{T}}}'\mathbf{Z} \tag{1}$$

defines the aggregation of the double centred empirical frequencies in $\mathbf{T}$, according to *external* information both on *fragments* and *words*.

3.2 The analysis

Aiming at analysing the relationships between the two kinds of information in $\mathbf{X}$ and $\mathbf{Z}$, or better, aiming at understanding how they influence the frequencies in $\mathbf{T}$, we refer to regression models with qualitative predictors, with $\mathbf{X}$ and $\mathbf{Z}$ the two different sets of predictors. Therefore, we look for a set of parameters $\boldsymbol{\Theta}$ which allows the best approximation (in a least square sense) of $\mathbf{T}$ by $\mathbf{X}$ and $\mathbf{Z}$ (Figure 1). From an analytical viewpoint, the following minimisation problem has to be solved:

Figure 1: *The data structure*

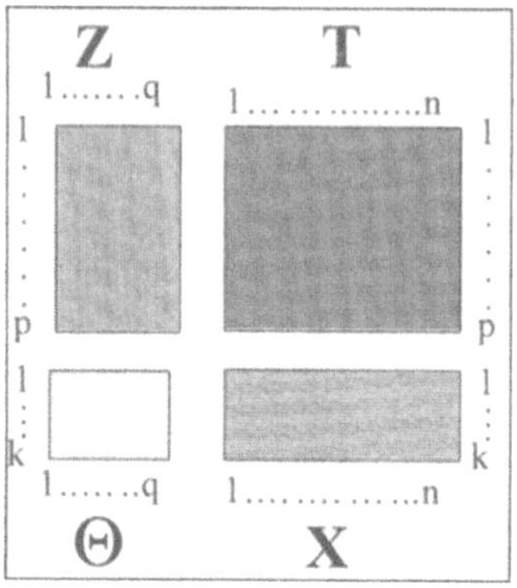

$$\|\mathbf{T} - \mathbf{Z}\Theta'\mathbf{X}\|^2 = \min_{\Theta}! \tag{2}$$

The solution is given by:

$$\Theta = (\mathbf{XX}')^{-}\mathbf{XT}'\mathbf{Z}(\mathbf{Z}'\mathbf{Z})^{-1} = (\mathbf{XX}')^{-}\mathbf{M}(\mathbf{Z}'\mathbf{Z})^{-1} \tag{3}$$

where $(\mathbf{XX}')^{-}$ is the Moore-Penrose generalised inverse of $(\mathbf{XX}')$.
Note that Θ can be derived by a two step multivariate regression:

1. in the first step, we consider the model that explains the use of the j-th word in the n observed *fragments*, through the context in which it appears. The related model assumes the contents of the n-th fragment (column of **T**) as depending on the columns of **Z**:

$$\mathbf{T} = \mathbf{ZB} + err, \tag{4}$$

with the usual least squares solution given by:

$$\mathbf{B} = (\mathbf{Z}'\mathbf{Z})^{-1}\mathbf{Z}'\mathbf{T} \tag{5}$$

2. in the second step, we try to understand how the external information on *fragments* influences the coefficients in **B** through the model:

$$\mathbf{B}' = \mathbf{X}'\Theta + err \tag{6}$$

By least squares estimation of the model, we easily obtain the matrix **Θ** defined in (3) (Giordano, Scepi, 1998).
Thus, the coefficients in **Θ** join the q categories that characterise words (subspace spanned by the columns of **Z**) with the k categories, which stratify the *fragments* (subspace spanned by the rows of **X**).
Each generic element of **Θ**, $\theta_{q,k}$, can be interpreted as a measure of the association between the q-th category for *words* and the k-th category for *fragments*. It is also possible to interpret the values in **Θ** as *inter-reference* coefficients, because they allow to show the relationship between the two reference subspaces **X** and **Z**, as shown in (3).
In order to graphically represent such association between the two informative structures, a singular value decomposition of the *inter-reference* matrix **Θ** can be performed:

$$(\mathbf{XX}')^{-}\,\mathbf{M}\,(\mathbf{Z}'\mathbf{Z})^{-1} = \mathbf{U}\,\Lambda\mathbf{V}' \tag{7}$$

under the usual normalising constraints $\mathbf{U}'\mathbf{U} = \mathbf{V}'\mathbf{V} = \mathbf{I}$.
The factorial decomposition allows to represent in a low-dimensional subspace the association structure. The display, known as *biplot* (Gabriel, 1981), allows a 2-rank joint representation of rows and columns of a matrix. Thus, we can visualise vector-points related to both rows and columns of **Θ** on a plane, by the following formulas ($\alpha = 1, 2$):

$$\psi_\alpha = \sqrt{\lambda}_\alpha v_\alpha \qquad (8)$$
$$\varphi_\alpha = \sqrt{\lambda}_\alpha u_\alpha$$

According to the interpretative rules of biplot, the symmetric co-ordinates (8) represent interrelations between the two sets of variables **X**' and **Z**.
The main aim of our proposal consists in providing tools for better understanding the role of words in the identified interrelation structure.
Consequently, we are interested in visualising, as *supplementary points*, the rows of the matrix **Z**, in order to show the characterisation of each word, according to its own external information.
At this aim we have to consider the proper geometry for representing columns of **Θ**. Thus, we rescale the co-ordinates (8), according to classic asymmetric biplot (Gabriel, 1971) in the following way:

$$\psi^*_\alpha = v_\alpha \qquad (9)$$
$$\varphi^*_\alpha = \lambda_\alpha u_\alpha$$

The co-ordinates of the words are:

$$\phi^{\bullet}_{\alpha} = \lambda_{\alpha} \mathbf{Z} u_{\alpha} = \mathbf{Z} \varphi^{\bullet}_{\alpha} \tag{10}$$

While the symmetric biplot represents the association between the two sets of external criteria in order to detect any interesting interrelations, the second (asymmetric) biplot has to be read by looking at clouds of words whose direction tends towards particular external criteria.

4. An Italian wines advertisement campaign

The aim of the analysis is to understand how Italian wine producers present their product when they want to encourage sales in a specific market segment, i. e. not actually experts, but people aware and interested in knowing what they drink. This market segment has been identified in the buyers of *Gambero rosso*, "*the monthly magazine for gourmets and curious consumers*". Thus, advertisements for wines from the beginning of 1992 to the whole 1994 have been recorded and analysed. Together with the text of the message, other information on the advertisements, like the presence of rhetorical effects, was collected. The analysis of the lexical table built cross-classifying textual units and the region where the wine is produced is in Balbi (1995), both by symmetrical and non-symmetrical correspondence analysis. The proposed method allows enriching the analysis with further characteristic inherent to the context in which words are used.
Here we are interested in introducing the information relative to the textual units, connected with textual expedients. Specifically, we consider if the use of each word was: *evocative*, *technical*, *naturalistic*, *gastronomical*, or *other*. Each message is considered according to the geographical area producing the corresponding wine (*Northwest*, *Northeast*, *Centre*, *and South*). Thus the data structure is the following: **T** is the lexical table crossing advertisements and textual forms; **X** is the indicator matrix with 1 in the cell corresponding to the wine origin area, and 0 elsewhere; **Z** is the matrix with general element the number of times each textual form has been used in the corresponding expedient.
In the matrix Θ, see table 1, we can read the association structure between areas and expedients.

Table 1: *The inter-reference matrix* Θ *for* ***areas*** *and* ***expedients***

	Northwest	*Northeast*	*Centre*	*South*
Technical	-0.05	-0.02	0.05	-0.03
Gastronomical	0.04	-0.02	-0.01	0.05
Evocative	0.01	0.01	-0.02	0.01
Naturalistic	-0.07	-0.00	0.04	-0.05
Other	0.01	0.01	-0.01	-0.01

In Figure 2, the association structure between *area* and *expedient* is visualised in a lower dimensional principal subspace, by a symmetric biplot, with co-ordinates given by (8).

Figure 2: *Biplot for areas (in italics) by context (in* **bold***)*

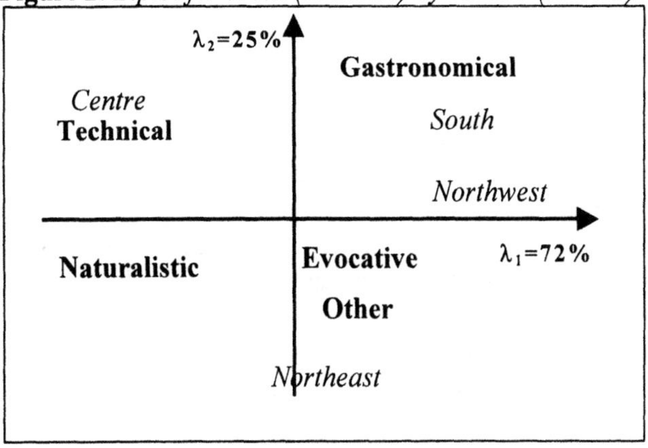

The 2 rank approximation takes into account almost the whole information in table 1 ($\lambda_1 + \lambda_2$=97%).

Thus it is possible to enhance how the wine produced in central regions (particularly Tuscany, with the noble tradition of Chianti) use advertisement for describing the technical characteristics of their production, while the Northeast, area with an old but less famous history, prefer evocative descriptions.

Figure 3: *Words as supplementary elements (textual forms in lower case)*

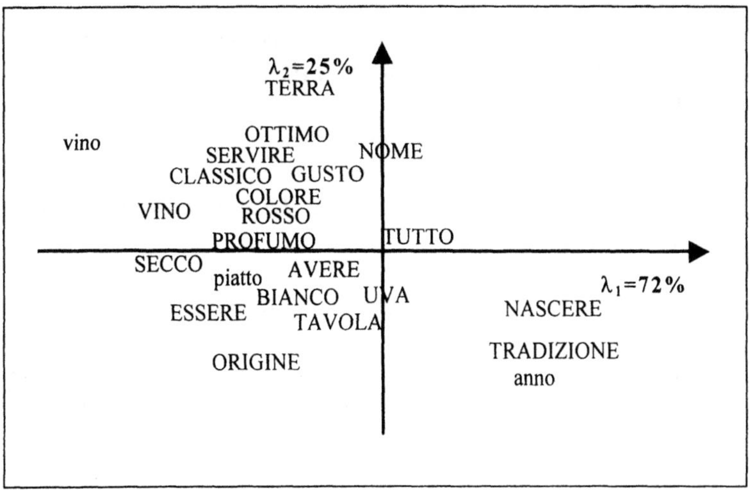

The analysis can be enriched by considering, as supplementary points, the textual units, related to their context, by projecting them on the same factorial plane.

Figure 3 shows the textual forms more interesting in interpreting the association in

Figure 2. The word "classico" (*classic*) is a typical reference for Chianti, while the evocative style is characterised by words like "nascere" (*to be born*), or "tradizione" (*tradition*).

References

Balbi, S. (1995), Non symmetrical correspondence analysis of textual data and confidence regions for graphical forms, in S. Bolasco, L. Lebart, A. Salem (eds.), *JADT'95*, vol. 2, Roma, CISU, 5-12.

Becue M., Peiro R. (1993), Les quasi-segments pour une classification automatique des reponses ouvertes, in S. J. Anastex (ed.), *JADT' 93*, Paris, Telecom, 310-325.

Bolasco S. (1993), Choix de lemmatisation en vue de reconstructions syntagmatiques du texte par l'analyse des correspondance, in S. J. Anastex (ed.), *JADT'93*, Paris, Telecom, 399-414.

Gabriel K. R. (1971), The biplot graphical display of matrices with application to principal component analysis, *Biometrika*, 58, 453-467.

Gabriel K. R. (1981), Biplot display of multivariate matrices for inspection of data and diagnosis, in V. Barnett (ed.), *Interpreting Multivariate Data*, Chichester, Wiley, 147-174.

Giordano G., Scepi G. (1998), La progettazione della qualità attraverso l'analisi di strutture informative differenti, *Atti della XXXIX Riunione Scientifica SIS*, vol. II (CD-ROM), Sorrento, 707-714.

Greenacre, M. (1984), *Theory and Applications of Correspondence Analysis*, London, Academic Press.

Lauro N.C., D'Ambra L. (1984), L'analyse non symétrique des correspondances, in *Data Analysis and Informatics, III* (E. Diday *et al.* eds.), Amsterdam, North-Holland, 433-446.

Lebart L. (1981), Vers l'analyse automatique des textes: le traitement des réponses libres aux questions ouvertes d'une enquête, in: J. P. Benzécri & collaborateurs, *Pratique de l'Analyse des Données. Linguistique & Lexicologie*, Paris, Dunod, 414-419.

Lebart L., Salem A. (1994), *Statistique Textuelle*, Paris, Dunod.

Lebart, L., Salem A., Berry, L. (1998), *Exploring Textual Data*, Dordrecht, Kluwer Academic Publishers.

Salem A. (1984), La typologie des segmentes répétés dans un corpus, fondée sur l'analyse d'un tableau croisant mots et textes, *Les Cahiers de l'Analyse des données*, 9, 4, 489-500.

Subjects on Using Open and Closed-Ended Questions

Arjuna Tuzzi
Department of Statistical Sciences of Padua
via San Francesco, 33
35121 Padova (Italy)
e-mail: arjtuzzi@stat.unipd.it

Abstract: In order to go on with the study of differences between open and closed-ended answers we want to overcome the simple univariate comparison to arrive at the study of relations between open and closed-ended alternatives. The coding procedure of the open answers texts is made by means of the manual post-coding and the textual analysis techniques. Since there is the possibility of observing open and closed-ended information taken from the same respondents, the aim of this work is to understand differences in the frequencies collected with the two data collecting tools and new methods of analysis are proposed.

Keywords: open interview, closed-ended questionnaire, textual analysis, manual post-coding, correspondence analysis

1. Introduction

Open and closed-ended answers produce different kinds of information and, in the strict sense, not comparable in a direct and simple way. Anyway in the past literature we can find a few examples where open and closed-ended answers are compared by authors such as: Belson and Duncan (1962), Bradburn and Sudman (1979), Schuman and Presser (1981), Sudman and Bradburn (1982). These authors submitted an open interview to a *first sample* of respondents and a closed-ended questionnaire, concerning the same topics, to a *second sample* of respondents. They observed wide differences in the two univariate distributions of the alternatives. The main advantage of the work I'm going to present is that we comment answers given from the *same* respondents to *both* the two different data collecting tools, so that we can work with the bivariate distribution.

The above mentioned authors compared results taken from open and closed-ended questions by means of a manual post-coding of the information contained in the open answers texts into new structured items (most of the content analysis techniques are based on new manual post coded closed-ended items) and in this work we try to exploit the manual post coding and the automatic numerical labelling of the word-tokens that occur in the texts (most of the modern textual analysis techniques are based on the direct numeric labelling of the text).

The closed-ended answers are taken from the results of a survey on the professional profiles of the graduates of the Faculty of Political Sciences in Padua (self compiled postal closed-ended questionnaires) and the open interviews are made by a group of students interviewers. We analyse only the questions occurring on the open interview and on the closed-ended questionnaire and cross results.

To save space we shall abbreviate the term "open interview question" and refer to it hereafter as OIQ, and "closed-ended questionnaire question" as CEQ.

2. CEQ alternatives *versus* OIQ manual post-coded alternatives

The first step is to draw a contingency table crossing alternatives born from the manual post-coding of the answers the OIQ (e.g. the rows of the table) with items answered by the same respondents on the CEQ (e.g. the columns of the table). Generally this matrix is rectangular because the alternatives on rows and columns are not all the same, nevertheless the manual post-coding has been directed in order to obtain as many alternatives in common as possible. In fact if we have two questions differing in the form but concerning the same topic, the possible alternatives are the same, but we can have some alternatives that are mentioned in the OIQ answers and are not proposed in the CEQ or *vice versa*. These are common researcher's faults.

The cell crossing the same alternative occurring on row and columns represents the coherence cases (the same answer has been observed on the open and on the closed-ended form). All the other cells represent incoherence cases (the answer given on the OIQ differs from the one given on the CEQ). The analysis of the association of the table can be done by means of techniques like χ^2 test and simple correspondence analysis or the analysis of concordance can be done on the square matrix obtained from the original contingency table keeping only the alternatives in common to both the forms. Unfortunately all these methods reveal only the obvious underlying coherence pattern between the two forms of the same question. By this way we always find the presence of significant association and concordance between the two data collecting tools. Reasonably, this result could not be surprising, because there is a definite prevalence of coherence cases, but we have not a good instrument to evaluate the incoherence cases that are surely the most useful ones in our study.

In order to obtain an instrument for analysing incoherence cases, we split the analysis of the contingency table into the analysis of as many 2×2 tables as many the alternatives of interest are. For each alternative we have two dichotomies, the first concerning its presence on the OIQ answers (variable I) and the second concerning its presence on the CEQ (variable Q). The symbol I denotes that the alternative has been mentioned on the OIQ and $\bar{I}$ otherwise, Q denotes that the alternative has been chosen on the CEQ and $\bar{Q}$ otherwise. Crossing the two variables we obtain a 2×2 table (tab.1) in which the coherence cases are expressed by cells a and d and incoherence ones by b and c.

This analysis of a selected alternative is based on the hypothesis that the answer the OIQ in the natural language is the natural way to express opinions and attitudes, while the CEQ is able to deviate the instinctive answer, for instance towards other more attractive or less threatening alternatives. Given the total number of valid responses, we are in particular interested in inequalities concerning the three free parameters (a, b, c) and in relations between conditional frequencies.

Table 1: *The 2×2 table for the analysis of a selected CEQ alternative.*

	Q	$\bar{Q}$	
I	a	b	$a+b$
$\bar{I}$	c	d	$c+d$
	$a+c$	$b+d$	

$$a<b \Leftrightarrow Q\,|\,I<\bar{Q}\,|\,I$$
$$a<c \Leftrightarrow I\,|\,Q<\bar{I}\,|\,Q$$
$$b<c \Leftrightarrow \bar{Q}\,|\,I<\bar{I}\,|\,Q$$

Starting from these simple elements and working with inequalities, we are able to understand more about the reasons of the incoherent cases (tab.2).

If $b<c$ we have an *overloaded* CEQ alternative. Despite the alternative is hardly mentioned during the open interview, it is very often chosen on the CEQ. We can have an overloaded CEQ alternative because it is considered far more *attractive* then others ($a>b$, $a>c$) or because it is considered either an *expedient* to avoid a delicate issue (the questionnaire is taking from the interview the threatening answers and makes them deviate towards this alternative) or a *substitute* of the real answer, when the respondent cannot find the alternative perfectly suitable to his mind ($a>b$, $a<c$). But the overloading may be due to a *misunderstanding* of either (or both) open or closed-ended forms ($a<b$, $a<c$).

If $b>c$ we have a *lightened* CEQ alternative. Despite the alternative is mentioned during the open interview, it is often omitted in the closed-ended questionnaire. We can have a lightened CEQ alternative because it is not considered interesting enough, so it is *penalised* by the presence of a more attractive one ($a>b$, $a>c$) or it is avoided either because concerns a *delicate* issue or because it is considered *unsuitable* ($a<b$, $a>c$). Finally, the CEQ alternative may be lightened because of a *misunderstanding* too ($a<b$, $a<c$).

Table 2: *Interpretation model for the CEQ selected alternative.*

	$a<b$ $a<c$	**$a>b$ $a<c$**	**$a<b$ $a>c$**	**$a>b$ $a>c$**	
$b<c$	*misunderstanding*	*expedient/substitute*	-	*attractive*	***overloaded***
$b>c$	*misunderstanding*	-	*delicate/unsuitable*	*penalised*	***lightened***

A pair of examples could be on purpose.

In the first example we show a part of the analysis of the questions about distrust. During the Seventies the Paduan Faculty of Political Sciences had serious ideological contrasts with the Italian Government, so that its graduates are still often labelled as the Bad Teachers' students. In the analysis of the question about the perceived distrust when the graduates were looking for a job, the alternative "*there is distrust*" of Political Sciences graduates occurs 188 times in the OIQ and only 82 times in the CEQ (tab.3 – 239 on 256 non missing data). Clearly, this is a delicate issue and in front of the CEQ choice most of the respondents prefer to deny the existence of distrust. Following our criteria this alternative being in the OIQ answer, on the contrary is in the CEQ one gone towards other alternatives. It is well known that in presence of delicate issues the OIQ is able to collect more cases than the CEQ (Bradburn and Sudman 1979) and results are more reliable.

Table 3: *Distrust of Political Sciences graduates. Alternative "there is distrust".*

	Q	$\bar{Q}$		
I	75	113	*188*	$b>c$
$\bar{I}$	7	44	*51*	$a<b$
	82	*157*	*239*	$a>c$
				avoided delicate issue

In the second example we show a part of the analysis of the questions about employment condition at graduation. The possible alternatives are "*unemployed*", "*employed*", and "*military service*". The alternative "*employed*" is an attractive one, in fact 41 respondents that in the OIQ turn out to be unemployed, in the CEQ turn out to be employed (tab.4 – 255 on 256 non missing data).

Table 4: *Employment condition. Alternative "employed at graduation".*

	Q	$\bar{Q}$		
I	71	8	*79*	$b<c$
$\bar{I}$	41	135	*176*	$a>b$
	112	*143*	*255*	$a>c$
				attractive overloaded alternative

It is well known that there is the problem of setting correctly up events in time (Sudman and Bradburn 1974) and normally the answers given during the open interview are more reliable. Following our criteria this CEQ alternative attracts the respondents in a very unexpected way. This phenomenon can be explained this way: many graduates at degree were not able to declare themselves unemployed because for instance they had already done a previous job interview, or had the possibility of working in their father's company, or they had precise promises, etc. They preferred to declare on the CEQ to have already a job firstly because they prefer to pride and secondly because they prefer to be respectful of their less lucky companions.

The 2×2 table allows a punctual analysis of the non coherence cases and underlines the limits of the alternatives proposed on the CEQ. Shortly, the items proposed on the closed-ended questionnaire may deviate the instinctive answer towards another more attractive or less threatening alternative, which is quickly assumed by the respondent (the questionnaire is suggesting a more interesting answer or is giving the possibility of avoiding a delicate issue). But these are not the only problems. For instance, because of the ambiguity of the phrasing, the question-form presented on the closed-ended questionnaire may be not properly understood by the respondent (and it is important to underline that a certain amount of ambiguity may occur in each non banal verbal expression). Further the respondent's choice on the CEQ may be an alternative that is collecting different meanings together. In fact either the respondent is giving a value which is different from the estimated one (the verbal expression cannot enter into details) or the respondent cannot find among the proposed items the one perfectly suitable to his mind, so he is obliged to chose another one (similar, but more generic). However, the researcher is not able to come back to the original sense of the collected information.

The manual ex-post coding technique has serious problems too, such as:
(a) coders effect (instability of manual post-coded structured items);
(b) information become poor (loosing of the original expressions and dictionary peculiarities);
(c) creation of too heterogeneous residual categories;
(d) growth of the time taken for generating structured data (the longer and the more complex the interviews are, the more time this operation takes);
(e) growth of missing data because of the incompleteness of the single interview towards all the topics comprehensively found.

3. CEQ alternatives *versus* OIQ texts words occurrences

Considered the above mentioned limits of the manual post-coding technique, we want to compare open and closed-ended answers by means of the textual analysis methods. Using a direct statistical analysis of the texts (Lebart Salem e Berry 1998; Bolasco 1995, 1997 and 1999) we do not need the tables crossing open and closed items, but a table crossing the word-types occurring in the responses concerning the selected OIQ together with the alternatives presented on the CEQ. We use the correspondence analysis of this new contingency table in order to observe if word-types are coherent or not with those CEQ alternatives that are closed to them on the correspondence analysis factorial plan.

Using again the example concerning the employment condition at graduation we can see on the correspondence analysis first two factors graph (fig.1 – 83% in. expl.) a latent coherent pattern (e.g. *lavoravo-working*, *già-already* and *inserito-hired* are closed to the alternative *OCCUPATO-employed*) but it is more interesting observing other patterns shown. For example we can see that words concerning post degree training (e.g. *specializzazione-specialisation*, *Inglese-English*, *conoscenze-knowledge*) are closed to unemployed graduates (*DISOCCUPATO*) and words concerning job (e.g. *carriera-carrier*, *stipendio-salary*, *lavoravo-working*), and in particular job in the public sector (*PA-Public Administration*, *insegnare-to teach*), are closed to employed ones (*OCCUPATO*). In other words the textual analysis techniques underline the coherent pattern, but other new unexpected interesting patterns are shown, for example the prevalence of employment in the public sector.

With textual analysis techniques we can avoid the above mentioned problems of the manual post coding procedures, but these methods involve some limitations too, in particular in the choice of the word types to be submitted or not to the analysis and in the interpretation of graphical results.

Figure 1. *Employment condition at graduation. Textual analysis results.*

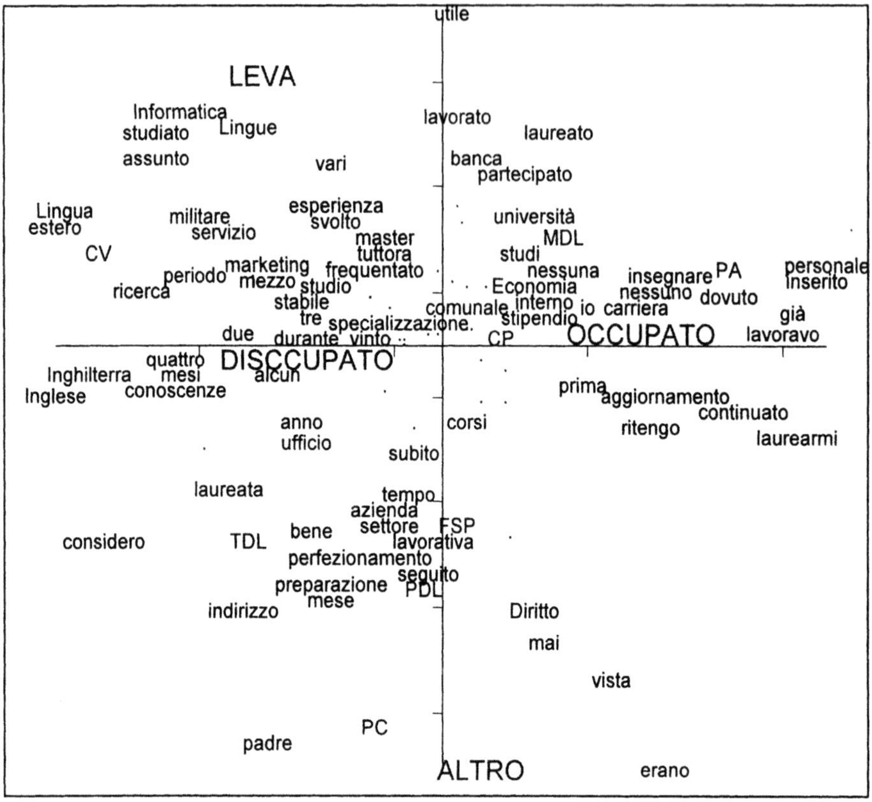

4. OIQ manual post-coded alternatives *versus* OIQ texts word occurrences

A crossed comparison between the manual post-coding results and the textual analysis word-types may be useful to understand more about the problems of textual information analysis. Furthermore, this kind of investigation may be exploited in order to check the goodness of the manual post-coded alternatives and/or the coders' ability. For example the OIQ about the reasons why there is distrust of Political Sciences graduates, has been manual post-coded into four cathegories: "*SPstoriaPD*" (because of the past of the Paduan Faculty of Political Sciences), "*Preparazione*" (because of the poor and generic preparation of the graduates), "*ignoraNE*" (because of the ignorance of the typical small businessman from the North East of Italy) and "*nonso-NoDiff*" (don't know–no distrust). On the correspondence analysis first two factors graph (fig.2 - 73% in. expl.), in the neighbourhood of the former cathegory, we can find words like *Padova-Padua*, *passato-past*, *periodo-period*, *tempi-times*, *Settanta-Sixties*; closed to the second words like *preparazione-preparation*, *conoscenza-*

knowledge, *specifica-specific*, *facile-easy* and so on. All these considerations support the goodness of the coding procedure done.

Figure 2: *OIQ about distrust. Checking the manual post coded alternatives.*

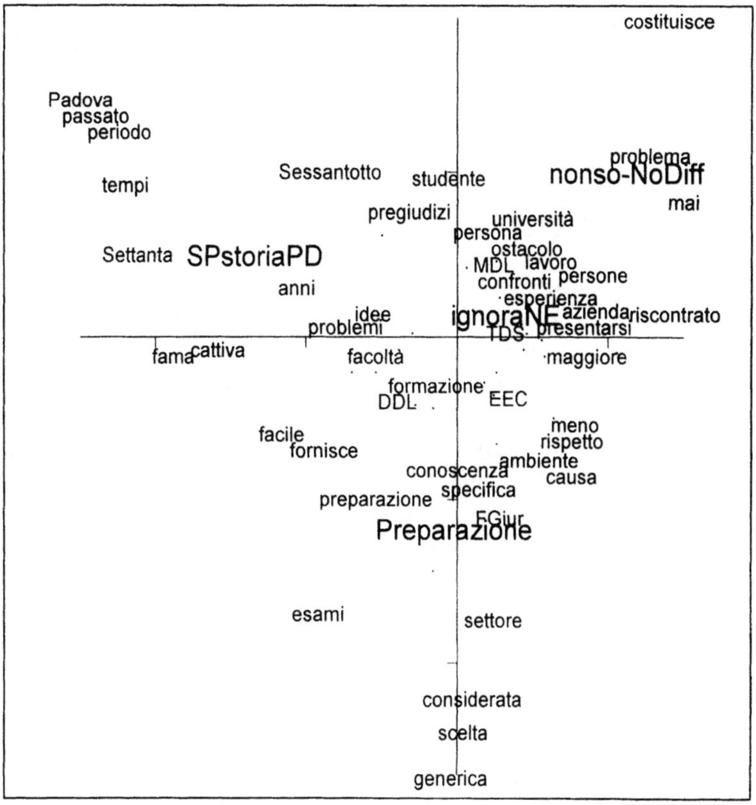

5. Conclusions

In order to explain easily the path followed in this work, a summarising scheme (tab.5) seems to be suitable. The table shows the main goal of each comparison proposed. The analysis of a selected alternative occurring on the open and on the closed-ended form by means of 2×2 contingency tables seems to be a useful instrument to evaluate incoherent cases. Furthermore, under the hypothesis that the open answer represents the respondent's real attitude, this method suitably investigates the limits of the CEQ. The interpretation model needs more research to assess the strength of the relations. Textual analysis seems not to be sensible in evaluating incoherence patterns because the correspondence analysis of the lessical table (crossing OIQ texts word occurrences and CEQ alternatives) shows only the natural coherence pattern between the two different data collecting tools. However this kind of graphical representations are very useful to discover new patterns and to formulate new interpretation hypothesis. The peculiar characteristics of this analysis make it suitable for another purpose: evaluating the goodness of the coding procedure and the coders' ability by means of the lessical table crossing the OIQ texts word occurrences with the OIQ manual post coded alternatives.

Table 5: *Summarising scheme of the comparisons proposed.*

Comparison	***Method***	***Main Goal***
OIQ post coded alternatives *versus* CEQ alternatives	2×2 tables	analysis of incoherent cases
OIQ texts word occurrences *versus* CEQ alternatives	correspondence analysis	analysis of new latent patterns
OIQ texts word occurrences *versus* OIQ post-coded alternatives	correspondence analysis	checking manual post-coding procedures and coders' ability

The possibility of a joint use of different techniques and different data collecting tools (open and closed-ended) might improve the results of the surveys. In Lazarsfeld's opinion the open interview ought to be always used in a preliminary survey in order to obtain a better closed-ended questionnaire (Lazarsfeld 1944) and this means *a priori*. Exploiting the methods proposed, the open interview could be useful for *a posteriori* quality control too. Finally we would like to state that the joint use of different methods always allows the researcher to be more conscious of the main limits of the interpretations often easily extracted from data analysis, even if it can not help the problem-solving of each peculiar data collecting tool.

References

Belson W. A., Duncan J. A., (1962), A comparison of the check-list and the open response questioning system, *Applied Statistics*, 1962, 2, 120-132

Bolasco S. (1995) Criteri di lemmatizzazione per l'individuazione di coordinate semantiche, in *Ricerca qualitativa e computer: teorie metodi e applicazioni*, Bolasco S. & Cipriani R. (Eds), Franco Angeli, Milano, 87-111

Bolasco S. (1997) L'analisi informatica dei testi, in *La ricerca qualitativa*, Ricolfi L. (Eds.), NIS, 165–202

Bolasco S. (1999) *Analisi multidimensionale dei dati*, Carocci editore, Roma

Bradburn N.M., Sudman S. (1979), *Improving Interview Method and Questionnaire Design*, Jossey-Bass Publisher, San Francisco

Lazarsfeld P. L. (1944) The controversy over detailed interviews - an offer for negotiation, *The Public Opinion Quarterly*, 8, 38-60

Lebart L., Salem A., Berry L. (1998) *Exploring textual data*, Kluwer-Academic

Schuman H., Presser F. (1981) *Question and Answers in Attitude Surveys*, Academic Press, New York

Sudman S., Bradburn N. M. (1982) *Asking Questions. A Practical Guide to Questionnaire Design*, Jossey-Bass Publishers, San Francisco

Sudman S., Bradburn N. (1974) *Response Effect in Survey*, Aldine, Chicago

Alternative Error Term Specifications in the Log-Tobit Model [1]

Rosa Bernardini Papalia
Dipartimento di Scienze Statistiche, Università di Perugia, bernard@stat.unipg.it

Francesca Di Iorio
Istituto Nazionale di Statistica, DCII/Pre, Roma, diiorio@istat.it

Abstract: In this paper a logarithmic transformation of the standard Tobit model is proposed. The model represents an interesting tool in order to specify alternative form of data heterosckedasticity. The properties of applied estimators are compared by a set of Monte Carlo experiments.

Keywords: Log-Tobit model, heterosckedasticity, EM algorithm.

1. Introduction

In many studies the consistent estimation of censored regression models under the assumption of a non-autocorrelated, homosckedastic, normally distributed error term has been explored (Amemiya 1984, Heckman 1976).
Hurd (1979) and Maddala and Nelson (1982) showed that ignoring conditional heterosckedasticity in a Tobit model leads to inconsistent estimates. The same result arises when the Gaussian quasi-maximum likelihood Tobit estimator is employed under non normality of error terms (Arabmazar and Schmidt 1982). Kalbfleisch and Prentice (1980) present a procedure that includes several distributions such as the exponential, lognormal and Weibull. Relating the conditional heteroskedasticiy problem a Tobit model with GARCH error terms was recently proposed by Calzolari and Fiorentini (1997).
The biasedness and inconsistency of OLS estimator in the presence of censoring and the non-robustness of Tobit ML estimator to incorrect specification of error distribution have motivated the development of semiparametric estimators (Powell 1984).
In this paper we propose a logarithmic transformation of the standard Tobit model (Tobin 1958), which seems to be useful to investigate the error term assumptions. The model also represents an interesting tool in order to specify alternative form of data heterosckedasticity. The properties of the ML and semiparametric estimators applied for the proposed model are also compared by a set of Monte Carlo experiments.
This paper is organised as follows: in section 2 we briefly present the LOG-TOBIT

[1] An early version of this paper was presented at the COMPSTAT'98, Bristol, UK.

model; in section 3 the results of the Monte Carlo experiments are given and discussed; an empirical application of the Log-Tobit model is described in section 4. Conclusion and final remarks are presented in section 5.

2. The model

It is well known that for limited dependent variables models estimation methods such as the Maximum Likelihood may not be robust against departures from misspecification in distributions. So, it is desirable to devise alternative specification of the Tobit model. In this paper we propose a log-transformation of Tobit model where an exponential error term distribution can represent the heterosckedasticity of data.

Assuming that y_i is an heteroschedastic observed censored dependent variable and x_i is a vector of exogenous explanatory variables. The heterosckedasticity is modelled by the multiplicative structure of the error term. Using the logarithmic transformation of y_i, $y_i^\star = ln(1+y_i)$, we define the LOG-TOBIT model as follows:

$$y_i = \begin{cases} exp(x_i'\beta + u_i) - 1 & y_i^\star > 0 \\ 0 & otherwise \end{cases} \tag{1}$$

where u_i are iid $N(0,\sigma^2)$, β is the vector of unknown coefficients and:

$$y_i^\star = x_i'\beta + u_i \qquad i = 1, ...n \tag{2}$$

The assumption of a particular type of heterosckedasticity and a logarithmic rather than linear relationship implies the specification of a truncated model. On the contrary, our model presents the advantage to impose such hypotheses maintaining the structure of a censored model.

Because of $y_i^\star = ln(1+y_i)$, the moments of this censored model can be easy derived from the moments of a censored normal variable. In particular the marginal effects are:

$$\frac{\partial}{\partial x} E[ln(1+y_i)|x_i] = \beta\Phi\left(\frac{x_i'\beta}{\sigma}\right).$$

A first application of Tobit models with logistic and Weibull distribution of the error term and a logarithmic transformation of the dependent variable has been proposed in Filippucci et al. (1996) to estimate Engel's models for the italian consumption.

The log likelihood function for the LOG-TOBIT model is:

$$lnL(\beta;x) = -\frac{n}{2}ln(\sigma^2) - \frac{1}{2}\sum_{y_i^*>0}\left(\frac{ln(1+y_i) - x_i'\beta}{\sigma}\right)^2 + \sum_{y_i^*\leq 0} ln\left[1-\Phi\left(\frac{x_i'\beta}{\sigma}\right)\right]$$

The ML estimates was computed by the BFGS optimisation algorithm and by the EM-algorithm, useful in contexts ranging from incomplete data problems.
We also consider the Least Absolute Deviation (LAD) estimator proposed by Powell (1984) for censored and truncated regression models, which is consistent under general distributions. The LAD estimator is the solution of:

$$\hat{\beta} = argmin_\beta \sum_{i=1}^{n} |ln(1+y_i) - max(0, x_i\beta)| \tag{3}$$

The LAD estimates were computed by the Nelder-Mead (1965) simplex method. This is a direct search method, which is less efficient then a gradient based algorithm, but it is more robust with respect to numerical problems involved by absolute deviation criterium. From a starting point, the simplex method builds a politope and evaluates the objective function in the politope vertices. By a sequence of expansions and contractions the minimum is reached.

3. The Monte Carlo Study

A Monte Carlo experiment was designed to investigate the properties of the parametric estimators in the LOG-TOBIT model discussed above. Specifically, we are interested in evaluating the performance of the ML estimator respect to: *i*) the percentage of censoring observation in the sample; *ii*) the sample length.
The simulated model is:

$$y_i = max(0; [exp(\beta_0 + x_i'\beta + u_i) - 1])$$

where $u_i \sim N(0, \sigma^2)$.
Three exogenous variables x, held fixed over all Monte Carlo replications, were generated as in Calzolari and Fiorentini (1997) setting the covariance matrix of the error terms equal to the identity matrix.
Two different average percentage of censored observations are considered. This different percentage is controlled by setting the intercept coefficient β_0 equal to 0.1 and -0.6 obtaining 40% and 75% levels of censoring respectively.
The value of the slope coefficient are (β_1=0.2, β_2=0.3, β_3=0.4), and σ^2=0.4. For each experiment we performed 1000 Monte Carlo replications.
Finally, the sample lengths are set equal to $100, 400, 800, 1500, 10000$. Following Amemiya (1984) the EM-algorithm for the LOG-TOBIT model proceeds by maximising with respect $\theta = (\beta, \sigma^2)$

$$\begin{aligned}\mathbf{E}[lnf(y_i|\theta)|x,\theta_1] &= -\frac{n}{2}ln(\sigma^2) - \frac{1}{2\sigma^2}\sum_{y^*>0}(ln(1+y_i) - x_i'\beta)^2 \\ &\quad -\frac{1}{2\sigma^2}\sum_{y^*\le 0}[\mathbf{E}(y_i^*|\theta_1) - x_i'\beta]^2 - \frac{1}{2\sigma^2}\sum_{y^*\le 0}\mathbf{V}(y_i^*|\theta_1)\end{aligned}$$

where θ_1 is the parameter estimates at the previous M step, and

$$\mathbf{E}(y_i^*|\theta_1) = x_i'\beta_1 - \frac{\sigma_1\phi_{1i}}{1-\Phi_{1i}}$$

and

$$\mathbf{V}(y_i^*|\theta_1) = \sigma_1^2 + x_i'\beta_1\frac{\sigma_1\phi_{1i}}{1-\Phi_{1i}} - \left[\frac{\sigma_1\phi_{1i}}{1-\Phi_{1i}}\right]^2$$

where $\phi_{1i} = \phi(\frac{x_i'\beta_1}{\sigma_1})$, $\Phi_{1i} = \Phi(\frac{x_i'\beta_1}{\sigma_1})$ where ϕ and Φ are the standard normal pdf and cdf respectively. The experiment involves 10 cases with variation in sample size and censoring degrees.

For each experiment we obtain the Monte Carlo mean and the Monte Carlo standard error of the estimated parameters calculated by ML, LAD and EM-algorithm; the Monte Carlo means of the standard errors of the parameter estimates obtained by the appropriate covariance matrices.

Some results are reported in Table 1 and Table 2.

Table 1: Mean estim. parameters, (std.err), and [Monte Carlo std.err]

Par	*True*	Log-Tobit Max. Lik.	Log-Tobit EM-alg	Log-Tobit LAD	Log-Tobit Max. Lik.	Log-Tobit EM-alg	Log-Tobit LAD
β_0	.1	.1008	.1001	.1609	.1001	.1001	.1509
		(.0769) [.0767]	(.0771) [.0719]	(.1281) [.0715]	(.0290) [.0291]	(.0290) [.0248]	(.0447) [.0679]
β_1	.2	.2015	.2019	.2067	.1994	.1994	.1968
		(.1209) [.1290]	(.1216) [.1285]	(.1314) [.0379]	(.0377) [.0383]	(.0377) [.0376]	(.0443) [.0283]
β_2	.3	.3016	.3020	.2765	.2996	.2996	.2776
		(.1013) [.1078]	(.1014) [.1068]	(.1165) [.0252]	(.0320) [.0319]	(.0320) [.0309]	(.0379) [.0161]
β_3	.4	.4003	.4008	.4259	.4008	.4008	.4190
		(.0819) [.0823]	(.0819) [.0801]	(.0862) [.0707]	(.0232) [.0228]	(.0232) [.0212]	(.0260) [.0503]
σ^2	.4	.3782	.3787	...	.3961	.3961	...
		(.0722) [.0757]	(.0725) [.0664]		(.0274) [.0271]	(.0274) [.0226]	
Cens=40 %			n=100			n=800	

The ML and EM parameter estimates are very close to the real values.
As expected, for a given sample length, the Monte Carlo mean of standard error and the Monte Carlo Standard error are higher with 75% degree of censoring.
For 75% degree of censoring and a given sample length the EM-estimates seems to be higher than ML estimates. This evidence could be explained by the convergence rule adopted in EM-algorithm. For the same reason the Monte Carlo standard error in these cases shows some divergence from the Monte Carlo mean of the standard error of the estimated parameter.
As expected, the LAD estimates are not so close to the true values, especially for 75% degree for censoring, and they show some discrepancies between the Monte Carlo standard error and the Monte Carlo mean of the standard error.
The LAD has the smaller Monte Carlo standard error than the EM and ML parametric estimators for 40% censoring and sample size n=100. Those results depend on the optimisation procedure involved by the simplex method, especially the step sizes related to every objective function parameter in the politope construction.

Table 2: Mean estim. parameters, (std.err), and [Monte Carlo std.err]

Par	*True*	Log-Tobit Max. Lik.	Log-Tobit EM-alg	Log-Tobit LAD	Log-Tobit Max. Lik.	Log-Tobit EM-alg	Log-Tobit LAD
β_0	-.6	-.5605	-.5843	-.5158	-.5946	-.5982	-.5135
		(.1411) [.1499]	(.1720) [.0673]	(.1510) [.0738]	(.0539) [.0541]	(.0548) [.0220]	(.0447) [.0706]
β_1	.2	.1900	.1946	.2190	.2006	.2010	.2132
		(.1452) [.1578]	(.1504) [.1471]	(.0397) [.4901]	(.0468) [.0487]	(.0470) [.0420]	(.0312) [.0443]
β_2	.3	.2905	.2980	.2781	.2972	.2985	.2796
		(.1219) [.1329]	(.1278) [.1185]	(.0164) [.4762]	(.0406) [.0396]	(.0408) [.0335]	(.0121) [.0379]
β_3	.4	.3857	.3956	.4200	.3984	.3997	.3816
		(.1024) [.1099]	(.1112) [.0872]	(.1033) [.2544]	(.0313) [.0312]	(.0316) [.0223]	(.0601) [.0260]
σ^2	.4	.3563	.3738	...	.3930	.3963	...
		(.1070) [.1243]	(.1348) [.0619]		(.0408) [.0405]	(.0418) [.0216]	
Cens= 75 %			n=100			n=800	

4. An empirical application

An empirical application of the Log-Tobit model was conducted by using data from the survey "Sistema dei conti delle imprese" conducted by ISTAT in 1995 on the

balance items of 35000 firm with more then 20 employed.
The Research and Development (R&D) behaviour seems an interesting field because of many studies on industrial organization are devoted to testing the hypotheses advanced by Schumpter about the innovation and the industrial market structure. In this application the R&D expenditures are related to the global sales proceeds (GSP).
We consider the sector of medical appliance producers (104 firms considered). Two model in y_i^* are estimated:

model A: $ln(1 + R\&D) = \beta_0 + \beta_1\sqrt{GSP} + u_i$

model B: $ln(1 + \frac{R\&D}{GSP}) = \beta_0 + \beta_1 GSP + u_i$

Model A is suggested by the data, model B is a classical "Budget share" model for Household budget survey studies. The results from parametric and semiparametric estimation methods are reported in Table 3.

Table 3: Mean estim. parameters, (std.err)

Par	Log-Tobit Max. Lik.	Log-Tobit EM-alg	Log-Tobit LAD	Log-Tobit Max. Lik.	Log-Tobit EM-alg	Log-Tobit LAD
		Model A			*Model B*	
β_0	2.252	2.252	2.735	0.019	0.019	0.0018
	(0.319)	(0.319)	(0.260)	(0.002)	(0.002)	(0.007)
β_1	0.022	0.022	0.020	1e-7	1e-7	1e-8
	(0.002)	(0.002)	(0.002)	(1e-7)	(1e-7)	(1e-7)
σ	2.357	2.357	...	5e-3	5e-3	...
	(0.584)	(0.584)		(1e-3)	(0.001)	
Cens=7.7%			n=104			

The main result is for this sector incentive activities increase less than proportionately than increase in firm size measured by GSP. The hypothesis of a positive relations between the R&D activities and the firm size is supported in our case.

5. Conclusion

In this paper we propose a log transformation of Tobit model where an exponential error term distribution is assumed. The performance of the ML and LAD estimator

is analysed in Monte Carlo experiments with respect to sample length and censoring degrees. Our simulation results confirm the importance of the censoring degree in the estimation performance. The presented estimators confirm the expected results relating to the sample length.
An Hausman test (Hausman, 1978) can be used as specification test for alternative distribution assumptions; or it is also possible to derive a score-type test following for example Bera et alii (1984), Ruud (1984) and Peters ans Smith (1991).
An empirical application of the Log-Tobit model was conducted by using data from the survey "Sistema dei conti delle imprese" conducted by ISTAT in 1995. A Research and Development (R&D) expenditure model was estimated and the results from parametric and semiparametric estimation methods was presented. The empirical results do not support the Schumpter's hypothesis that inventive activities increase faster or more than proportionately than increase in firm size.

References

Amemiya T. (1984) Tobit models: a survey, *Journal of Econometrics*, 24, 3-61.

Arabmazar A., Schmidt, P. (1982) An investigation of the robustness of the Tobit estimators to non-normality, *Econometrica*, 50, 1055-1069.

Bera A. K., Jarque, C. M., Lee, L.F. (1984) Testing for the normality assumption in limited dependet variable models, *International Economic Review*, 25, 563-578.

Calzolari G., Fiorentini, G. (1997) A Tobit model with GARCH errors, *WP-AD* n.97-13, Instituto Valenciano de Investigaciones Economicas.

Filippucci C., Drudi, I., Bernardini Papalia, R. (1996) Testing the relevance of Tobin's approach for modelling consumption, *Economic Notes*, 2, 225-248.

Hausman J. (1978) Specification tests in econometrics, *Econometrica*, 46, 1251-1272.

Heckman J.J. (1976) The common structure of statistical models of truncation, sample selection and limited dependent variables and a simple estimator for such models. *Annals of Economic and Social Measurement*, 15, 475-492.

Hurd M. (1979) Estimation in truncated samples when there is heterosckedasticity, *Journal of Econometrics*, 11, 247-258.

Kalbfleisch J., Prentice, R. (1979) *The statistical analysis of failure time data*, New York, Wiley and Sons.

ISTAT (1998) *Rilevazione sul sistema dei conti delle imprese con 20 addetti ed oltre. Panel 1991-1994*, Informazioni n.78.

Maddala G. S., Nelson, F. (1975) Switching regression models with exogenous and endogenous switching, *Proceedings of the American Statistical Association* (Business and Economic Section), 423-426.

Nelder J. A., Mead R., (1965) A simplex method for function minimization, *Computer Journal*, 7, 308-313.

Peters S., Smith, R. J. (1991) Distributional specification tests against semiparametric alternatives, *Journal of Econometrics*, 47, 175-194.

Powell J. L. (1984) Least absolute deviation estimation for censored and truncated regression models, *Journal of Econometrics*, 25, 303-325.

Ruud P. A. (1984) Tests of specifications in econometrics, *Econometric Reviews*, 3, 211-242.

Tobin J. (1958) Estimation relationship for limited dependent variables, *Econometrica*, 26, 24-36.

A Customer Satisfaction Approach for User-oriented Comparative Evalutations of Services

Giulio D'Epifanio

Dip. di Sc. Statistiche, Univ. di Perugia, via A.Pascoli, 06100 Perugia, Italy

Abstract: A goal-driven multi-criteria approach to evaluation is proposed to help a potential new client to choose among competing suppliers of a type of service (or product). The actor of evaluation is an independent agency, perhaps a magazine, which interprets the point of view of a type of potential client. Evaluations are grounded on the surveyed perceived satisfaction of old clients, but on the dimensions of interest for a type of potential client accounted for by the criterion he adopts. An approach which uses structured stochastic models is presented, but we emphasize a new interpretation of the role of modeling in evaluations. In the our setup evaluations are fully conditioned on the accounting criterion - even subjective - which the actor of evaluations would adopt as grounds for its evaluations.

Keywords: Comparative Evaluation, Customer Satisfaction, Logit Normal Model

1. Introduction

Over the last decade there has been an increasing interest in the development of quantitative methods for comparative evaluations of suppliers of a type of service. Mainly from an institutional perspective, full model approaches to comparative evaluations of non-profit organizations were presented in Goldstein and Spiegehalter (1996), Gori and Vittadini (1999). Recently in non-profit sectors (Stiefel, 1997), attention has increased also towards the point of view of customers, even in the public sector where specific legislative norms have been approved. Thus several approaches (for a brief review see Vedaldi, 1997; Lauro, Balbi and Scepi, 1997; Magagnoli,1997; Zanella, 1999; Clark *et al.*, 1999) to the customer satisfaction have been recently developed although in different directions. Differences among approaches essentially concern the role of actors of evaluations (management, internal client, external client, independent agency, etc..) and subsequently the aims which these actors pursue (to improve the internal organization, allocate resources to sub-agent, choose suppliers, etc..). Thus, to avoid misinterpretation we need to make clear from the outset our setup. We exclusively consider **the problem** of comparative evaluations of competing suppliers. Suppliers of a type of service are actual objects (universities, schools, hospitals, etc.) in a given collection of objects of interest, typically imposed by normative, geographic, even political considerations. **The actor** of evaluations is an independent agency, perhaps a magazine. Calibrated on a specific type of potential client, the main **actor' s aim** is to help a potential client in choosing a supplier. The role of the agency is that of interpreting (through focus group, preliminary interviews, pilot surveys) the point of view of a type of potential client, identifying the questions of interest (the latent profile of interest) and the relevant goals. Subsequent, elaborating and communicating types of scores to be adapted for personalized multi-criteria and synthetic evaluations.

Of course, the actual interest of a potential client is not in explaining differences[1] among suppliers ("why they differ"). But, the potential client is interested in evaluating suppliers in respect to the goals he pursue (Gori *at al.* 1999, pp. 128). These goals depend on his personal preferences and are typically constrained by his budget. But the agency interprets the prevalent requirements of a well identified target-population of potential clients, assuming the role of an **inter-subjective actor** of evaluations. The evaluation system we propose is structured in two sub-system. The first. **The observational apparatus** is the (possibly) objective component, constructed to identify statistically from responses the latent profile of interest, accounted for by context conditions. The second. **The targeting-system** is the (inter-)subjective[2] component, grounded on the criterion - even personal - which the actor would adopt to assign reference-values (targets), over the latent profile of interest. To be operative, this criterion - perhaps latent and informal - should induce coherent rules to assign the targets. Although founded on facts, the rationale behind comparative evaluations is intrinsic to the assumed specific criterion. The rationality concerns coherency between the criterion and the goals's actor. To be clear, the agency could adopt the surveyed (from old clients) customer-satisfaction to predict, for each supplier, performances over the latent profile of characteristic of interest for a client-type. Evaluations measure then the degree the statistical predictions adhere to the targets which the actor of evaluation were pretended. Thus evaluations depend on the specific criterion used to assign targets. But this criterion could adopt, for instance, the costs[3] (the potential client should weigh on to receive services), to assign the targets to be pretended, *irrespective of whether or not, in the collection of interest, the responses (levels of satisfaction) are statistically predictable by costs.* Thus in the our setup, the rules to assign targets have essentially prescriptive valences: they serve to assess reference-values (base-lines) which the actor retain "fair", accounted for by the criterion he adopts.

Using probabilistic structured models, for each objects, predictions - accounted for by context conditions - on the latent profile of interest might be automatically performed in a formal Bayesian setup. But, different from an actual

[1]This contrast with other approaches, typically management-oriented, where the interest is essentially focused just on relationships between perceived customer satisfaction and factors of quality.

[2]Certainly it is commonly accepted that, between two services (products) at the same level of performances, a rational client chooses that less expensive. Thus, he operatively has evaluated the more the less expensive object, *coeteris paribus*. Therefore, implicitly his evaluation system is grounded on a criterion which would pretend the performances to be positively correlated to the prices, irrespective of whether or not statistically this actually occurs. Unfortunately, in the given collection of interest, relationships might suffer from the absence of statistical significance (for detail see Gori *et al.*, 1999, pp. 218-219). Thus, pure econometric approaches to evaluations may fail. In contrast with full econometric approaches (like that of Goldstein *et al.*, 1996), deliberately we will consider statistical estimating and evaluating as *different tasks which concern different aims*, although the latter - to have a certain scientific ground - should incorporate statistical inferential tools as specialized preliminary sub-tasks.

[3]sometime, they coincide with the prices

bayesian approach, in the our setup the priors are working models to describe the residual uncertainty of the evaluator over the targets he pretends. To calibrate automatically the rules which assign targets coherent with the actor'goals, we propose an approach that uses "reference collections" of objects, actual o even virtual. Thus a reference- empirical-Bayes setup (Carlin *et al.*, 1996) constitutes a powerful framework to elicit automatically rules for targeting. Subsequent, these rules may be translated over the actual collection of interest for targeting their objects, accounting for by their actual characteristics.

To outline a more general approach to comparative evaluation, we use a concrete, perhaps provoking example. This paper consist of three parts. In Chap. 2, we present a concrete model and introduce modeling. In Chap. 3, we introduce targeting, scoring and briefly discuss applications.

2. Examples and modeling

A concrete evaluation problem. The actor of evaluation: an independent agency. The concrete problem: to support a new potential client in choosing one among N competing universities. Focusing the attention on a target-population of potential client, from preliminary inquires the agency has identified: (a) the dimensions of the service which the clients consider relevant, (b) the prevalent criterion which the clients, at least implicitly, would adopt in evaluating performances, (c) the available responses (outputs, outcomes) of performances. In this example, the ingredients are the following:

(i) *The dimensions of the interest*[4]: teaching-methods(1), technology(2), entertainment(3), accommodation(4).

(ii) *The accounting criterion*[5]: "for each dimension of interest, the more services costs the client, the higher satisfaction the client should pretend."

(iii) *The available data set*:

University	Item-1	Item-2	...	Item-p	N. interviews	fee	costs
1	x_{11}	x_{12}	...	x_{1p}	n_1	z_{11}	z_{12}
2	x_{21}	x_{22}	...	x_{2p}	n_2	z_{21}	z_{22}
...	...	...	...	...	...	...	...
k	x_{k1}	x_{k2}	...	x_{kp}	n_k	z_{k1}	z_{k2}
...	...	...	...	...	...	...	...
N	x_{N1}	x_{N2}	...	x_{Np}	n_N	z_{N1}	z_{N2}

For each k in a collection of N competing universities, n_k old students (the customers surveyed from the university k) are interviewed. They respond Yes/No to

[4]Here, in contrast with other approaches, the dimension of interest are established by the clients, not imposed by normative-institutional directives.

[5]It seems naturally having an inter-subjective valency. See note 2.

each of p binary questions concerning perception of satisfaction from their university. Here, x_{ki} denotes the number of positive responses (satisfied? YES) to the binary item i. Furthermore, this table includes the costs (requested from the universities to supply services) which the agency-evaluator would account as grounds for its evaluations. For the university k, z_{k1} and z_{k2} denote, respectively, the requested fee for the dimensions of teaching-methods & technology and the estimated average cost for the dimensions of entertainment & accommodation.

3. Models

Notations. Let $x_k := (x_{k1}, \ldots, x_{kp})$ denote the realized customer satisfaction profile, $\eta_k := (\eta_{k1}, \ldots, \eta_{kp})$ the full latent customer satisfaction profile underlying x_k, $\tilde{\eta}_k := (\tilde{\eta}_{k1}, \ldots, \tilde{\eta}_{kR})$ the latent customer profile of interest.

The model we propose is structured in two sub-models. The first, the observational apparatus is the (possibly) objective component, constructed to identify statistically the latent profile of interest, from responses (eventually) accounted for by context conditions. The second, the targeting-model is the (inter-)subjective component, grounded on the criterion - in principle even personal - which the actor would adopt to assign the reference-values (targets), over the latent profile of characteristic of interest. This model constitutes a formal setup to make the assumed criterion operative.

(A) **The observational apparatus**. This model should have some substantive-predictive ground. It decomposes the full latent profile η_k to separate an item-specific effect (the general propensity to respond "Yes" to a specific question) from the effects due to the latent profile of interest $\tilde{\eta}_k$, which characterizes the performance of each university.

$$X_{ki} \mid \theta_{ki}, n_k \overset{indep.}{\sim} Bin(x_{ki}; \theta_{ki}, n_k), \; k := 1, \ldots, N, i := 1, \ldots, p, \tag{1}$$

$$\theta_{ki} := \exp(\eta_{ki})(1 + \exp(\eta_{ki}))^{-1},$$

$$\eta_{ki} := logit(\theta_{ki}) = \alpha_i + \sum_{r:=1}^{R} D_{ir} \cdot \tilde{\eta}_{kr}$$

Here, $Bin(x_{ki}; \theta_{ki}, n_k)$ denotes the binomial distribution with parameters θ_{ki} and size n_k; η_{ki} denotes the logit transform of θ_{ki}, D the boolean $(p \times R)$ design matrix which groups together all the items related to the same latent dimension r ($r := 1, \ldots, R$, where $R := 4$, $R < p$), α_i the parameter (constant across the universities) which characterizes the question i. This model[6] is assumed here adequate

[6]Of course, more complex multi-level models might here be considered to include specific client effects (Gori *et al.*, 1999) over satisfaction. But, to outline basic ideas this is irrelevant. However we have assumed here that, for each university, the surveyed individuals are drawn from samples calibrated on the same target-population.

and the parameters α_i assigned[7].

(B) **The targeting-model.** It serves to develop a formal framework to prescribe targets to the objects of interest.

$$\tilde{\eta}_k \mid m_k, \Sigma_k \overset{indep.}{\sim} \Phi_R(\tilde{\eta}_k; m_k, \Sigma_k), \; k := 1, \ldots, N, \tag{2}$$

$$m_k := \mathcal{Z}_k \cdot \beta, \tag{3}$$
$$\Sigma_k := \tilde{\Sigma}(\zeta)$$

Here, $\Phi_R(\tilde{\eta}; m_k, \Sigma_k)$ denotes the normal R-variate with expectation vector m_k and matrix of covariance Σ_k. For each object k, the vector m_k and the matrix Σ_k represent, respectively, the multi-dimensional target and the residual uncertainty - after the target has been assigned - of performances. The system of vectorial equations (3) assigns to each object the expected target and the residual variances of the performances, taking into account - through the design matrix Z_k and parameters β and ζ - the structure of the actual costs. Here, $\tilde{\Sigma}$ is a linear matrix function which structures the covariance through the vector ζ. To conform to the accounting criterion (ii) of Sec. 2.1, for instance we could set $\mathcal{Z}_k := \begin{bmatrix} 1 & z_{k1} & 0 & 0 & 0 & 0 & 0 & 0 \\ 0 & 0 & 1 & z_{k1} & 0 & 0 & 0 & 0 \\ 0 & 0 & 0 & 0 & 1 & z_{k2} & 0 & 0 \\ 0 & 0 & 0 & 0 & 0 & 0 & 1 & z_{k2} \end{bmatrix}$.

In contrast with other approach[8], in this setup the distribution Φ_R has not frequentist grounds. It is considered a working-reference model, useful to describe the virtual residual uncertainty which may be ascribed to the actor-evaluator, after he has prescribed the target, due to the intrinsic complexity of any mental process of evaluation. The relationships (3) might be virtual, but they need to be conformed to the accounting criterion (for instance that quoted in Sec. 2.1.) which the actor-evaluator would retain "fair", regardless of whether or not - in this example - the costs are statistical predictors of satisfaction, in the actual collection of interest. Thus, these relationships are rules - perhaps artificial - to prescribe expected values, on the latent profile of interest, up to residual uncertainty. Thus, they have a valency essentially **prescriptive** rather than predictive.

[7]The collection of objects of interest is typically imposed from normative considerations. Therefore it can not absolutely be considered a sample - from some (quasi-)experimental design - useful for inferences over the observational apparatus, to identify the latent profile of interest. As a consequence, these inferences should be grounded on samples, which typically are different from the actual collection of interest. In contrast with Goldstein *et al.*(1996), inferences on the observational apparatus should be specialized sub-tasks preliminary to evaluations. See also note 2.

[8]Since evaluations regard *the* actual objects in *the* specific collection of objects of interest - which is typically imposed by normative, even political considerations - we generally contrast approaches as that of Goldstein *et al.*(1996) which use this actual and specific collection as whether it was really a sample drawn from some super-population, in a pseudo experimental design. See also note 7.

4. Calibrating the rules for targeting, scoring and evaluations

4.1. Calibrating rules. The core problem of a goal-driven evaluation systems is to calibrate automatically the rules for assigning targets, conform to a criterion coherent with the actor's goals. To briefly delineate approaches we distinguish among purely subjective "ex-ante" approaches and data-driven "ex-post" approaches which contextualize - among the competing objects - the rules to assign targets, conditional on the assumed conceptual framework. To tackle this problem, we should devote attention again on the following points: (a) the actor of evaluations (the agency) represents common interests shared by more subjects, thus inter-subjective valences of evaluations are requested, (b) typically, in the actual collection of interest, "what the potential client would pursue" is not necessarily compatible with hypothetical "statistical means of what actually occurs". Therefore in practice, calibrating targets neither must be purely subjective[9] neither it can be a mere statistical task$^{(2)}$. Pragmatically, we propose an approach that uses "reference collections" of objects, actual o even virtual, which the actor (supported by expert teams, possibly publicly) select "ad hoc" to elicit the rules (3). The rationale behind this approach is that the selected "reference collection" should explicitly operationalize a criterion - perhaps implicit - coherent with the actor's goals, that is a reference-model of behaviors in respect to which he would judge performances of the actual objects. Using data - actual or perhaps virtual - from this reference collection, formally through model (1-3) the empirical-Bayes setup (Carlin *et al.*, 1996) constitutes then a powerful formal framework to elicit automatically rules for targeting, i.e. for determinating values $\gamma := (\beta, \zeta)$. Computationally very efficient procedure has been developed based on quasi-linearized estimating equations (D'Epifanio, 1998). Subsequent, these rules may be translated over the actual collection of interest for targeting their objects, accounting for by their specific costs.

4.2. Scoring. Let $\gamma^* := (\beta^*, \zeta^*)$ indicate the values determined by calibrating rules (3). Thus, $m_k^* = E(\tilde{\eta}_k \mid z_{k1}, z_{k2}; m_k(\gamma^*), \Sigma_k(\gamma^*)) = \mathcal{Z}_k\beta^*$ is the target expected by the evaluator and $E(\tilde{\eta}_k \mid x_k, z_{k1}, z_{k2}; m_k(\gamma^*), \Sigma_k(\gamma^*))$ is the actual prediction for university k, on the latent profile of interest. The point to be focused on is that the multi-dimensional differential (variation, discrepancy)

$$\Delta_{x_k}(m_k^*, \Sigma_k^*) := E(\tilde{\eta}_k \mid x_k, z_{k1}, z_{k2}; m_k^*, \Sigma_k^*) - m_k^*$$

measures the degree of disagreement of the actual performance with the expectation which the evaluator would pretend. The larger this differential is, the more distant the actual performance is from the target "fair". For each dimension, a large positive differential indicates a type of effectiveness. In fact, using well known - although predictive - formal identities of conditioned moment we have:

$$E_{m_k(\gamma^*),\Sigma_k(\gamma^*)}[\Delta_{x_k}(m_k^*, \Sigma_k^*)] = 0, \tag{4}$$

where $E_{(m_k(\gamma^*),\Sigma_k(\gamma^*))}[\cdot]$ denote the operator expectation over the space of the realizations of the responses, supposing they have been generated according to the

[9] A pure subjective approach seems not realistic, too much arbitrary to be rationally justified, communicated and therefore accepted by other actors

model (1-2), where γ^* has been assigned. Taking into account the uncertainty, due to evaluator, to standardize these differentials, finally a multi-dimensional scoring system may be obtained as the following:

$$Score_k := \left[diag(Cov_{m_k(\gamma^*),\Sigma_k(\gamma^*)}\Delta_{x_k}(m_k^*, \Sigma_k^*)\right]^{-1/2} \cdot \Delta_{x_k}(m_k^*, \Sigma_k^*).$$

4.3. Evaluations. Qualitative multi-criteria analysis - perhaps with the aid of dynamical graphic tools - can be carried out involving multi-dimensional score plots. The scores derived from the differentials, on each dimension of interest, may be interpreted as a type of inter-subjective measure of effectiveness. Of course, a client-specific type of effectiveness also depends on the client's personal preferences over dimensions and it is typically constrained by his budget. Using the inter-subjective target-equations (graphically furnished by the agency), the individual-client could determinate, on the space of the profile of interest, the zone of expected performances feasible with his budget. His personalized "ideal target" should be that point whose total cost coincides with his budget - therefore over the frontier of the feasible zone - for which the gradient of his benefit (utility) is orthogonal to the frontier of the feasible zone. This is sketched in figure 1. This ideal target may be now considered the base-line in respect to which he should measure discrepancies of predictions, to evaluate specific effectiveness, over objects whose targets is budget-compatible. Finally, the projections of these variations over the gradient of his benefit might give the his personalized synthetic measure of effectiveness. The evaluation of this gradient may be derived from the marginal benefit in respect to each items.

Figure 1: *Synthesizing client-oriented effectiveness scores.*

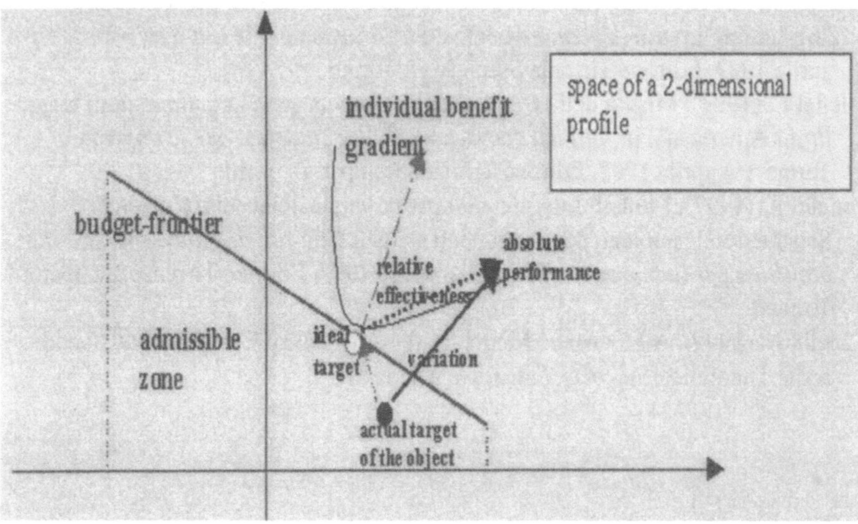

References

Carlin B., Louis T. (1996), *Bayes and Empirical Bayes Methods for Data Analysis*, Chapman Hall, London

Clark L.A., Cleveland W.S., Denby L., Liu C. (1999), Modeling Customer Survey Data, *Case Studies in Bayesian Statistics*, Springer, New York,

D'Epifanio G. (1998), An estimation Procedure for Multilevel Models with Binary Responses, in: *Data Science Classification and Related Methods, Book of Short Papers*, Roma, ISTAT

Fabris L., Corsi M., Franci A. (edit.) (1997), *La Valutazione della Qualita' dei Servizi Socio-Assistenziali*, CLEUP, Padova

Fornel C, Johnson M., Anderson E., Cha J., Everitt B. (1996), The American Customer Satisfaction Index:Nature, Purpose, and Findings, *Journal of Marketing*, n. 60, 7-18

Gori E, Vittadini G. (1999), *Qualita' e Valutazione nei Servizi di Pubblica Utilita'*, Etas, Milano

Gori E. (1992), La Valutazione dell' Efficienza e dell' Efficacia dell' Istruzione, in: *atti della XXXVI Riunione Scientifica SIS*, Pescara, CISU, Roma

Goldstein H., Spiegelhalter D.J. (1996), League Tables and Their Limitations: Statistical Issues in Comparisons of Institutional Performance, *Journal of the Royal Statistical Society*, A, 159, 385-443

Kristensen K, Kanji G.K., Dahlgaard J.J. (1992), On Measurement of Customer Satisfaction, *Total Quality Management 3*, pp. 123-128

Lauro C., Balbi S., Scepi G. (1997), L' Analisi Multidimensionale dei Dati per la Misurazione della Customer satisfaction nei Servizi, in: *atti del convegno SIS "La Statistica per le Imprese"*, Torino 1-4 aprile 1997, Editrice Tirrenia Stampatori, Torino

Magagnoli U. (1997), Indicatori di Prestazione Complessiva e della Customer Satisfaction, in: *atti del convegno SIS "La Statistica per le Imprese"*, Torino 1-4 aprile 1997, Editrice Tirrenia Stampatori, Torino,

Stiefiel L, (1997), Misura della Qualita' degli Outputs nelle Organizzazioni Non-Profit Americane, in: *atti del convegno SIS "La Statistica per le Imprese"*, Torino 1-4 aprile 1997, Editrice Tirrenia Stampatori, Torino

Vedaldi R. (1997), Modelli Interpretativi per la Valutazione della Customer Satisfaction:l' impiego degli strumenti statistici, in: *atti del convegno SIS "La Statistica per le Imprese"*, Torino 1-4 aprile 1997, Editrice Tirrenia Stampatori, Torino

Zanella A. (1999), A Stochastic Model for the Analysis of Customer Satisfaction: some Theoretical Aspects, *Statistica*, n.1, 1-40

Mixture Models for Maximum Likelihood Estimation from Incomplete Values

Filippo Domma, Salvatore Ingrassia
Dipartimento di Economia Politica, Università della Calabria
87036 Arcavacata di Rende (CS), Italy
[f.domma, s.ingrassia]@unical.it

Abstract: In this paper we consider methods for the analysis of the relationship between input and output variables when missing values occur in the input data. In such situations the incomplete cases cannot be suppressed and then the missing values must be estimated on the basis of some suitable statistical model. This problem is here approached by means of mixture distributions in which the parameters are estimated using likelihood-based methods. Applications in neural network training from incomplete data are discussed and the results are compared with those obtained using the mean imputation method. These results lead to some practical criteria for the use of either method in the learning of neural network from incomplete data.

Keywords: Missing data, mixture models, likelihood, neural networks.

1. Introduction

Statistical analysis based on incomplete data may follow one of two directions, depending on the objectives and resources of the experimenter. One possibility is to suppress incomplete cases and analyze the remaining data set; the alternative is to estimate the missing values and then to analyze the original data set with these values inserted. When we consider problems of parameter estimation, often in data analysis it is possible to suppress cases with missing values; but when we have to face problems involving prediction, this is not possible. For example one might want to continue to use a predictor even when a sensor fails in giving readings or, in economic forecasting, when an input variable becomes meaningless (for example, due to political changes in a country). In such problems the question is if the observed values of an incomplete input vector contain sufficient information in order to predict the missing values and afterwards to make useful predictions. Here we consider this case in which the incomplete input vectors cannot be ignored and thus the missing values must be estimated and inserted in the original data set.

In this context, in this paper we focus on finite mixture distributions for modeling incomplete data, according to ideas proposed by Ghahramani and Jordan (1997). The paper aims at a first comparison between this method and the mean imputation one (which is the methodology usually implemented in many statistical packages for neural networks) both in discrimination and regression problems on the basis of some real data sets.

The rest of the paper is organised as follows. In the next section we outline main statistical properties of missing at random data, in Section 3 we introduce finite mixture distribution for modeling incomplete data; in Section 4 we consider parameter estimation in mixture models according to the maximum likelihood approach; in Section 5 we present results of some numerical studies; concluding remarks are given in Section 6.

2. Statistical properties of missing at random data

The statistical framework we present below is based on Little and Rubin (1987). Let $\mathcal{X}$ be the $N \times (p+1)$ data matrix that would occur in the absence of missing values and assume that $\mathcal{X}$ can be divided into an observed component $\mathcal{X}^{\mathrm{o}}$ and a missing component $\mathcal{X}^{\mathrm{m}}$, that is $\mathcal{X} = (\mathcal{X}^{\mathrm{o}}, \mathcal{X}^{\mathrm{m}})$. Each data vector $\mathbf{x}_i \in \mathcal{X}$, with $i = 1, \ldots, N$, may have different patterns of missing features. The missing data can be given by a $N \times (p+1)$ matrix called *missing data indicator matrix* $\mathbf{R} = (r_{ij})$ where $r_{ij} = 1$ if x_{ij} is observed and $r_{ij} = 0$ if x_{ij} is missing.

Usually one distinguishes between the *environment*, which is assumed to generate complete data, and the *missing data mechanism*, which renders some of the output of the environment unobservable to the experimenter. Both the data generation process and the missing data mechanism are considered to be random processes: the environment is modelled by a probability distribution parametrized by some $\gamma \in \Gamma$ and the missing data mechanism is modelled by a probability distribution parametrized by some $\phi \in \Phi$. The parameters γ and ϕ are assumed to be distinct, that is the joint parameter space of (γ, ϕ) is $\Gamma \times \Phi$. Then the joint probability of $\mathcal{X}$ and $\mathbf{R}$ can be specified as the product of the densities of $\mathcal{X}$ and the conditional distribution of $\mathbf{R}$ given $\mathcal{X}$:

$$f(\mathcal{X}, \mathbf{R}; \gamma, \phi) = f(\mathcal{X}; \gamma)\, f(\mathbf{R}|\mathcal{X}; \phi)\,. \tag{1}$$

We point out that the actual observed data consist of the values of the variables $(\mathcal{X}^{\mathrm{o}}, \mathbf{R})$, so that the distribution of the observed data $f(\mathcal{X}^{\mathrm{o}}, \mathbf{R}; \gamma, \phi)$ is obtained by integrating $\mathcal{X}^{\mathrm{m}}$ out of the joint density of $\mathcal{X} = (\mathcal{X}^{\mathrm{o}}, \mathcal{X}^{\mathrm{m}})$ and $\mathbf{R}$:

$$f(\mathcal{X}^{\mathrm{o}}, \mathbf{R}; \gamma, \phi) \;=\; \int f(\mathcal{X}^{\mathrm{o}}, \mathcal{X}^{\mathrm{m}}; \gamma) f(\mathbf{R}|\mathcal{X}^{\mathrm{o}}, \mathcal{X}^{\mathrm{m}}; \phi)\, \mathrm{d}\mathcal{X}^{\mathrm{m}}\,. \tag{2}$$

The second hypothesis assumed here is that the data are *missing at random* (MAR), that is $f(\mathbf{R}|\mathcal{X}^{\mathrm{o}}, \mathcal{X}^{\mathrm{m}}; \phi) = f(\mathbf{R}|\mathcal{X}^{\mathrm{o}}; \phi)$. This implies that in the likelihood approach the parameter ϕ of the missing data mechanism can be ignored for the purposes of estimating γ. As a matter of fact the likelihood of γ and ϕ is proportional to $f(\mathcal{X}^{\mathrm{o}}, \mathbf{R}; \gamma, \phi)$, that is $L(\gamma, \phi; \mathcal{X}^{\mathrm{o}}, \mathbf{R}) \propto f(\mathcal{X}^{\mathrm{o}}, \mathbf{R}; \gamma, \phi)$, moreover under the MAR hypothesis the relation (2) yields $f(\mathcal{X}^{\mathrm{o}}, \mathbf{R}; \gamma, \phi) = f(\mathcal{X}^{\mathrm{o}}; \gamma) f(\mathbf{R}|\mathcal{X}^{\mathrm{o}}; \phi)$, where

$$f(\mathcal{X}^{\mathrm{o}}; \gamma) \;=\; \int f(\mathcal{X}^{\mathrm{o}}, \mathcal{X}^{\mathrm{m}}; \gamma)\, \mathrm{d}\mathcal{X}^{\mathrm{m}}\,.$$

Since our interest is in the estimation of γ, then $f(\mathbf{R}|\mathcal{X}^{\mathrm{o}}; \phi)$ can be considered as a constant in $L(\gamma, \phi; \mathcal{X}^{\mathrm{o}}, \mathbf{R})$, thus we get $L(\gamma, \phi; \mathcal{X}^{\mathrm{o}}, \mathbf{R}) \propto L(\gamma; \mathcal{X}^{\mathrm{o}})$ where $L(\gamma; \mathcal{X}^{\mathrm{o}})$ is the likelihood based on the observed data ignoring the missing data mechanism defined as $L(\gamma; \mathcal{X}^{\mathrm{o}}) \propto f(\mathcal{X}^{\mathrm{o}}; \gamma)$, see section 5.3 in Rubin (1987) for details.

3. Modeling incomplete data by mixture distributions

Let $(\mathbf{x}_1, y_1), \ldots, (\mathbf{x}_N, y_N)$ be a set of observations and assume that the functional dependence between a p-dimensional input $\mathbf{x}$ and the response y is $g(\mathbf{x}; \theta)$. Here the data matrix $\mathcal{X}$ will be rewritten as $\mathcal{X} = [\mathbf{y}\ \mathbf{X}]$ where $\mathbf{y} = (y_1, \ldots, y_N)^t$ is the first column of $\mathcal{X}$ and $\mathbf{X} = (\mathbf{x}_1, \ldots, \mathbf{x}_N)^t$ are the columns $2, \ldots, p+1$ of $\mathcal{X}$; the superscript ^{t} denotes vector transposition. In many applications, the functional dependence $g(\mathbf{x}; \theta)$ can be assumed as linear in the θ_h's; in more complicated situations $g(\mathbf{x}; \theta)$ can be approximated by a linear combination of sigmoidal functions.

Now let us assume that the probability density of $\mathbf{x}$ is parametrized by some $\xi \in \Xi$, say $f(\mathbf{x}; \xi)$, thus the parameter of the probability distribution of the environment is given by $\gamma = (\theta, \xi)$. In addition let us assume that also θ and ξ are distinct so that it results $f(y, \mathbf{x}; \theta, \xi) = f(y|\mathbf{x}; \theta)\, f(\mathbf{x}; \xi)$ and then the loglikelihood of (θ, ξ) can be written as:

$$\mathcal{L}(\theta, \xi; \mathbf{y}, \mathbf{X}) = \sum_{i=1}^{N} \ln f(y_i|\mathbf{x}_i; \theta) + \sum_{i=1}^{N} \ln f(\mathbf{x}_i; \xi) = \mathcal{L}(\theta; \mathbf{y}, \mathbf{X}) + \mathcal{L}(\xi; \mathbf{X}).$$

When missing values occur in the input data, the likelihood function of θ, $L(\theta; \mathbf{y}, \mathbf{X}) = \prod_{i=1}^{N} f(y_i|\mathbf{x}_i; \theta)$ is not well defined since $f(y_i|\mathbf{x}_i; \theta)$ depends on the complete vector $\mathbf{x}_i$. Under the MAR hypothesis we can consider the *observed loglikelihood* $\mathcal{L}(\theta; \mathbf{y}, \mathbf{X}^{\mathrm{o}}) \propto \sum_{i=1}^{N} \ln f(y_i|\mathbf{x}_i^{\mathrm{o}}; \theta)$ where:

$$f(y_i|\mathbf{x}_i^{\mathrm{o}}; \theta) = \int f(y_i|\mathbf{x}_i^{\mathrm{o}}, \mathbf{x}^{\mathrm{m}}; \theta) f(\mathbf{x}^{\mathrm{m}}|\mathbf{x}_i^{\mathrm{o}}; \xi) \mathrm{d}\mathbf{x}^{\mathrm{m}} . \tag{3}$$

Equation (3) shows that integrating out over the missing data requires some assumptions for the distribution of $\mathbf{x}$ and algorithms for estimating the parameters of the model as well. Thus the maximum likelihood estimation of θ requires the following two steps: i) we should estimate the density $f(\mathbf{x}; \xi)$ on the basis of the data $\mathbf{X}^{\mathrm{o}}$ and afterwards ii) we can estimate the parameters θ. Below we shall consider continuous data and then we can model $f(\mathbf{x}; \xi)$ by a Gaussian mixture having density $f(\mathbf{x}; \xi) = \alpha_1 p_1(\mathbf{x}; \mu_1, \Sigma_1) + \cdots + \alpha_k p_k(\mathbf{x}; \mu_k, \Sigma_k)$ where the α_j's are the *weights* of the mixture, and the $p_j(\mathbf{x}; \mu_j, \Sigma_j)$'s are the probability density functions of the multinormal components of the mixture with vector mean μ_j and covariance matrix Σ_j. Thus $\xi = (\alpha_1, \mu_1, \Sigma_1, \ldots, \alpha_k, \mu_k, \Sigma_k)$. Mixture models look quite attractive as they combine much of the flexibility of nonparametric models with certain analytic advantages of parametric models.

4. MLE of mixture parameters from incomplete data

Let $\mathbf{x}_1, \ldots, \mathbf{x}_N$ be a sample of size N from a Gaussian mixture with density $f(\mathbf{x}; \xi)$. For our purposes, we can assume multinormal components with the same covariance matrix Σ. Each input vector can be written as $(\mathbf{x}_i, \mathbf{z}_i)$ where $\mathbf{z}_i \in \{0, 1\}^k$ is such that $z_{ij} = 1$ if $\mathbf{x}_i$ is generated by the j-th component of the mixture and $z_{ij} = 0$ otherwise. Afterwards we set $\mathbf{Z} = (\mathbf{z}_1, \ldots, \mathbf{z}_N)'$. The loglikelihood function corresponding to $(\mathbf{x}_1, \mathbf{z}_1), \ldots, (\mathbf{x}_N, \mathbf{z}_N)$ can be then written as:

$$\begin{aligned}\mathcal{L}(\xi; \mathbf{X}, \mathbf{Z}) &= -\frac{N}{2}(p \ln(2\pi) + \ln|\Sigma|) - \frac{1}{2}\mathrm{tr}\left(\Sigma^{-1}\sum_{i=1}^{N}\mathbf{x}_i\mathbf{x}_i^t\right) + \\ &+\mathrm{tr}\left\{\Sigma^{-1}\mathbf{M}\left(\sum_{i=1}^{N}\mathbf{z}_i\mathbf{x}_i^t\right)\right\} - \sum_{j=1}^{k}\left\{\left(\sum_{i=1}^{N}z_{ij}\right)(\ln\alpha_j - \frac{1}{2}\mu_j^t\Sigma^{-1}\mu_j)\right\}, \qquad (4)\end{aligned}$$

where $\mathbf{M}$ is the $p \times k$ matrix with columns $\mu_1, \ldots, \mu_k$ and $|\Sigma|$ denotes the determinant of Σ. Maximum likelihood estimates of the parameters can be obtained via the EM algorithm which is an iterative procedure based on an application of the Jensen inequality for maximizing a likelihood function in the context of partial information. Here the simplified form of the EM algorithm described in Little and Schluchter (1985) can be applied as the complete data model belongs to the regular exponential family with complete data sufficient statistics $\sum_i \mathbf{x}_i\mathbf{x}_i^t$, $\sum_i \mathbf{z}_i\mathbf{x}_i^t$ and $\sum_i \mathbf{z}_i$.

When missing values are present in the given data set, the E-step of the algorithm computes the quantities: $T_{1i} = \mathbb{E}(\mathbf{x}_i\mathbf{x}_i^t|\mathbf{x}_i^{\mathrm{o}}; \mathbf{M}, \Sigma, \alpha)$, $T_{2i} = \mathbb{E}(\mathbf{z}_i\mathbf{x}_i^t|\mathbf{x}_i^{\mathrm{o}}; \mathbf{M}, \Sigma, \alpha)$, and $T_{3i} = \mathbb{E}(\mathbf{z}_i|\mathbf{x}_i^{\mathrm{o}}; \mathbf{M}, \Sigma, \alpha)$ for each item $i = 1, \ldots, N$, where $\mathbf{M}$, Σ and α are the current estimates of the parameters at step t, and afterwards:

$$\begin{aligned}\alpha^{(t+1)} &= \frac{1}{N}\sum_i T_{3i}, \qquad \mathbf{M}^{(t+1)} = (\sum_i T_{2i})'\mathbf{A}^{-1} \\ \Sigma^{(t+1)} &= \frac{1}{N}\left\{\sum_i T_{1i} - (\sum T_{2i})'\mathbf{A}^{-1}(\sum T_{2i})\right\},\end{aligned}$$

where $\mathbf{A}$ is the diagonal matrix with elements $\sum T_{3i}$ along the diagonal and zeros elsewhere. Finally the missing values have been estimated by the least square linear regression between $\mathbf{x}_i^{\mathrm{m}}$ and $\mathbf{x}_i^{\mathrm{o}}$: $\hat{\mathbf{x}}_i^{\mathrm{m}} = \mathbf{E}(\mathbf{x}_i^{\mathrm{m}}|\mathbf{x}_i^{\mathrm{o}}, \hat{\xi})$.

5. Numerical comparisons

Properties of the mixture model (MM) methodology have been investigated considering both problems of discrimination and regression involving neural network learning from incomplete data. The obtained results have been compared with the ones obtained using the mean imputation (MI) method. We remark that neural network models can be set in the context described in the previous sections because, with suitable hypotheses, the error function of the multilayer perceptron can be viewed as proportional to the likelihood function of the network parameters (that are called *weights*). These parameters are estimated on the basis of a set called *learning set*; afterwards the performances of the networks are evaluated using another set of cases independent from the previous one called *test set*, see e.g. Bishop (1995).

As far as the discrimination problem is concerned, we have used well known Fisher's iris data set containing 4-dimensional points concerning three species of iris flower: *setosa, versicolor* and *virginica*. The data set contains 50 units for each species.

As for the regression problem, we have used data taken from a polymer test plant (from ftp.cis.upenn.edu in pub/ungar/chemdata). The data set contains 61 units of 10 predictors concerning measurements of controlled variables in a polymer process plant and 4 responses concerning the outputs of the plant; they are particularly good for testing the robustness of nonlinear modeling methods to irregularly spaced data. For this data, which lie in the interval $[0.1, 0.9]$, De Veaux *et al.* (1993) showed that neural networks are superior to multivariate adaptive splines regression and both are superior to linear regression.

a) Discrimination: Iris data. In order to compare the properties of the two methods for increasing number of missing values, starting from the complete data set, first of all we have generated eight groups of ten incomplete data sets eliminating at random some values (the whole data set contains 600 values): 10 data sets with 25 missing values, 10 data sets with 50 missing values, ..., 10 data sets with 200 missing values. For each data set we have considered 90 units as the learning set (30 for each species) and 60 units as the test set (20 for each species). For each of these 80 incomplete data sets we obtained a version with the missing values estimated according to the MM method and a version with inserted values obtained via the MI method. We point out that the missing data in the test set have been estimated using the model obtained from the learning set.

A first comparison concerned the distances between the complete data sets and the ones containing the estimated values on the basis of the following *standardized mean distances*:

$$\delta_L = \frac{1}{n_L} \sum_{(i,j)\in X_L} \frac{|\hat{x}_{ij} - x_{ij}|}{s_j} \qquad \delta_T = \frac{1}{n_T} \sum_{(i,j)\in X_T} \frac{|\hat{x}_{ij} - x_{ij}|}{s_j} \tag{5}$$

where X_L (X_T) denotes the learning (test) set containing n_L (n_T) points, $\hat{x}_{ij}$ denotes the estimated ij-th value and x_{ij} the corresponding value in the complete data set, and s_j is the standard deviation of the j-th variable. Representative results (one for each group of data sets with the same number of missing values) are given in Tables 1,2. We notice that the mean distances of the values estimated according to the MM method are about 3 times smaller than the distances of the values obtained using the MI method both for the learning and the test sets. This is quite important especially for the latter ones because the estimated values in the test set are based on the parameters estimated from the learning data. Finally we have approached the second part of the problem that is neural network learning from the reconstructed data sets with either the MM or MI method. We considered a perceptron having one hidden layer, with 4 inputs, 3 neurons in the hidden layer and 3 neurons in the output layer. Then we trained the neural network from the complete data, the obtained weights were utilized as the initial ones for all subsequent numerical studies. After many trials, we retained the weights in which the error function – computed on the test set – attained its minimum during the learning phase, as we are interested in models for prediction purposes. In the same Tables 1,2 we give the percentage of misclassifications vs. the number of inserted values in one data set for each group; in particular the value corresponding to zero missing values gives the percentage of misclassification in the original data set. In particular we remark that the MM method leads always to a smaller percentage of misclassifications in the test sets.

Table 1. *Results for different numbers of missing values in Iris data estimated using MM: distances from the original data and percentage of misclassifications.*

	Learning Set			*Test Set*		
# tot missing	# missing	δ_L	% errors	# missing	δ_T	% errors
0	0	0,000	2,22	0	0,000	0,00
25	14	0,300	2,22	11	0,138	1,67
50	30	0,273	3,33	20	0,292	1,67
75	50	0,258	1,11	25	0,367	0,00
100	62	0,274	5,56	38	0,219	1,67
125	68	0,341	2,22	57	0,265	1,67
150	86	0,330	5,56	64	0,240	6,67
175	108	0,325	1,11	67	0,352	3,33
200	122	0,320	3,33	78	0,346	6,67

Table 2. *Results for different numbers of missing values estimated in Iris data using MI: distances from the original data and percentage of misclassifications.*

	Learning Set			*Test Set*		
# tot missing	# missing	δ_L	% errors	# missing	δ_T	% errors
0	0	0,000	2,22	0	0,000	0,00
25	14	0,995	0,00	11	0,842	5,00
50	30	0,833	7,78	20	0,691	3,33
75	50	0,877	2,22	25	0,840	3,33
100	62	0,803	3,33	38	0,791	3,33
125	68	0,843	3,33	57	0,932	5,00
150	86	0,776	3,33	64	0,846	13,33
175	108	0,805	5,56	67	0,910	13,33
200	122	0,902	33,33	78	0,874	35,00

b) Regression: Polymer pilot plan data. Starting from the complete data set, we generated six groups of ten incomplete data sets eliminating at random some values from the variables X_4, X_5, X_6 and X_7: 10 data sets with 5 missing values, 10 data sets with 10 missing values, ..., 10 data sets with 30 missing values. As in DeVeaux *et al.*(1998), we have chosen the response variable X_{11} and we have considered 50 units as the learning set and 11 units as the test set. The analysis followed the same lines as in the above case. In Tables 3,4 we give the distances (5): in this case, the differences in the test sets between the MM and MI are considerably larger than in the previous case. Afterwards we considered a neural network with 10 inputs, 18 neurons in the hidden layer and one neuron in the output layer (the choice of this architecture is discussed in De Veaux *et al.*(1998)). Also in this case the mean square errors (MSE) in the test sets based on the MM method are always smaller than the ones obtained using MI. However the results in Table 4 show an irregular tendency as the number of missing values increases. This highlights a critical aspect of the mean imputation: as we remarked above, *polymer data* are irregularly spaced, in particular for most variables they are concentrated on the extremes of the interval $[0.1, 0.9]$; in this case the means are quite far from almost all values of the corresponding variables.

Table 3. *Results for different numbers of missing values in Polymer data estimated using MM: distances from the original data and mean square errors.*

	Learning Set			*Test Set*		
# tot missing	# missing	δ_L	MSE	# missing	δ_T	MSE
0	0	0,000	0,0179	0	0,000	0,0134
5	4	0,388	0,0132	1	0,156	0,0150
10	9	0,440	0,0131	1	0,139	0,0197
15	13	0,299	0,0142	2	0,096	0,0139
20	15	0,261	0,0236	5	0,302	0,0185
25	21	0,212	0,0251	4	0,178	0,0138
30	22	0,210	0,0428	8	0,129	0,0143

Table 4. *Results for different numbers of missing values in Polymer data estimated using MI: distances from the original data and mean square errors.*

	Learning Set			*Test Set*		
# tot missing	# missing	δ_L	MSE	# missing	δ_T	MSE
0	0	0,000	0,0179	0	0,000	0,0134
5	4	1,171	0,0145	1	2,192	0,0349
10	9	0,811	0,0146	1	2,023	0,0240
15	13	1,006	0,0231	2	1,387	0,0218
20	15	1,356	0,0216	5	0,968	0,0419
25	21	0,969	0,0199	4	1,209	0,0162
30	22	0,867	0,0490	8	2,022	0,0629

6. Concluding remarks

The obtained results can be analyzed from another point of view considering that the distances δ_L, δ_T and the percentage of misclassifications/MSE in the above tables concern respectively the input and the output data of the neural network. This analysis will lead to some practical criteria involving the use of either method in learning neural network from incomplete data.

From Tables 1,2 it emerges that the tendencies of the standardized distances and of the number of miclassifications are quite different. For a limited percentage of missing values, even if the differences between the normalized distances are relevant, the performance of the networks are comparable; this is due to the well known properties of robustness of the neural networks. On the contrary the differences between the two estimate procedures lead to quite different performances of the network as the percentage of missing values in the data set becomes larger and larger, in particular when mean imputation is considered the performances of the neural network get worse and worse. We point out that the results concerning the learning set can appear misleading but we recall that they have been obtained using the weights in which the error function has attained its minimum during the training phase.

Furthermore Tables 4 points out some troubles coming from the use of the mean imputation for irregularly spaced variables.

Finally we can draw the following conclusion. Even if the MM method always looks preferable from a theoretical point of view (bearing in mind the limits of the numerical experiments we carried out), in practical application of neural network involving small numbers of incomplete cases – provided that the data are regular in the sense said above – the properties of robustness of the neural networks allow the estimation of the missing values via the MI method. Indeed the MM method requires the implementation of the EM algorithm and this last requires some computational efforts which are not always justified. At present, more procedures are under investigation.

References

Bishop C.M. (1995) *Neural Networks for Patter Recognition*, Clarendon Press, Oxford.

DeVeaux R.D., Psichogios D.C. & Ungar L.H. (1993). A comparison of two non parametric estimation schemes: MARS and neural networks, *Computers and Chemical Engineering*, 17, n.8, 819-837.

DeVeaux R.D., Schumi J., Schweinsberg J. & Ungar L.H. (1998). Prediction intervals for neural networks via nonlinear regression, *Technometrics*, 40, n.4, 273-282.

Ghahramani Z. & Jordan, M.I. (1997) Learning from incomplete data, in *Computational Learning Theory and Natural Learning Systems*, Greiner, R., Petsche, T. & Hanson, S.J. (Eds.), vol. 4: Making Learning Systems Practical, MIT Press, Cambridge, MA, 67-85.

Little R.J.A. & Schluchter M.D. (1985) Maximum likelihood estimation for mixed continuous and categorial data with missing values, *Biometrika*, 72, 497-512.

Little R.J.A. & Rubin D.B. (1987) *Statistical Analysis with Missing Data*, John Wiley & Sons, New York.

Robust Inference in the Logistic Regression Model[1]

Michele La Rocca
Università degli Studi di Salerno
e-mail: larocca@unisa.it

Abstract: Empirical likelihood is extended to a class of robust estimators for the parameter vector of the logistic regression model so to improve on both the known inference procedures based on empirical likelihood, which are not robust, and the usual robust inference procedures based on the normal approximation.

Keywords: Logistic regression, Robust inference, Empirical likelihood.

1. Introduction

Empirical likelihood (EL) is a technique for forming confidence sets (or tests), based on nonparametric likelihood ratios (Owen, 1988). It provides inferences that are of comparable accuracy to those based on the delta method, the jackknife and the simpler bootstrap methods. But several advantages of this method have been highlighted: confidence regions have a data-driven shape, constraints among the parameters of interest can be easily imposed, confidence sets are range-preserving, analytical corrections to improve accuracy (such as Bartlett correction) are available.

The method has been studied extensively in the literature with applications to a large class of problems including regression models, GLM (Kolaczyk, 1994) and weakly dependent processes (Kitamura, 1997). But, besides an extension to *M*-estimators (Owen, 1991), very little has been done to study this technique in conjunction with robust procedures.

EL, in principle, can give inferences for any parameter defined implicitly through a set of unbiased estimating equations with finite higher order moments (Qin and Lawless, 1994). Hence, the procedure, which is not robust in general, can be made robust by a proper choice of its set of estimating functions. In this paper we consider a binary-response logistic regression model and we focus on the construction, based on EL, of approximate confidence regions for the parameters of the model with low coverage error and not sensitive to the presence of outlying observations.

In section 2, the proposed procedure is introduced and discussed. In section 3, some results of a simulation study are presented. Some concluding remarks will close the paper.

---

[1] The paper is supported by MURST 98 grant ex 60% 98.

2. The empirical likelihood method

Let $Y_1,\ldots,Y_n$ be a random sample from a non-degenerate distribution F. EL is defined as $L(F)=\prod_{i=1}^{n} F\{y_i\}$ where y_i is the observed value of Y_i and $F\{y_i\}$ is the probability, under F, of the set $\{y_i\}$ .The function $L(F)$ is maximised, over the space of all distribution function, by $F_n(y)=n^{-1}\sum_{i=1}^{n} I(y_i \le y)$ where $I(\cdot)$ is the indicator function. Hence, the empirical likelihood ratio (ELR) is defined as $R(F)=L(F)/L(F_n)$. If the y_i are all distinct, we may write $R(F)=\prod_{i=1}^{n} np_i$ with $p_i=F\{y_i\}$. If ties among the data are present, definition of $R(\cdot)$ is still appropriate with the modification $\sum_{j:y_j=y_i} p_j = F\{y_i\}$ (Owen, 1988). Given the d-dimensional parameter $\theta=T(F)$ the profile ELR function is

$$R(\theta)=\sup_{p_1,\cdots p_n}\left\{\prod_{i=1}^{n} np_i \middle| p_i \ge 0, \sum_{i=1}^{n} p_i = 1, T(F_{np})=\theta\right\} \tag{1}$$

where $F_{np}(y)=\sum_{i=1}^{n} p_i\, I(y_i \le y)$. Thus, empirical likelihood is the profile likelihood of a multinomial distribution putting masses $p_1,\ldots,p_n$ on $y_1,\ldots,y_n$. In many cases, under mild conditions, it can be shown that, if θ_0 is the true value of the parameter, $r(\theta_0)=-2\log(R(\theta_0))\xrightarrow{d}\chi^2_{(d)}$, for $n\to\infty$. Therefore, confidence regions and tests, with an asymptotically justified calibration, can be based on contours of likelihood ratio functions, by using an approach analogous to that used for parametric models.

In this paper, we consider a Bernoulli variable Y along with q covariates $\mathbf{X}=(X_1,\ldots,X_q)$. Under the logistic regression model, for any $\mathbf{x}\in\Re^q$ and some $\beta_0\in\Re^q$, we have $\Pr(Y=1|\,\mathbf{x})=P(\mathbf{x}^T\beta_0)$ where $P(t)=\exp(t)/(1+\exp(t))$. Given a sample, $(\mathbf{x}_i^T, Y_i)$, $i=1,\ldots,n$, of n independent observations, classical inference for the unknown parameter vector are usually based on maximum likelihood. However, it is well known that outliers can have a strong impact on this procedure and several robust alternatives have been proposed (see Bianco and Yohai, 1996 *inter alia*). A large class of them solves equations which generalise maximum likelihood of the form

$$\sum_{i=1}^{n} g(\mathbf{x}_i, Y_i, \beta)=0 \tag{2}$$

where $g(\mathbf{x}_i, Y_i, \beta)=w_i\,\mathbf{x}_i\left\{Y_i - P(\mathbf{x}_i^T\beta)\right\}$ is q vector with $\{w_i = w(\mathbf{x}_i, \mathbf{x}_i^T\beta)\}$ weights built to reduce the impact of unusual observations.

The estimating functions in the system (2) are unbiased and so, if $\beta = \beta_0$, they are such that $E(g(\mathbf{x}_i, Y_i, \beta)) = 0$. In this framework, we propose to construct the ELR function as

$$R(\beta) = \sup_{p_1, \cdots p_n} \left\{ \prod_{i=1}^n np_i \middle| p_i \geq 0, \sum_{i=1}^n p_i = 1, \sum_{i=1}^n p_i g(\mathbf{x}_i, Y_i, \beta) = 0 \right\} \quad (3)$$

If the hull of the points $g(\mathbf{x}_i, Y_i, \beta)$, $i = 1, \ldots, n$ contains the null vector, for each β, a unique maximum exists and it can be determined through a Lagrange multiplier approach. In this way we obtain

$$r(\beta) = 2\sum_{i=1}^n \log\left(1 + \lambda^T g(\mathbf{x}_i, Y_i, \beta)\right)$$

where $\lambda^T = \{\lambda_1, \ldots, \lambda_q\}$ satisfies

$$\frac{1}{n}\sum_{i=1}^n \frac{g(\mathbf{x}_i, Y_i, \beta)}{1 + \lambda^T g(\mathbf{x}_i, Y_i, \beta)} = 0 .$$

The following result can be shown.

PROPOSITION. Assume that:

1. $\Pr(ch(P) \cap ch(N) \neq \varnothing) \to 1$ as $n \to \infty$ where $P = \{\mathbf{x}_i | Y_i - P(\mathbf{x}_i^T \beta_0) > 0\}$ and $N = \{\mathbf{x}_i | Y_i - P(\mathbf{x}_i^T \beta_0) < 0\}$, with $ch(\cdot)$ denoting the convex hull,
2. $0 < w_i = w(\mathbf{x}_i, \mathbf{x}_i^T \beta_0) \leq 1, \forall i$,
3. $n^{-2} \sum_{i=1}^n \|x_i\|^4 \mu_4(\mathbf{x}_i) \to 0$ as $n \to \infty$, where $\mu_4(\mathbf{x}_i) = w_i^4 \int (Y_i - P(\mathbf{x}_i^T \beta_0))^4 dF_{x_i}$,
4. $a < \min eig(n^{-1} X^T X)$ with $a > 0$,
5. $n^{-1} \sum_{i=1}^n \|\mathbf{x}_i\|^2 < \infty$.

Then, for $n \to \infty$, $r(\beta_0) \xrightarrow{d} \chi^2_{(q)}$.

Proof. We simply remark that conditions 1-5 are sufficient conditions for the theorem on Empirical likelihood for triangular arrays in Owen (1991). First, we observe that condition 1 is a sufficient condition to have $0 \in ch\{g(x_i, Y_i, \beta_0), i = 1, \ldots, n\}$. Moreover our condition 3 is a sufficient condition for (3.3b) in Owen (1991).

Now, observe that $V_{in} = \mathrm{var}(g(\mathbf{x}_i, Y_i, \beta_0)) = w_i^2 P(\mathbf{x}_i^T \beta_0)(1 - P(\mathbf{x}_i^T \beta_0))\mathbf{x}_i \mathbf{x}_i^T$ so we have $V_n = n^{-1} \sum_{i=1}^n w_i^2 P(\mathbf{x}_i^T \beta_0)(1 - P(\mathbf{x}_i^T \beta_0))\mathbf{x}_i \mathbf{x}_i^T$. Simple inequalities show that

$$\max eig(V_n) \le \max_i \{w_i^2 P(\mathbf{x}_i^T \beta_0)(1 - P(\mathbf{x}_i^T \beta_0))\} \max eig(n^{-1} X^T X)$$
$$\min eig(V_n) \ge \min_i \{w_i^2 P(\mathbf{x}_i^T \beta_0)(1 - P(\mathbf{x}_i^T \beta_0))\} \min eig(n^{-1} X^T X).$$

Condition (3.3c) of the theorem on Empirical likelihood for triangular arrays in Owen (1991) is ensured if $w_i^2 P(\mathbf{x}_i^T \beta_0)(1 - P(\mathbf{x}_i^T \beta_0))$ is bounded away from zero and infinity and also the eigenvalues of $n^{-1} X^T X$ are bounded away from zero and infinity. Observe that, from condition 2 we have $0 < w_i^2 P(\mathbf{x}_i^T \beta_0)(1 - P(\mathbf{x}_i^T \beta_0)) \le 1/4$. Now, denote with $tr(A)$ the trace of the matrix A and observe that $tr(n^{-1} X^T X) = n^{-1} tr(\sum_{i=1}^n \mathbf{x}_i \mathbf{x}_i^T) = n^{-1} \sum_{i=1}^n \|x_i\|^2$. Moreover, denote with $\{\psi_i, i = 1, \ldots, q\}$ the eigenvalues of $n^{-1} X^T X$, from condition 4, we have that $tr(n^{-1} X^T X) = \sum_{i=1}^n \psi_i \ge \max eig(n^{-1} X^T X)$. So, condition 5 (which is obviously satisfied when the independent variables are bounded or they are sampled from a normal distribution) implies that $\max eig(n^{-1} X^T X) < \infty$. This complete the proof.

An approximated confidence region of nominal level $1-\alpha$ for β can be constructed as $I_{1-\alpha} = \{\beta \,|\, r(\beta) \le \chi^2_{(q;1-\alpha)}\}$. If inferences on a subset of k ($k<q$) coefficients are of interest, the remaining q-k coefficients can be treated as nuisance parameters and profiled out, referring then to a chi-square distribution with k degrees of freedom for the calibration.

The ELR depends on the data only through the functions $g(\cdot)$ and so it inherits all the robustness properties of the estimator obtained through equations (2). The weights are only function of the design and the vector parameter, so robustness is achieved by downweighting points with high leverage (and so potentially dangerous). Two sub-cases should be mentioned. First, the weights depend only on the design, $\{w_i = w(\mathbf{x}_i, \mathbf{x}_i^T \beta) = w(\mathbf{x}_i)\}$ and all those points not near the centre of the design itself will be downweighted. Of course, robustness properties depend on how well one can measure the leverage of each individual point, a task that is not trivial due to masking effect. Alternatively, the weights depend only on the probabilities, setting $\{w_i = w(\mathbf{x}_i^T \beta)\}$, so that extreme fitted probabilities are downweighted.

As weighting schemes we will refer to Carroll and Pederson (1993), who introduced and discussed some of them in the classical *M*-estimation framework, but other choices are possible as well (see Bianco and Yohai, 1996 *inter alia*). For the leverage downweighting (LD) case the suggested procedure runs as follows. Let $\mathbf{x}_i^T = \{1, \mathbf{z}_i^T\}$, let $(\hat{\mu}, \hat{\Sigma})$ be a robust estimate of the centre and covariance matrix of $\mathbf{z}_i$; estimates of (μ, Σ) that are highly robust can be based on *S*-

estimators (Ruppert, 1992). Define $d_i = (\mathbf{z}_i - \hat{\mu})^T \hat{\Sigma}^{-1} (\mathbf{z}_i - \hat{\mu})$ and the weights $w_i = w(\mathbf{x}_i) = h\left[(d_i/(q-1))^{1/2}\right]$ where $h(t) = \psi(t)/t$ with $\psi(t)$ any odd function.

For the prediction downweighting (PD) case one can consider,

$$w_i = w(\mathbf{x}_i^T \beta) = \left[P(\mathbf{x}_i^T \beta)\{1 - P(\mathbf{x}_i^T \beta)\}\right]^c \left[P(\mathbf{x}_i^T \beta)^\lambda + \{1 - P(\mathbf{x}_i^T \beta)\}^\lambda\right] \quad (4)$$

for constants (c, λ). The choices $c = 0$ and $\lambda = 0$ or $c = 1$ and $\lambda = -1$ yield to the usual logistic estimates. The choice $c = 1$ and $\lambda = 0$ weights the observations according to the variance of Y given x and will handle outliers at very low or very high predicted probabilities. Alternatively, one can consider the form

$$w_i = w(\mathbf{x}_i^T \beta) = (1 - 2\gamma)\left[P(\mathbf{x}_i^T \beta)\{1 - P(\mathbf{x}_i^T \beta)\}\right]\left[G(\mathbf{x}_i^T \beta, \gamma)\{1 - G(\mathbf{x}_i^T \beta, \gamma)\}\right]^{-1} \quad (5)$$

where $G(t, \gamma) = P(t) + \gamma(1 - 2P(t))$. In both cases tuning constants have to be fixed, with some advantages for the last proposal where γ can be interpreted as the nominal amount of misclassification assumed in constructing the weights.

We finally remark that the maximum EL estimate of β will be the same as the point estimate obtained by solving the system (2). Nevertheless the proposed procedure will give some clear advantages over the inferences for β based on the normal approximation to the sampling distribution of robust estimator. First, it is possible to form confidence regions without imposing any predetermined shape since EL implicitly takes into account higher order effects (asymmetries and/or tails heavier than those of the normal) possibly present in the data. Second, no explicit estimate of the covariance matrix of the estimators involved is required.

3. Some simulation results

To assess the performance of the proposed procedure a small simulation study was conducted. We considered a simple logistic regression model, with the intercept and one covariate, distributed as a standard normal. Four different models have been considered, with parameters chosen so that $\Pr(Y = 1) = 0.5$ (Model 1), $\Pr(Y = 1) = 0.4$ (Model 2), $\Pr(Y = 1) = 0.3$ (Model 3) and $\Pr(Y = 1) = 0.2$ (Model 4). We considered only the most influential mass contamination, that is at points (x_c, y_c) with $y_c = 1$ if $\Pr(\beta_{01} + \beta_{02} x_c) < 0.5$ and $y_c = 0$ in other case, and x_c chosen so that it varies on a grid of 80 points in the interval [-8, 8]. We considered one-wild outlier when the sample size is equal to 30, and two-wild outliers when the sample size is equal to 60.

In the first scheme, the weights depend only on the design and all those points not near the centre of the design itself will be downweighted (LD). As an alternative scheme, we consider weights as defined in the equation (4) with $c = 1$

and $\lambda = 0$, so that extreme fitted probabilities will be downweighted (PD).

The observed coverages corresponding to 5000 simulation runs for each contaminating point, for a sample size equal to 30 are shown in Figure 1. As expected, the observed coverage of the ELR based on the ML functions is extremely sensitive to outliers contamination. On the other hand, ELR based on the robust functions have a much more stable behaviour. In the case considered the PD scheme and the LD scheme have a similar behaviour when $\Pr(Y=1)$ is close to 0.5 with differences increasing when $\Pr(Y=1)$ gets close to zero. By symmetry the case of $\Pr(Y=1)$ close to one should behave in a similar manner.

Figure 1: *Observed coverage for EL based on ML, prediction downweighting (PD) and leverage downweighting (LD). 5000 simulation runs on samples of size 30 with one outlier. Nominal coverage equal to 0.90.*

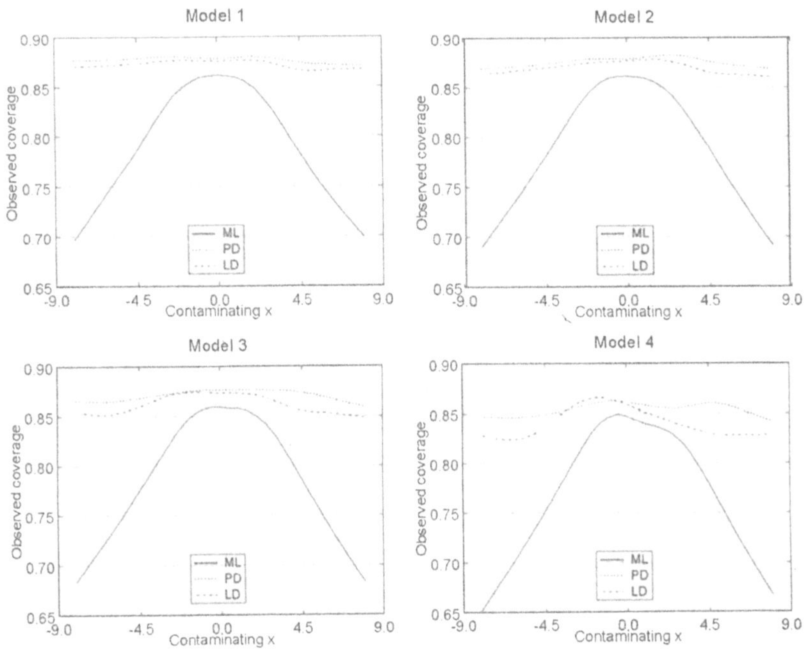

In Fig. 2 we reported the results of 5000 runs on samples of size 60 with two contaminating points and it is clear the above remarks still apply.

We also considered the weights as defined in the equation (5). In this case we compared ML with the observed coverage obtained with three different choices of the parameter γ, namely $\gamma = 0.02$, $\gamma = 0.04$ and $\gamma = 0.06$ (respectively termed G2, G4 and G6). The results are reported in Fig. 3 where, again, there is a clear improvement of the robust approach over the classical one.

In all cases considered, the actual coverage, for small n, can be quite different from the nominal one, especially for models where $\Pr(Y=1)$ gets close to zero (or by symmetry close to one). Therefore, as usual when using EL, corrections or adjustment to the statistics or to its approximating distribution would be useful. In any case, as expected from the asymptotic result shown in the previous section, coverage accuracy improves as n increases (as evident from Fig. 2 and Fig. 3).

Similar results can be obtained if more than one parameter is considered. In any case, due to the curse of dimensionality, the accuracy of the coverage worsen.

Figure 2: *Observed coverage for EL based on ML, prediction downweighting (PD) and leverage downweighting (LD). 5000 simulation runs on samples of size 60 with two outliers. Nominal coverage equal to 0.90.*

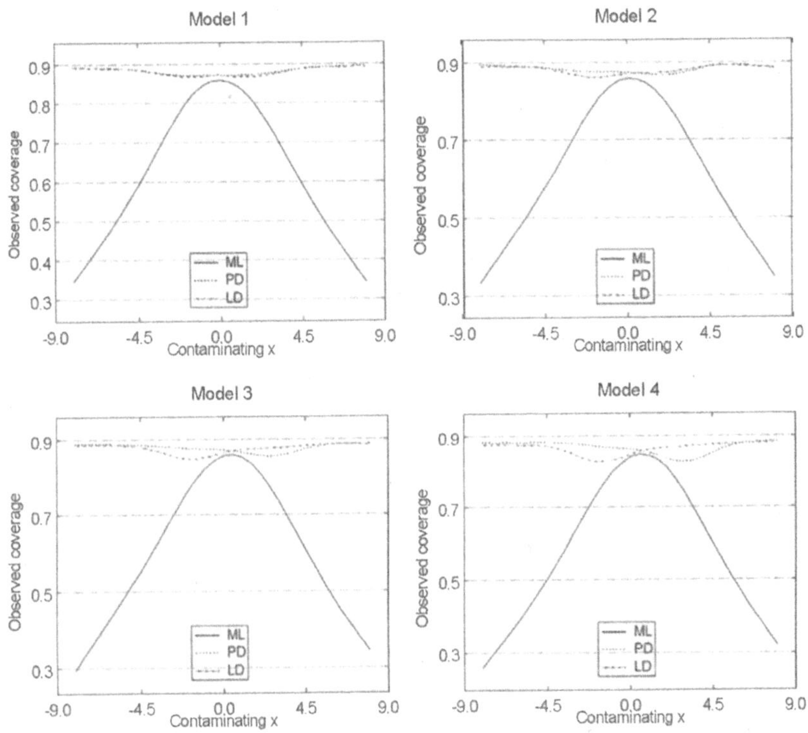

4. Concluding remarks

In this paper a new procedure to make inference in the logistic regression model, when M-type robust estimates are used, has been presented. It is based on the empirical likelihood method and it improves upon normal approximation in that confidence regions have a data-driven shape, constraints among the parameters of interest can be easily imposed, confidence sets are range-preserving, no variance covariance matrix estimate is needed. The proposed procedure has an asymptotic justification and performs quite well in finite samples, especially if sample size is not too small and if the probability of success is not too close to zero or one. A better accuracy could be obtained by using analytical correction to the empirical likelihood ratio statistics (such as the Bartlett correction) or by improving the accuracy of the approximating distribution, by using a bootstrap approach. These aspects are still under investigation.

Figure 3: *Observed coverage for EL based on ML and PD. 5000 simulation runs on samples with one outliers. Nominal coverage equal to 0.90.*

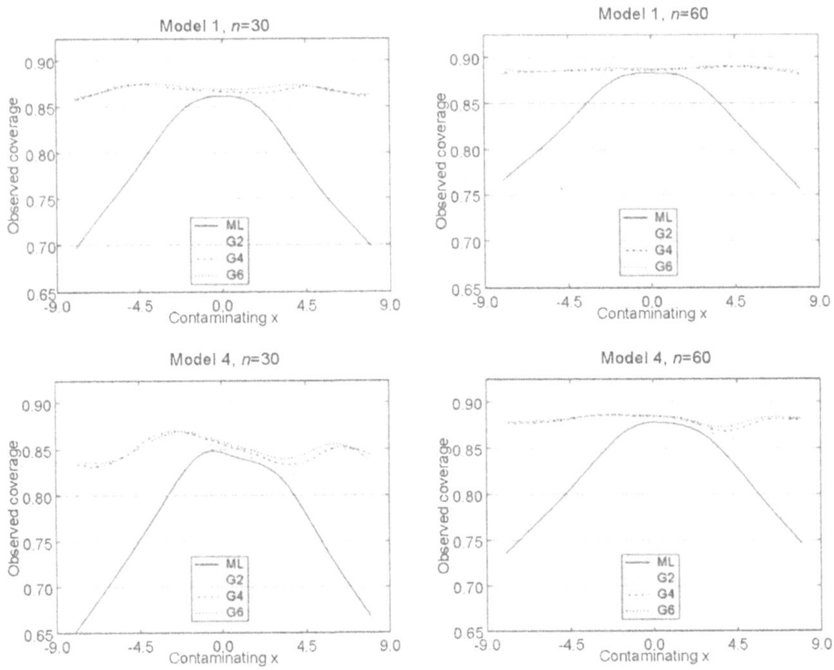

References

Bianco, A. M. and Yohai, V. J. (1996). Robust estimation in the logistic regression model in: *Robust statistics, data analysis and computer intensive methods*, H. Rieder (ed.), Lecture Notes in Statistics – Springer, 17-34.

Carrol R. J. and Pederson, S. (1993). On robustness in the logistic regression model. *Journal of Royal Statistical Society* B, 55, 693-706.

Kitamura, Y. (1997). Empirical likelihood methods with weakly dependent processes, , *Annals of Statistics*, 25, 2084-2102.

Kolaczyk, E. D. (1994). Empirical likelihood for generalized linear models, *Statistica Sinica*, 4, 199-218.

Owen, A. B. (1988). Empirical likelihood ratio confidence intervals for a single functional, *Biometrika*, 75, 237-249.

Owen, A. B. (1991). Empirical likelihood for linear models, *Annals of Statistics*, 19, 1725-1747.

Ruppert, D. (1992). Computing *S*-estimators for regression and multivariate location/dispersion. *Journal of Computational and Graphical Statistics*, 1, 253-270

A Plot for Submodel Selection in Generalized Linear Models

Giovanni C. Porzio[1]
Università degli Studi di Cassino
via Mazzaroppi, Cassino
g.porzio@unicas.it

Abstract: In applied regression analysis, model selection criteria are usually used to identify a set of submodels for further study. In this paper, we present a method for a graphical comparison of models that helps in selecting among submodels. The method, based on comparisons of fitted functions projected on two-dimensional surfaces, is offered in a generalized linear models framework, and it is explored in the binomial regression case.

Keywords: Binomial regression, dynamic graphics, logistic regression, linear projections.

1. Introduction

Submodel selection is a typical problem in most applied regression studies. Assuming a model that fits well the data is already available (the full model), submodel selection consists of selecting few covariates, among many, that could adequately describe the data as well.

As no tests are available to compare nonnested models, submodel selection is performed through some criteria, many of which are found in the literature. Mallows' C_p statistic is probably the most widely used in linear regression, while other measures can be applied within generalized linear models (see e.g. Jovanovic and Hosmer, 1997). However, in practice, such model selection criteria are commonly used to identify a *subset* of models for further study (Bedrick and Tsai, 1994), and tools are required to compare the selected models in greater depth.

With this aim, we present a new kind of plot, that we will call *submodel plot*, that is based on the representation of marginal features of the models. The plot can be used both to compare a submodel with the full model, and to compare submodels. Looking at the plot, the user will be able not only to judge wheter two models are essentially equivalent, but also to understand what reasons, if any, lie behind the selection of a particular submodel.

The basis for our proposal is given in Section 2, while the proposed *submodel plot* is described in Section 3. In Section 4 we apply our method to the binomial

[1]Research supported by MURST and CNR funds.

regression case. An illustrative example in the medical field follows in Section 5, and the last section contains some further remarks.

2. Submodel selection through projections

A model for regression analysis, that is for the study of the relationship between a response variable Y and some covariates X, is completely determined by a conditional cdf $F(y|x) = Pr(Y \leq y|X = x)$. When using parametric models, we consider cdf's $F_\theta(y|x)$ that are exactly identified for some value of parameters θ. In the generalized linear models framework, $F_\theta(y|x)$ is

$$F_\theta(y|X = x) = \int_{-\infty}^{y} f_\theta(t|x)\, dt = \int_{-\infty}^{y} \exp\{[t\,\theta(x) - b(\theta(x))] + c(t)\}\, dt$$

if Y is continuous, or with the sum replaced for the integral if discrete, and where we have omitted the dependence on the dispersion parameter ϕ for simplicity. In such a framework, the conditional mean function is $E(Y|X = x) = b'(\theta(x))$, the first derivative of b evaluated at $\theta(x)$. Parameter θ is modeled as a known function of a linear combination of x through some parameters β to be estimated. With a canonical link function, we have:

$$\theta(x) = x^T \beta.$$

Hence, for a given x, a value of θ is equivalent to a value for β.

Submodel selection consists of looking for parsimonious models to describe the relationship between the response and the covariates excluding some of the latter. In other words, some elements β_j of β are constrained to equal zero, and we test $H_0 : \beta = \beta_0$ against a general alternative (where β_0 denotes the constrained parameters vector). Equivalently, we can test:

$$H_0 : F_\theta(y|X = x) = F_{\theta_0}(y|X = x) \quad \forall x, \tag{1}$$

where $\theta_0 = x^T \beta_0$.

A submodel selection problem is then a comparison between two cdf's $F_\theta(y|X = x)$ and $F_{\theta_0}(y|X = x)$, where the former describes the full unrestricted model, and the latter the restricted model. To perform such a comparison graphically, as we are unable to give graphical representations of a multidimensional object like $F_\theta(y|X = x)$, we propose to consider its projections. Projections underlie many graphical tools in regression analysis, and are discussed extensively by Cook (1998), and Cook and Weisberg (1999).

It is indeed well known that $F_{\theta_1}(y|X = x) = F_{\theta_2}(y|X = x)$ for all values of x in the sample space Ω_x iff $F_{\theta_1}(y|Z = x^T a) = F_{\theta_2}(y|Z = x^T a)$ for all values of $x^T a$ in$\{x^T a | a \in \Re^k, ||a|| = 1, x \in \Omega_x\}$. And we can use this result to compare

$F_\theta(y|Z = x^T a)$ with $F_{\theta_0}(y|Z = x^T a)$ to infer about H_0 in (1). If they are not equal for some $Z = x^T a$ (i.e. for some projection directions a), we will have evidence against their global equality.

As in a regression setting, we focus on the mean function and we consider the comparison of $E_\theta(Y|Z = x^T a)$ with $E_{\theta_0}(Y|Z = x^T a)$. To obtain a graphical representation, we note that

$$\begin{aligned} E_\theta(Y|Z = x^T a) &= E[E_\theta(Y|Z = x^T a)|X = x] \\ &= E[E_\theta(Y|X = x)|Z = x^T a]. \end{aligned} \tag{2}$$

We can then treat $E_\theta(Y|Z = x^T a)$ as the regression problem for the dependence of $E_\theta(Y|X = x)$ on a single covariate $Z = x^T a$, and therefore we can represent it graphically on a two-dimensional plot.

Should we know $E_\theta(Y|X = x)$, the following simple method could be used in order to evaluate a submodel: *i)* select an a, and draw a plot with both an estimate for $E_\theta(Y|Z = x^T a)$ and $E_{\theta_0}(Y|Z = x^T a)$ on it; *ii)* if they differ, infer against H_0 in (1); *iii)* repeat the procedure, drawing plots for many a's.

3. Submodel plot

To perform a submodel selection through the method sketched above, we need an estimate for $E_\theta(Y|Z = x^T a)$ and $E_{\theta_0}(Y|Z = x^T a)$, which we will obtain in two steps. First, consider an estimate for $E_\theta(Y|X = x) = b'(\theta(x))$ and $E_{\theta_0}(Y|X = x) = b'(\theta_0(x))$. For our procedure to work, we can obtain them through any estimate $\hat{\theta}(x)$ under the full model, and any estimate $\hat{\theta}_0(x)$ under H_0 (we can rely on the maximum likelihood estimate, say).

Secondly, note that $E_{\hat{\theta}}(Y|X = x_i) = b'(\hat{\theta}(x_i)) = \hat{y}_i$, the predicted values from the model, where we have evaluated the expected value at $\hat{\theta}$ and at the observed values x_i. Therefore, according to eqn. (2),

$$E_{\hat{\theta}}(Y|Z = x_i^T a) = E[\hat{y}_i|Z = x_i^T a],$$

a single covariate regression problem, for the dependence of $\hat{y}_i$ on $x_i^T a$.

We can then estimate $E_{\hat{\theta}}(Y|X = x_i^T a)$ nonparametrically through a smoother superimposed on the plot of the predicted values $\hat{y}_i$ against $x_i^T a$. Similarly, an estimate under H_0, of course, will be given by a smoother superimposed on the plot of the predicted values under H_0 against $x_i^T a$.

A *submodel plot* will be a scatterplot of the data with these two smoothers, one according to the unrestricted model and the other to the submodel, superimposed. From their comparison, we can infer about H_0 in (1). The comparison will be performed for many projection directions. If for some direction a the two smoothers differ, we will have evidence that the two models do not carry the same information.

In theory, the choice of the directions along which to project the data is crucial. However, it seems that in practice looking at many directions works well in most cases. Yet, the directions of the covariates included in the full model and excluded from the submodel must be examined. In addition, it is worth looking at the estimated linear predictor direction ($a = \hat{\beta}$), the other covariate directions, and even some random directions.

If we are dealing with both continuous and categorical covariates, *submodel plots* are still meaningful. However, as for any graphical methods relying on projections, they are not very useful when data are projected along a categorical covariate (see also Cook and Weisberg, 1999, p.405).

The use of *submodel plots* does not need to be restricted to the comparison of a full model with a submodel. Yet, it can supply a tool to compare submodels. Write eqn. (1) as: $H_0 : F_{\theta_1}(y|X = x) = F_{\theta_2}(y|X = x)$, with θ_1 and θ_2 being respectively the value of $\theta(x)$ under submodels 1 and 2. We can still apply our results to get a plot with an estimate for $E_{\theta_1}(y|Z = x^T a)$ and $E_{\theta_2}(y|Z = x^T a)$ on it. Evidence against H_0 will be given if for some direction they substantially differ. Obviously, in this particular case, the projection directions set must contain the covariates not shared by the submodels.

4. The plot in binomial regression

To exemplify the proposed method we consider the construction of the *submodel plot* in binomial regression, a case within the generalized linear models framework. Suppose we have n binomial observations of the form y_i/n_i $(i = 1, ..., n)$, where $E(y_i|X = x_i) = n_i p_i$ and p_i is the success probability corresponding to the i - th observation. Assuming the canonical link holds, the regression model for the dependence of p_i on the values of k covariates, $x_{1i}, ..., x_{ki}$, associated with that observation, is: $logit(p_i) = log(p_i/(1 - p_i)) = x_i^T \beta$.

Let us denote with $p_i(\hat{\theta}) = logit^{-1}(x_i^T \hat{\beta})$ the predicted probabilites (i.e. the predicted values) under the full model, and with $p_i(\hat{\theta}_0)$ the predicted probabilities under the submodel of interest. Consider also a set A of projection directions ($||a_j|| = 1$). A *submodel plot* for a given projection direction a_j will be given by the following steps.

i) Fit the full model and the submodel of interest and save their predicted probabilities $p_i(\hat{\theta})$ and $p_i(\hat{\theta}_0)$;

ii) Compute a nonparametric estimate on the scatterplot of $p_i(\hat{\theta})$ versus $x_i^T a$;

iii) Compute a nonparametric estimate on the scatterplot of $p_i(\hat{\theta}_0)$ versus $x_i^T a$;

iv) Plot the two smoothers on the same graph, perhaps on the plot of y_i/n_i versus $x_i^T a$;

v) Compare the two smoothed curves: if they substantially agree then we have no evidence to consider that in the direction a_j the submodel is less informative than the full model.

This procedure must be repeated for all the projection directions a_j in A. If for some direction the two curves are substantially dissimilar, we conclude that the submodel is not adequate to substitute the full model.

5. Using submodel plots

To illustrate how *submodel plots* can be used and their advantages, we will compare submodels for the Prostate Cancer data (Brown, 1980), a data set already considered in submodel selection in binomial regression (Collett, 1991, Chap.3). This data set was used to examine which of five preoperative variables are predictive of nodal involvement. The point is in evaluating, without performing surgery, if the prostate cancer has spread to the neighbouring lymph nodes. Data refer to 53 patients, the binary response ($nodes$) presents the finding at surgery (a value of one denotes nodal involvement, while zero denotes no nodal involvement), while the covariates are age, $grade$ (grade of the tumor determined by a biopsy), $size$ (a rough measure of the size and location of the tumor obtained by palpation), $X - ray$ reading, and *acid* (serum acid phosphatase level).

Before performing any further analysis, Collett (1991) looking at the values of *acid* for patients with and without nodal involvment, noted an unusual high value for patient number 24. In the absence of any medical reason for excluding this observation from the sample, he retained this individual in the data set. Rather, to accommodate for the data point in question, he decided to take as covariate the natural logarithm of *acid*.

Unlike Collett, we retain the values of *acid* in their original scale, postponing any action to take into account this anomalous observation.

Our analysis starts by considering a full model that includes all the covariates and a quadratic term in *acid*

$$logit(p) = \beta_0 + \beta_1 age + \beta_2 grade + \beta_3 size + \beta_4 X - ray + \beta_5 acid + \beta_6 acid^2.$$

This model does not show lack-of-fit, according to the Hosmer and Lemeshow goodness-of-fit test (p-value=0.8151) and the marginal model plots.

A backward elimination procedure is performed, and the Akaike information criterion reaches its minimum (4.952) when age and $grade$ are omitted, and we have

submodel 1: $$logit(p) = \beta_0 + \beta_1 size + \beta_2 X - ray + \beta_3 acid + \beta_4 acid^2.$$

However, because of possible overfitting, we consider the subsequent model in the elimination procedure, that has the term $acid^2$ dropped, obtaining

submodel 2: $$logit(p) = \beta_0 + \beta_1 size + \beta_2 X - ray + \beta_3 acid.$$

A likelihood ratio test for the significance of $acid^2$ in submodel 1 yields indeed a p-value of 0.0443.

We rely then on our method for further investigation. In Fig. 1 we present two *submodel plots* for the submodels under analysis. The figure shows the jittered data set projected on the estimated linear predictor and the *acid* covariate directions: the two curves on the plots correspond to the smoothed predicted probabilities for submodel 1 (solid line), and for the submodel without $acid^2$ (submodel 2, dashed line). These plots reveal that submodel 2 is not adequate to fully substitute submodel 1, as the two nonparametric curves do not coincide for low values of the linear predictor, and clearly differ in the higher values of the *acid* covariate.

Figure 1: *Submodel plots for Prostate data. The data have been projected along the linear predictor direction (a), and the acid covariate direction (b). The solid line corresponds to the smoothed probabilities according to submodel 1, while the dashed line accounts for submodel 2. Any difference between the smoothers highlights differences between the submodels in describing the data.*

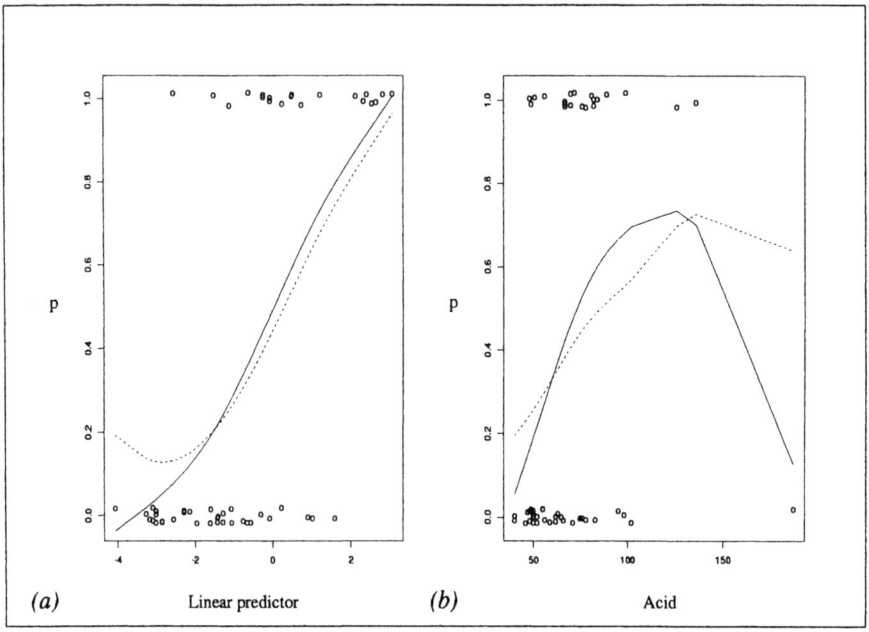

Indeed, a closer look at the plot (Fig.1 (b)) shows that the quadratic term could be required only to fit the single extreme value along the *acid* axis with *nodes*=0 (i.e. obs. # 24). Should this observation be omitted, we will see that *submodel plots* for the same models do not show any considerable differences. In other words, observation 24 is influential with respect to the choice between the above submodels, and to take into account such observation the quadratic term $acid^2$ has to be retained into the model.

Finally, we use *submodel plots* to compare our submodel 1 with the findings of Collett (1991, p. 86). For illustrative purposes, he examined a number of possible models looking at deviances along with the corresponding degrees of freedom. He eventually considered a final model that includes four covariates plus two interactions:

$$logit(p) = \beta_0 + \beta_1 grade + \beta_2 size + \beta_3 X - ray + \beta_4 log(acid) \\ + \beta_5 size * grade + \beta_6 grade * log(acid).$$

In Fig. 2 we offer *submodel plots* to compare the final model from Collett (dashed line) with submodel 1 (solid line) we considered above. We present plots for the same projection directions as in Fig. 1. The smoothers in the plots essentially agree, and then the two models can be considered equivalent for this data set. We can hence choose either of them, and we would prefer submodel 1 because of its smaller number of terms.

Figure 2: *Submodel plots for Prostate data. Data are projected along the linear predictor direction (a), and the acid covariate direction (b). The solid line corresponds to the smoothed probabilities according to submodel 1, while the dashed line accounts for the model from Collett.*

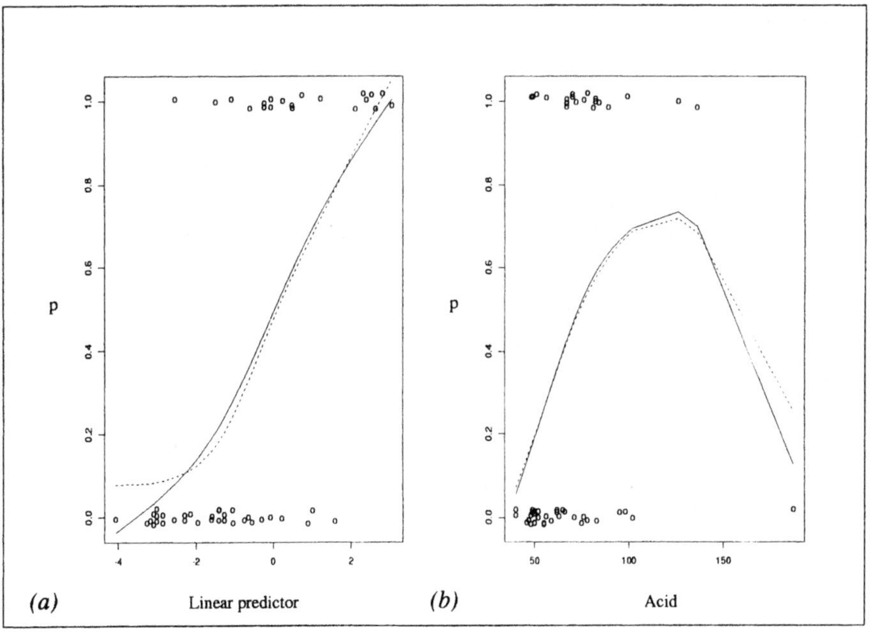

6. Discussion

We presented *submodel plots*, a graphical technique which is well suited to provide relevant information when performing submodel selection. It consists of looking at many plots where two estimated functions have to be compared. If they differ, the submodel is not adequate to substitute the full model.

A possible drawback for the method could appear the choice of the smoother and of the bandwidth parameter, a critical point in nonparametric estimation. Notwithstanding, this choice is somewhat less important for *submodel plots*. In the comparison of two nonparametric estimates, if the same estimator with the same bandwidth parameter is used for both curves, then the pointwise bias in the estimates will be canceled. In a different setting, Bowman and Young (1996) showed that the comparison of nonparametric curves yields a remarkable stability of results over a wide (reasonable) range of the bandwidth parameters.

In addition, we note that the procedure could be time consuming in the presence of many covariates. However, the method can be strongly enhanced by a dynamic graphics device that allows for an easy and quick change of the projection directions.

Further work will involve a general use of the proposed method to compare complex multivariate regression models, not necessarily restricted to parametric models (e.g. generalized additive models, projection pursuit regression models).

References

Bedrick E.J., Tsai C.L. (1994). Model Selection for Multivariate Regression in Small Samples. *Biometrics*, 50, pp. 226-231.

Bowman A., Young S. (1996). Graphical comparison of nonparametric curves. *Applied Statistics*, 45, pp. 83-98.

Brown, B.W. (1980). Prediction analysis for binary data, *Biostatistics Casebook*, Miller, R.J., Efron, B., Brown, B.W. & Moses, L.E. (Eds.), Wiley, New York.

Collett D. (1991). *Modelling binary data.* Chapman and Hall, 1991.

Cook R.D. (1998). *Regression Graphics.* Wiley, New York.

Cook R.D., Weisberg, S. (1999). *Applied Regression Including Computing and Graphics.* Wiley, New York.

Jovanovic, B.D. and Hosmer, D.W. (1997) A simulation of the performance of Cp in model selection for logistic and Poisson regression, *Computational Statistics and Data Analysis*, 23, pp. 373-379.

On the Use of Multivariate Regression Models in the Context of Multilevel Analysis

Giorgio Vittadini
Department of Statistical Sciences - University of Milan-Bicocca. E-mail: giorgio.vittadini@unimib.it

Abstract: The use of Multivariate Regression Models with mixed data to evaluate and decompose relative effectiveness of different social agencies presents numerous problems. The solution proposed is to use the Seemingly Unrelated Equations Models (SURE) in the framework of Multilevel Analysis, following quantification of the response variables by means of simultaneous Multidimensional Scaling methods. An example is provided.

Keywords: Multivariate Regression Analysis, Mixed Data, Multilevel Analysis, Interaction Parameters, SURE Analysis, Relative Effectiveness.

1. Study with mixed indicators

In the studies about the effectiveness evaluation of a social service, the outcome is defined as a long-term result of the output onto a particular aspect of the user. Such an outcome can be measured by an appropriate set of indicators.

Of particular relevance is the relative effectiveness which evaluates different social agencies (hospitals, schools, and etc...). In such cases, the outcome depends on the explicative variables connected with the users and on other variables concerning the agencies. Outcomes can be described by a set of mixed indicators or by latent variables, obtained with measurement errors from those indicators[1].

To this aim the following models have been utilized:

- Multivariate Models with mixed variables that take into account associations and correlations among outcomes.
- Multilevel Models which clearly reveal the relative effectiveness of every social agency.

This paper proposes an integration of the two families of instruments.

2. Multivariate Models and relative effectiveness

From the interpretative point of view, it is reasonable to calculate simultaneously the value of outcomes which describe different but correlated aspects of the "state of well being" of the user. The Multivariate Models are suitable for this purpose.

[1] For a systematic explanation of these concepts see Gori and Vittadini (1999), pp. 135-146.

There have been numerous attempts in the literature to utilize Multivariate Models with mixed variables for outcome studies[2]. The most sophisticated model is Fitzmaurice-Laird (1997), a Multivariate Regression Model that analyzes the linear dependence of a set of mixed response variables (binary, politomic, and continuous) from explicative variables[3]. In the Fitzmaurice-Laird model (1997) we can consider the conditional expected values of every dependent variable with respect to a set of explicative variables, by means of Marginal Regression Models (linear functions for the quantitative dependent variables and logit functions for categorical and binary responses are used). Using the likelihood methods applied to the joint density of the quantitative and qualitative dependent variables, one gets a solution, which considers the associations between dependent and explicative variables[4].

There are still some unresolved problems in this approach:

a) First, the Gaussian multivariate distribution hypothesis for the dependent quantitative variables or the use of binomial distribution hypothesis for the qualitative variables can give unsatisfactory results with respect to the distributions of the indicators.

b) Secondly, direct use of mixed indicators may be inappropriate to describe the outcomes. Most importantly in many cases we have ordinal indicators obtained from surveys or experiments. Data obtained must be transformed in quantitative objective measures independent upon type of sample, difficulty of the trials and ability of the individuals (Rasch 1960, Vittadini 1999). Secondly, from the statistical point of view latent variables are more indicative to express the outcome (Gori and Vittadini 1999, Vittadini 1999).

c) In order to measure the relative effectiveness of every single social agency (effectiveness A) and decompose it into effectiveness due to the observable resources (effectiveness B) and to non-observable capacity of the management (effectiveness C), it is necessary to utilize a Multilevel approach. In fact, in a case in which the effectiveness C was identified with dummy variables instead of

---

[2] For example: using Marginal Multivariate Regression Models with discrete variables, taking into account the associations between outcomes, the relationship between vision loss and demographic characteristics, educational level, and access to medical care was studied (the sample of 5000 people aged 40 years and above was collected by the *Baltimore Eye Survey)*; the relationship between diarretic and respiratory infections and the lack of *vitamin A* was also studied (the sample of 3000 children was collected by the *Indonesian Children Study*); the genetic association between asthmatic parents and children was studied in a sample of 100 people [Liang Zeger Qaqish (1992)]. Using an Alternate Logistic Regression Model the positive effects of a particular therapy on epileptic patients were analyzed in a sample of 52 individuals, collected by *John Hopkins Hospital* [Carey-Zeger-Diggle (1993)]. Using a Marginal Regression Model with mixed variables for Clustered Ordinal Measurement, the post-operative neurological improvement of 144 patients, followed for ten years, was analyzed [Heagerty and Zeger (1996)].

[3] Using this model the health consequences of air pollution were studied in a sample of 7243 children between 9 and 10 years old. Data were compiled by *the Harvard Six Cities Study*. The effects of parents' psychological disorders on the development of their children was studied in a sample of 56 children. Data were compiled by *the St. Louis Research Project*.

[4] The solutions are characterized by the relative asymptotic efficiency with respect to the case in which the equations were considered one by one. They are robust with respect to misspecification of association pairs between binary variables. They solve the problem of the missing values.

random variables (used in Multilevel analysis), there was collinearity with the variables inherent to the effectiveness B (Gori and Vittadini 1999).

d) In covariance models, where the dependent variable is measured after a treatment at the time t=1, this variable is introduced as an explicative variable covariate at the time t=0. When more treatments are implemented in order to take into account the effects of interaction between these covariates and the treatments, interaction parameters are also introduced (Cox and McCullagh, 1982). In relative effectiveness studies, interactions between agencies and the initial state of the user also exist. Therefore the variation of the outcome with respect of the initial state of the user is not identical for every initial state and every social agency (Fleiss 1986, Rice and Leyland 1996). Thus interaction parameters should be introduced also in Multivariate Regression Models with mixed variables. However, there are difficulties in creating adequate maximum likelihood functions, in finding computational solutions and in proposing the interpretation of the results.

3. The SURE Model in the framework of multilevel analysis

In order to resolve the problems illustrated above, we propose the utilization of Multivariate Models with quantitative dependent variables characterized by stochastic links between different equations, in a Multilevel contest (Hox 1995). In order to quantify all the observed indicators and simultaneously obtain the outcomes as their linear transformations, we utilize the appropriate Multidimensional Scaling instruments and the Vectorial Space Decomposition Method named Restricted Regression Component Decomposition (RRCD) (Haagen and Vittadini 1998, Vittadini 1999). In this way we resolve the problem reported in section 2b. Among the quantitative dependent variable models with stochastic links between different equations, we chose the Seemingly Unrelated Equation Models (SURE). Furthermore, the introduction of a Multilevel framework hypothesizes that "there is a hierarchical data set, with one single dependent variable that is measured at the lowest level and explanatory variables at all existing levels..." and that in every equation "the intercept and slope coefficients are assumed to vary across the schools; for that reason they are often referred to as random coefficients" (Hox 1995),(Aitkin and Longford 1986). This fact allows studying completely the problem. Therefore, relatively to j-th outcome ($j=1,\dots,h$) at the time t_1 (after the service allocation) drawn on the i-th ($i=1,\dots\dots,n_\ell$) user to whom the service is allocated by the ℓ-th social agency ($\ell = 1,\dots,q$; $n = n_1 + \dots + n_\ell + \dots + n_q$), the SURE model in a Multilevel context equation has the following structure:

$$y_{ji\ell} = \alpha_{j\ell} + \sum_{k=1}^{p} \beta_{kj} x_{kjil} + e_{jil} \tag{1}$$

$$\alpha_{j\ell} = \delta_{j\ell} + \sum_{w=1}^{t} \delta_{wj\ell} \cdot f_{wj\ell} + m_{j\ell}; \quad \beta_{kj} = v_{kj\ell} + \sum_{w=1}^{t} v_{wkj\ell} \cdot f_{wkj\ell} + \chi_{kj\ell}; \tag{1a}$$

with: $y_{ji\lambda}$ outcome; $x_{kji\lambda}$ explicative k-th ($k=1,\ldots,p$) variable belonging to the set of outcome explicative variables $X_{kj\lambda}$; $\alpha_{j\lambda}$ and β_{kj} random coefficients which identify the effectiveness A; $e_{ji\lambda}$ random residual; $\delta_{j\lambda}$, $\delta_{wj\lambda}$, $v_{kj\lambda}$, $v_{wkj\lambda}$ parameters; $f_{wj\lambda}$ ($w=1,\ldots,t$) specific observable w-th characteristic of the social agency; $\sum_{w=1}^{t}\delta_{wj\lambda}\cdot f_{wj\lambda}$ and $\sum_{w=1}^{t}v_{wkj\lambda}\cdot f_{wkj\lambda}$ technological relationships which identify the effectiveness B; $m_{j\lambda}$ and $\chi_{kj\lambda}$ random residuals which identify effectiveness C. From (1a), (1) we obtain:

$$y_{ji\lambda} = \delta_{j\lambda} + \sum_{w=1}^{t}\delta_{wj\lambda}\cdot f_{wj\lambda} + \sum_{k=1}^{p}v_{kj\lambda}\cdot x_{kji\lambda} + \sum_{k=1}^{p}\left(\sum_{w=1}^{t}v_{wkj\lambda}\cdot f_{wkj\lambda}\right)x_{kji\lambda} +$$
$$+ m_{j\lambda} + \sum_{k=1}^{p}\chi_{kj\lambda}\cdot x_{kji\lambda} + e_{ji\lambda} \qquad (2)$$

Defined the $g_{j\lambda} = \sum_{k=1}^{p}\chi_{kj\lambda}\cdot x_{kji\lambda}$ random part $\pi_{ji\lambda} = [m_{j\lambda} + g_{j\lambda} + e_{ji\lambda}$ contains the random error structure of the equation (Hox 1994). The matrix model of the explicative variables is of full rank and non-stochastic; the expected values of the residuals are null and their variances are finite; the contemporaneous covariances between the diverse residuals are constant; the intertemporal covariances are null for every couple of residuals. The contemporaneous covariances among residuals postulate stochastic relationships between the equations of the model (Srivastava and Giles, 1987).

The variances and covariances of the random error structure $\pi_{j\lambda}$ have the following structure:

$$\sigma^2_{\pi_{j\lambda}} = \sigma^2_{m_{j\lambda}} + \sigma^2_{g_{j\lambda}} + \sigma^2_{e_{j\lambda}} + 2\sigma_{m_{j\lambda}g_{j\lambda}} + 2\sigma_{m_{j\lambda}e_{j\lambda}} + 2\sigma_{g_{j\lambda}e_{j\lambda}} \quad (j=1,\ldots,h;\ \ =1,\ldots,q) \quad (3)$$

Then there are: the $\sigma_{\pi_{j\lambda}\pi_{jm}}$ covariances inherent to the same outcome, for different social agencies λ and m ($j=1,\ldots,h$; $\lambda,m=1,\ldots,q$, $\lambda \neq m$); the $\sigma_{\pi_{j\lambda}\pi_{z\lambda}}$ covariances inherent to different outcome j and z for equal social agency λ (j, $z=1,\ldots,h$, $j \neq z$; $=1,\ldots,q$); the $\sigma_{\pi_{j\lambda}\pi_{zm}}$ covariances between different outcomes j and z and different social services λ and m. Introducing in every equation of the model (1) the initial state of the user we obtain:

$$y_{ji\lambda} = \alpha_{j\lambda} + \sum_{k=1}^{p}\beta_{kj}x_{kji\lambda} + \gamma_{j\lambda}\cdot y_{ji\lambda(t_0)} + e_{ji\lambda};\quad \gamma_{j\lambda} = \tau_{j\lambda} + \lambda_{j\lambda}\cdot s_{j\lambda} + u_{j\lambda} \qquad (4)$$

with $\gamma_{j\lambda}$ random coefficient inherent to initial state of the user, $\tau_{j\lambda}, \lambda_{j\lambda}$ parameters; $s_{j\lambda}$ explicative variable, $u_{j\lambda}$ random residual. Expressing (4) in compact terms we obtain:

$$y_{ji\lambda} = \delta_{j\lambda} + \sum_{w=1}^{t} \delta_{wj\lambda} \cdot f_{wj\lambda} + \sum_{k=1}^{p} v_{kj\lambda} \cdot x_{kji\lambda} + \sum_{k=1}^{p} \left(\sum_{w=1}^{t} v_{wkj\lambda} \cdot f_{wkj\lambda} \right) x_{kji\lambda} +$$
$$+ \tau_{j\lambda} \cdot y_{ji\lambda(t_0)} + \lambda_{j\lambda} \cdot s_{j\lambda} \cdot y_{ji\lambda(t_0)} + m_{j\lambda} + g_{j\lambda} + u_{j\lambda} \cdot y_{ji\lambda(t_0)} + e_{ji\lambda} \quad (5)$$

with $\tau_{j\lambda} \cdot y_{ji\lambda(t_0)}$ the effects of interaction between the observable variables inherent to the social service λ-th and the initial state j-th of the user i-th. From the random error structure of the equation $\pi^*_{ji\lambda} = \left[m_{j\lambda} + g_{j\lambda} + u_{j\lambda} \cdot y_{ji\lambda(t_0)} + e_{ji\lambda}\right]$ contains the random error structure of the equation; from that we obtain the variances and covariances of the random error structure which take into account the interactions.

4. Possible further developments

We observed that the integration between the two models resolves the problems reported in paragraph 3. Some problems still remain open:

1) The models postulate linear relationships, whereas non-linear relationships can exist between explicative and dependent variables.

2) In the case of numerous parameters there is difficulty in constructing robust and consistent estimations of the parameters.

3) The normality hypothesis of the errors is often inadequate. In fact, "specific distributional assumption regarding the disturbances of the SURE model it has been one of normality" and "its violation may have serious consequences for our inferences" (Judge, Griffiths, Hill and Lee, 1980). The problem is partially resolvable if the observed distributions are symmetrical (conservative leptocurtic or liberal platicocurtic). In this case, using the Generalized Linear Models, such distributions may be studied as an extension of Gaussian distribution. The problem becomes more serious when the distribution is not symmetrical, especially in the presence of numerous parameters. In this case, the methods based on Multivariate Ranks and the bootstrap methods (which have been used for the SURE models) appear inadequate (Rilstone and Veall, 1995), because if the starting distribution is not normal the new distribution will also be not normal. The method for constructing inferences based on iterative simulation and Bayesian inference for models with not normal distribution of errors appears more proficient (for example using Monte Carlo Markov Chain methods by means of algorithms such as Metropolis Hastings and Gibbs sampling) (Gelman and Rubin, 1992, Geyer, 1992). Such methods are commonly employed, but they have never been used in the case of SURE models.

4) The problem of the selection bias is unresolved in simpler models[5] as in multivariate models.

5. An example of relative effectiveness study

In this example the relative effectiveness of 5 retirement homes (RSA) of different sizes in Lombardy[6], is studied. The sample was made up of 225 residents. By means of the Functional Independent Measure (FIM) scale we measured motor, cognitive, sfinteric skills in two instances; simultaneously we collected other explicative variables.

Table 1: *The variables of the model*

DEPENDENT VARIABLES (t=1 2° draft)	EXPLICATIVE VARIABLES (t=0 1° draft)
Qualitative: Latent Variables 11 indicators of motor skills → FM1; 5 indicators of cognitive skills → FC1; 2 indicators of sfinteric skills → FS1	Qualitative: 11 indicators of motor skills → FM0; 5 indicators of cognitive skills → FC0; 2 indicators of sfinteric skills → FS0; Quantitative: Nosological Macroclassification = MCN. Qualitative: Provenience = PRO; Dummy: Vision (2 modalities / 1 dummy) = V_1, Hearing (3/2) = U_1,U_2; Social Services (5/4)=RSAI (I=1,2,3,4), Interaction FQ0 (Q=M,C,S)-RSAI (I=1,2,3,4)

By means of the ALS Multidimensional Scaling methodology (Young, de Leeuw, Takane, 1976) the ordinal modalities of the motor, cognitive, sfinteric skills indicators are transformed into quantitative values and simultaneously 6 latent variables are obtained by means of RCDR. Therefore we have a model in which the dependent variables are the latent variables FM1, FC1, FS1. In table 3 we observe that the results obtained with the SURE method compared with those obtained with the Ordinary Least Squares (OLS) method present significant differences. Moreover, using the fifth RSA as a reference point we observe that in the first equation no RSA has more effectiveness, in the second we find significant differences between the RSA2 and the RSA4, in the third the effectiveness of the RSA1 and RSA4, RSA2 and RSA3 are equivalent. Finally we observe (Fig.1) that distributions of FM1, FC1, FS1 are almost symmetrical and transformable into normal distributions.

Table 2: *Explained variance of the latent variables*

	FM	FC	FS
t=1	78.14	79.71	98.96
t=0	78.89	83.09	99.36
ρ^2	68.70	60.10	60.20

[5] Copas and Li (1997) examine a case in which two independent subsamples of equal numerousity of patients are not subdivided among the medical treatments in the random mode; also in this simple case the robustness and validity of estimates is heavily influenced by the non randomness of the subsamples. Gori (1992) proposes a study of this problem in two cases of two social agencies, but an extension to the case of more than two social agencies is still lacking at this point.

[6] Having considered the universe of the social agencies, the parameters inherent to the social agencies are fixed effects. Indeed we do not decompose the effectiveness as postulated in section 2c. Therefore there was no need to utilize the Multilevel Analysis.

Figure 1: *Cumulative Distributions of the latent variables*

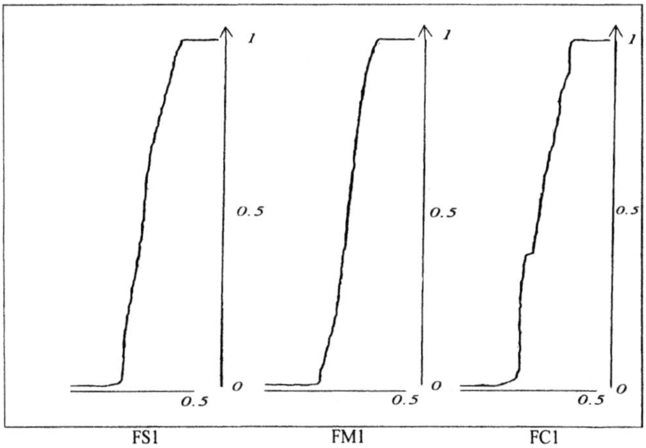

Table 3: *Estimations and standard errors of the parameters*

Variables	FM1 Estimation	FM1 Standard error	FC1 Estimation	FC1 Standard error	FS1 Estimation	FS1 Standard error	
INT	0.179013	0.42328	-0.293252	0.32140	-0.035074	0.22631	
F0	0.702301	0.05986	0.567080	0.07067	0.708070	0.07376	
V1	-0.316453	0.37974	0.111068	0.29559	-0.027746	0.20211	
U1	0.00904327	0.34456	0.153666	0.26650	-0.140215	0.18541	
U2	-0.171813	0.44592	-0.670021	0.34592	-0.069402	0.23843	
MCN	-0.427725	0.27051	-0.284346	0.21095	0.096989	0.14511	
PR0	-0.214866	0.27587	-0.187318	0.21251	-0.210229	0.14605	
RSA1	0.368478	0.27334	0.371480	0.13613	-0.037974	0.17453	OLS
RSA2	0.195361	0.18386	0.319548	0.14581	0.036307	0.17500	
RSA3	0.090738	0.08220	0.134334	0.09586	-0.011763	0.10565	
RSA4	0.183009	0.07963	0.185559	0.09942	0.092408	0.09613	
I1	0.895490	0.86308	0.665183	0.39349	-0.254697	0.31801	
I2	0.800830	0.48131	0.752619	0.41328	0.219688	0.08680	
I3	0.244861	0.33049	0.867803	0.25647	0.219688	0.08680	
I4	0.509220	0.29733	0.657681	0.22866	0.196530	0.15638	
INT	0.281961	0.42152	-0.301730	0.32210	-0.031446	0.22452	
F0	0.601513	0.07562	0.548824	0.08614	0.598249	0.08639	
V1	-0.286105	0.37513	0.182117	0.29680	-0.00573215	0.19915	
U1	-0.064152	0.34047	0.126007	0.26819	-0.157545	0.18363	
U2	0.067896	0.44814	-0.600076	0.35046	-0.00624369	0.23609	
MCN	-0.381401	0.26656	-0.229918	0.21223	0.104402	0.14313	
PR0	-0.141028	0.27578	-0.121240	0.21459	-0.184113	0.14494	
RSA1	0.365700	0.34861	0.365283	0.16760	0.024663	0.20496	SURE
RSA2	0.131445	0.24659	0.287541	0.18086	0.145475	0.19391	
RSA3	0.285164	0.10458	0.219554	0.11782	0.222087	0.12381	
RSA4	0.117353	0.10384	0.101879	0.12355	0.081229	0.11511	
I1	0.468137	1.02470	0.602479	0.39918	-0.313396	0.30692	
I2	-0.053030	0.75421	0.414974	0.48207	0.195712	0.08617	
I3	0.046121	0.33419	0.778766	0.25960	0.195713	0.08617	
I4	0.501007	0.29661	0.653265	0.22941	0.221006	0.15801	

References

Aitkin M., Longford N. (1986) Statistical modelling issues in school effectiveness studies, *Journal of the Royal Statistical Society*, 149, 1, 1-43.

Carey V., Zeger S. L., Diggle P. (1993) Modelling multivariate binary data with alternating logistic regressions, *Biometrika*, 80, 3, 517-526.

Copas J. B., Li H.G. (1997) Inference for non-random samples, *Journal of Royal Statistical Society, B,* 59, 1, 55-95.

Cox D. R., McCullagh P. (1982) Some Aspects of Analysis of Covariance, *Biometrics*, 38, 541-561.

Fitzmaurice G. M., Laird N. M. (1997) Regression models for mixed discrete and continuous responses with potentially missing values, *Biometrics,* 53, 110-122.

Fleiss J. L. (1986) *The design and analysis of clinical experiments*, Ed. Wiley.

Gelman A., Rubin D. B. (1992) Inference from Iterative Simulation using Multiple Sequences, *Statistical Science*, 7, 4, 457-472.

Geyer J. C. (1992) Practical Markov Chain Monte Carlo, *Statistical Science*, 7, 4, 473-483.

Gori E. (1992) La valutazione dell'efficienza ed efficacia dell'istruzione, *Comunicazione invitata alla XXXVI Riunione Scientifica della S.I.S.*, Pescara, 21-24 Aprile 1992, 219-229.

Gori E., Vittadini G. (1999) La valutazione dell'efficienza ed efficacia dei servizi alla persona. Impostazione e metodi, in *Qualità e valutazione nei servizi di pubblica utilità*, Etas, 121-241.

Haagen K. L., Vittadini G. (1998) Restricted Regression Component Decomposition, *Metron*, 56, 1-2, 53-75.

Heagerty P. J., Zeger S. L. (1996) Marginal regression models for clustered ordinal measurements, *Journal of the american statistical association*, 91, 435, 1024-1036.

Hox J.J. (1994) Hierarchical regression models for interviewer and respondent effects, *Sociological methods and research*, 22, 3, 300-318.

Hox J.J. (1995) *Applied Multilevel Analysis*, TT-Publikaties, Amsterdam.

Judge G. G., Griffiths W. E., Hill R. C., Lee T. S., Lütkepohl H. (1980) *The Theory and Practice of Econometrics*, Wiley, New York.

Liang K. Y., Zeger S. L., Qaqish B. (1992) Multivariate regression analyses for categorical data, *Journal of Royal Statistical Society,* B, 54, 1, 3-40.

Rasch G. (1960), *Probabilistic Models for Some Intelligence and Attainment Tests,* Danmarks Paedagogiske Istitut, Copenaghen.

Rice N., Leyland A. (1996), Multilevel models: applications to health data, *Journal of health services research and policy*, 1, 3, 154-164.

Rilstone P., Veall M. (1996), Using bootstrapped confidence intervals for improved inferences with seemingly unrelated regression equations, *Econometric Theory*, 12, Dec 1996, 569-580.

Srivastava V. K., Giles D. E. A. (1987) *Seemingly Unrelated Regression Equations Models*, Marcel Dekker, New York

Vittadini G. (1997) Una metodologia statistica per la performance evaluation dei servizi alla persona di pubblica utilità, Atti del convegno «La statistica per le imprese» organizzato dal SIS, Torino, 2-4 IV-1997, I, 387-401

Vittadini G. (1999) "Analysis of qualitative variables in structural models with unique solutions", in *Classification and Data Analysis – Theory and Application*, a cura di M. Vichi e O. Opitz, Springer (1999), 203-210.

Young F. W., de Leeuw J., Takane Y (1976) Regression with qualitative and quantitative variables: an alternating least squares method with optimal scaling features, *Psychometrika*, 41, 4, 505-529.

Nonparametric Estimation Methods for Sparse Contingency Tables

Riccardo Borgoni
Attuariato Rami Danni RAS SpA
Corso Italia, 23 - 20122 Milano, Italy

Corrado Provasi
Dipartimento di Scienze Statistiche, Università di Padova
Via S. Francesco, 33 – 35121 Padova, Italy

Abstract: The problems related with multinomial sparse data analysis have been widely underlined in statistical literature in recent years. Concerning the estimation of the mass distribution, it has been widely spread the usage of nonparametric methods, particularly in the framework of ordinal variables. The aim of this paper is to evaluate the performance of kernel estimators in the framework of sparse contingency tables with ordinal variables comparing them with alternative methodologies. Moreover, an approach to estimate the mass distribution nominal variables based on a kernel estimator is proposed. At the end a case study in actuarial field is presented.

Keywords: Ordinal and nominal variables, Kernel estimator, Sparse data.

1. Introduction

Data in the form of counts occur often in statistical practice. The empirical mass function is a good estimator of the true distribution under usual asymptotic conditions, but it is very poor with small or even moderate cell counts, when sample zeros can appear. In such situation, Bishop, Fiemberg and Holland (1975) introduced the idea of *sparse asymptotic conditions* to give a more realistic asymptotic framework in which the number of classes and the sample size increase together. They also introduced a well known estimator (hereafter *BFH estimator*) obtained by summing a small value to each cell frequency. Even if this estimator is not consistent, it has been proved that it has a better behaviour than empirical distribution in a sparse context. In a sparse asymptotic context kernel estimator has good asymptotic properties (Hall and Titterington, 1987).

In this paper we make a review of the kernel estimator for univariate and multivariate contingency tables with ordered marginals and an evaluation of its performance. Moreover, we make an attempt to apply this approach to nominal variables. In the end we try to underline the utility of smoothing by an application. In section 2 we briefly review some results on the kernel estimator for ordinal data in the unidimensional and multidimensional cases and some useful corrections to improve it. In the multidimensional case, a simulation study is performed to evaluate the behaviour of an improved version of the kernel estimator. In section 3 we introduce a kernel-type estimator for nominal data and present a simulation study for its evaluation while, in section 4, an application of the methodology in an actuarial case study is performed. Conclusions and final remarks end the paper.

2. The Kernel Estimator for Ordinal Variables

Kernel estimators for multinomial probabilities adapt kernel estimators for densities to the discrete case. The aim is to reduce the harshness of the empirical distribution supposing that the probabilities associated to neighbouring cells are similar, which is reasonable when we refer to ordinal variables.
Let $\boldsymbol{s}=\{X_1,\ldots,X_n\}$ be a random sample drawn from a multinomial distribution $\boldsymbol{p}=\{p(1),\ldots,p(K)\}$. The kernel estimator of $p(i)$ is defined as

$$\tilde{p}(i)=\tilde{p}(i|h,\mathbf{s})=\frac{h}{n}\sum_{j=1}^{n}W_h\{h(i-X_j)\}=\frac{h}{n}\sum_{l=1}^{K}n_l W_h\{h(i-l)\},$$

where $W_h(xh)$ is a kernel function and h is the smoothing parameter (Hall and Titterington, 1987). Under some regularity conditions this estimator has good properties both in standard and sparse asymptotic conditions. In the last case, Hall and Titterington (1987) derived the optimal convergence rate for any estimator of a probability law $\boldsymbol{p}$ with respect to the metric given by the Mean Summed Squared Error (MSSE) defined as $E\left\{\sum_{i=1}^{K}(\tilde{p}_i-p_i)^2\right\}$.
Thinking the vector $\boldsymbol{p}$ generated by an underlying density function $f(x)$ with s-bounded continuous derivatives on a compact support through the relations $p_i=\int_{(i-1)/K}^{i/K}f(u)du$, they showed this rate is $O(n^{-2s/(2s+1)}\delta)$ if $n^{-1/(2s+1)}\delta\to 0$ as $n\to\infty$, where δ is a sequence such that $\delta\to 0$ as $n\to\infty$. Kernel estimator achieves this optimal rate. The previous boundary conditions are needed to face the well known bias problem of smoothing when we refer to distributions with a limited support. In order to avoid the need of these conditions, the kernel estimator can suitably be corrected on the extremes of the support introducing a boundary kernel constructed by integrating a second order boundary kernel function (Dong and Simonoff, 1994).
Obviously, the sparseness of data is a case widely present in multivariate analyses. Kernel estimators can be generalised to the multidimensional case immediately (Grund, 1993). The probability estimator of the multicell $\boldsymbol{i}$ is defined as

$$\tilde{p}(\mathbf{i})=\tilde{p}(\mathbf{i}|\mathbf{h},\mathbf{s})=\sum_{\mathbf{l}}\hat{p}_{\mathbf{l}}W_{\mathbf{i}}(\mathbf{l},\mathbf{h}),$$

where $\mathbf{h} \in [0,1]^d$ is the vector of smoothing parameters and d is the dimension of the table. Usually the multidimensional kernel function is determined as a product of univariate kernel functions (kernel product), each of which can be obtained from a density with a fairly regular compact support.

Dong and Simonoff (1995) proposed a geometric combination of kernel estimators with different smoothing parameters for d-dimensional tables. The MSSE convergence rate of this estimator is $O(K^{-1}n^{-8/(d+8)})$ (with $d \leq 4$) assuming appropriate boundary conditions, where K is the whole number of cells in the table. The authors suggested using the following simple form for applications:

$$p^{\circ}(\boldsymbol{i} \mid \boldsymbol{h}) = \tilde{p}(\boldsymbol{i} \mid \boldsymbol{h})^{4/3}\, \tilde{p}(\boldsymbol{i} \mid 2\boldsymbol{h})^{-1/3}.$$

Without requiring any restrictive boundary condition on the probability matrix to avoid bias near the frontier, they introduced a d-dimensional boundary-corrected estimator with kernel function of Epanechnikov's type. Its construction is quite formal (see the references for details). However if this corrected version is introduced in the geometric combination, the resultant estimator has a Summed Squared Error of order $O_p(K^{-1}n^{-8/(d+8)})$ for all d.

2.1 An Empirical Evaluation

In this section we perform a Monte Carlo experiment to evaluate the estimator described before. We also compared it with the well-known BFH estimator defined as $p_i' = (n_i + \alpha)/(n + \alpha K)$.

The comparison is made in terms of the MSSE calculated among the estimated distribution and the true distribution used to generate the data. In the simulations, we put α=0.5 and we consider different sparseness conditions, i.e. tables with an increasing mean cell frequency. We considered tables in two and three dimensions, obtained by clustering data generated from a normal distribution with independent components. We referred to tables with the same number of classes for each factor of classification for simplicity, but the procedures can be extended to general contingency tables. Previous simulations were made to select optimal values (which minimise a quadratic loss function) of smoothing parameter used in the kernel estimator for data of such a structure. Results are shown in the table 1. They show that kernel estimator performs better (the not shaded cells in columns titled Kernel MSSE of the tables) than BFH one in almost every case.

3. A Kernel Estimator for Nominal Variables

In this section we introduce a smoothing approach for nominal variables based on an estimator of the same kind of Titterington and Hall's one. It is based on the introduction of a fictitious ordering on the empirical probabilities of a table.

Table 1: *Monte Carlo experiment for ordered variables results: the performance of kernel and BFH estimator in the case of bivariate and trivariate tables.*

Cells	Mean frequency	Number of iterations	Kernel MSSE		BFH MSSE	
			Mean	Stand. Err	Mean	Stand. Err
16x16	1.5	1000	3.55E-09	1.07E-08	4.83E-09	1.09E-08
	3	1000	1.78E-09	4.01E-09	2.33E-09	5.44E-09
	10	1000	4.66E-10	8.26E-10	5.45E-10	1.16E-09
18x18	1.5	1000	1.54E-09	3.25E-09	3.18E-09	6.74E-09
	3	1000	1.03E-09	1.25E-09	1.71E-09	3.86E-09
	10	1000	4.35E-10	4.31E-10	5.27E-10	9.18E-10
22x22	1.5	1000	4.84E-10	5.77E-10	1.22E-09	2.85E-09
	3	1000	2.87E-10	2.62E-10	7.31E-10	1.46E-09
	10	1000	3.09E-10	2.16E-10	2.13E-10	3.42E-10
14x14x14	1.5	400	1.04E-12	5.03E-12	2.66E-12	2.83E-19
	3	400	5.51E-13	1.77E-12	7.84E-13	8.36E-20
	10	400	1.49E-13	3.18E-13	4.95E-14	6.23E-21
16x16x16	3	400	9.83E-14	3.43E-13	3.39E-13	5.88E-13
	10	400	3.09E-14	6.43E-14	5.44E-14	0

The biggest probability is associated to the central cell (or to one of the two central cells if the cardinality is not odd), the second biggest probability to the adjacent right cell, the third one to the left adjacent cell of the central cell and so on, so that data are reorganised in a vector whose frequencies present a bell aspect. The smoothing procedure is applied to reordered data and, in the end, the estimated probabilities are reallocated to their original positions. The considered kernel function $W_h(x)$ is a local triangular kernel function on a compact support. This function changes according to the currently smoothed cell in such a way that it is symmetric for inner cells and becomes more and more asymmetric towards the boundary. On the extreme right and left cells of the support it is a squared triangle so that a positive weight is given to cells belonging only to the support . The kernel support depends on the distance d of the current cell i from the modal one i_0 thus $d=d(i,i_0)$. We actually used an index of divergence between cells, whose values belong to [-1,1], rather than a real norm, defined as $d(i_{0,}i) = (i - i_0)/(i_{0-1})$ if K is odd and, $d(i_0,i) = \dfrac{i - i_0}{i_0 - 1}$ if $i \le i_0$ and $d(i_0,i) = \dfrac{i - (i_0 + 1)}{i_0 - 1}$ if $i \ge i_0 + 1$ if K is even. We considered a support of the form $[-p(d), q(d)]$ where $p(d)$ and $q(d)$ are functions $[-1,1] \rightarrow [0,2]$ and such that $p(d)+q(d)=2$ thus the support has a fixed width equal to 2. We choose the simple form: $p(d)=1+d$. Imposing the kernel function is a density, so that the triangle has unit area, it is easy to prove that the kernel function is specified as

$$W(u,d)=\left(1+\frac{1}{1+d}u\right)I\{u:-p(d)\leq u\leq 0\}+\left(1-\frac{1}{1-d}u\right)I\{u:0\leq u<q(d)\}$$

where $I(A)$ is the characteristic function of the set A. Defining a system of weights as $W_h(uh)=W(hu)\times s(h)$ with $s(h)=\left[h\sum_l W(hl)\right]^{-1}$ where h is the smoothing parameter, we obtain the following final form for the estimator

$$\tilde{p}(i)=h\sum_{l-1}^{K}\hat{p}_l W_h\{h(l-i)\}$$

$$=hs(h)\sum_{\{l:-p(d)\leq h(l-i)\leq 0\}}\hat{p}_l\left[1+\frac{h}{1+d}(l-i)\right]+hs(h)\sum_{\{l:0\leq h(l-i)\leq q(d)\}}\hat{p}_l\left[1-\frac{h}{1-d}(l-i)\right].$$

3.1 An Empirical Evaluation

The proposed estimator has been compared with the BFH estimator. Simulations are based on 1000 samples from a bivariate 5×6 table using a Monte Carlo experiment. In order to consider contexts characterised by different degrees of sparseness, the sample size n has been put equal to 30, 60, 150, moving from a very sparse situation (average size equal to 1) to a standard situation (average size equal to 5). The parameter h has been evaluated at each iteration using cross validation. The computations have been done using *S-plus* (Chambers and Hastie, 1991) and *Fortran 77* routines used to build the estimator. The indicators used have been: the Euclidean norm in $\Re^K$ and a weighted form of it calculated as $\sum_{l=1}^{K}\left[p_i-\tilde{p}_i\right]^2/p_i$, the Kullback and Leibler divergence, the norm of sup, the Matusita distance, the distance $\sup_i\left|\hat{p}_i-p_i\right|$ (in the following inappropriately indicated as Kolmogorov distance). The results can be found in table 2. We show the proposed estimator is characterised by a greater bias in average and in median and a greater variability in every context of sparseness compared to the other (the fifth and the last column titled IR contain the Interquartile Range).

Some simulations have been done also in the case of presence of structural zeros (Borgoni, 1998) concerning three indicators among these mentioned above and considering several cases in which the number of zeros increases compared to the number of classes. In particular, we have considered a number of null probabilities equal to 11, 16 and 21 for each case. The values of average frequencies, equal to 1, 2 and 5 respectively, are to be referred to cells not necessarily empty. The results indicate an improvement of the estimator proposed when the level of sparseness and of the number of structural zeros increases.

Table 2: *Indicators of kernel and BFH estimator performance: simulation results for nominal variables.*

Indicator	Kernel estimator				BFH Estimator			
	Median	Mean	Variance	IR	Median	Mean	Variance	IR
N=30								
L2	0.020	0.020	5.4e-05	0.009	0.014	0.015	1.89e-05	0.005
Weighted L2	0.646	0.637	0.051	0.274	0.426	0.442	0.0142	0.159
Kullback	0.491	0.480	0.0284	0.195	0.289	0.294	0.0042	0.087
Kolmogorov	0.061	0.064	0.0002	0.013	0.057	0.06	0.00023	0.020
Sup	2.132	2.124	0.3765	0.711	1.914	1.92	0.3433	0.889
Matusita	0.212	0.211	0.0079	0.103	0.102	0.104	0.0005	0.029
N=60								
L2	0.011	0.011	1.13e-05	0.004	0.01	0.010	9.4e-06	0.004
Weighted L2	0.336	0.347	0.00974	0.118	0.30	0.310	0.00696	0.112
Kullback	0.234	0.234	0.005	0.083	0.216	0.222	0.00272	0.069
Kolmogorov	0.047	0.049	0.00012	0.014	0.046	0.047	0.00015	0.020
Sup	1.612	1.632	0.151	0.47	1.469	1.515	0.18647	0.527
Matusita	0.084	0.090	0.001	0.033	0.079	0.081	0.00036	0.0004
N=150								
L^2	0.005	0.006	1.92e-06	0.002	0.005	0.005	2.33e-06	0.002
Weighted L^2	1.66	0.171	0.0018	0.055	0.155	0.159	0.0018	0.055
Kullback	0.115	0.118	0.00084	0.036	0.115	0.117	0.0009	0.040
Kolmogorov	0.034	0.035	5.52e-05	0.009	0.032	0.033	7.19e-05	0.012
Sup	1.129	1.17	0.0876	0.376	0.972	1.037	0.00650	0.267
Matusita	0.040	0.041	0.0001	0.013	0.041	0.042	0.00014	0.015

4. An Actuarial Application

A technical evaluation of the performance of an insurance company's portfolio is made by some indicators based on the number and the value of claims. One of these is called *frequency of claims* (Daboni, 1993) calculated as the ratio between the number of claims, n_i, and the temporal exposition, r_i, of a client, thus the frequency of claims of the *i*-th unit is calculated as $f_i = n_i / r_i$. This estimate can be evaluated both in the case of a single client and on sets of clients.

We have considered the problem of estimating the frequency of claims in 103 geographical areas of Italy of interest for an Italian insurance company. We considered this classification as an ordinal categorical variable since these regions are sorted by the company on the basis of an internal risk index. The data used in this application are a sample of 10000 items taken from the company's portfolio. In this sample we had 627 claims and 34 sample zeros. Let *n* be the total number of claims and *r* the whole period of exposition. To estimate the frequency of claims we decomposed it as

$$f_i = \frac{n_i}{r_i} = \frac{n_i}{n} \times \frac{r}{r_i} \times \frac{n}{r}.$$

Table 3: *The considered sample: exposure, claims and frequency of claims for kind of fuel. Last two columns show the case study results.*

Kind of fuel	Exposure	Claims	N° of zeros	Frequency of claims	Cases over upper level	Cases under lower level
Petrol	21840	649	34	3%	0	28
Diesel	14890	652	26	4,4%	22	13
Total	36731	1301	6	3,5%	22	41

The ratio n_i / n can be considered as the probability that a claim might happen in the i-th province. We referred to it as *probability of claim*. We estimated it via kernel with smoothing parameter calculated by cross validation and we rebuilt the value of the frequency of claims by putting this value in the previous formula for fixed values of the other two ratios. At the end only 5 zeros remained. Of course a smoother estimate can be obtained by increasing the width of smoothing window. To evaluate the performance of this estimation we made a smoothed bootstrap from the smoothed probability of claims (Efron and Tibshirani, 1993).

In estimating functionals of a density or mass function, smoothed bootstrap substitutes the smoothed distribution for the empirical one to which usual bootstrap refer to. Unlike the ordinary bootstrap sample, the smoothed bootstrap one will not have repeated values and they are less likely to be dominated by unusual values (Simonoff, 1996). Theoretical attention to the smoothed bootstrap has focused on continuous densities (Hall, Di Ciccio and Romano, 1989, Silverman and Young, 1987). This resampling technique is quite useful in a sparse data context in the place of standard bootstrap. As a matter of fact, the standard bootstrap cannot differentiate between structural and sample zero in a cell. On the other hand, samples generated by smoothed bootstrap, coming from a smoothed probability distribution which has generally positive values in cells with sample zeros, do not present this problem.

We made 1000 replications obtaining a coverage, at a 95% confidence level, of 75% of the provinces' values observed in the year in the whole company portfolio.

A second application regarded the evaluation of the effect of different kinds of fuel (petrol or diesel) on the frequency of claims. Data are described in the first five columns of table 3.

We smoothed the frequency of claims on geographical areas and performed a smoothed bootstrap resampling 2000 times from the smoothed mass distribution for the whole group, as shown in the previous example. Thus we calculated a 95% percentile interval of the smoothed frequency of claims. We then smoothed the frequency of claims conditioned to the type of fuel. We considered an effect as significant when the smoothed conditioned frequency was out of the constructed interval. Results are shown in the last two columns of table 3. We can see that in many geographical areas there is an effect of fuel on the frequency of claims.

5. Final Remarks

In this paper we made a review of the kernel estimator in a suitably corrected form for multidimensional contingency tables and we tried to introduce a kernel-type estimator for nominal variable. We noticed smoothing is very useful for ordered data. The solution proposed for nominal ones is not totally satisfactory even if it seemed to become more and more useful when structural zeros appear in data.
We underlined the usefulness of smoothing for sparse data above all jointly with resampling methods, by performing an application taken from an actuarial case study. Possible further directions of research could concern not only aspects tied with estimation problems, i.e. other non-parametric estimation methods suitable for sparse data (Simonoff, 1995) or the selection of smoothing parameters in multidimensional case, but also their application in modelling the association or the dependence among categorical variables.

References

Bishop Y. M. M., Fiemberg S. E., Holland P. W. (1975) *Discrete Multivariate Analysis*, MIT Press, Cambridge, MA.

Borgoni R. (1998) Uno stimatore a nucleo per dati sconnessi nel caso di frequenze sparse, *Working Papers 1998.2*, Dipartimento Scienze Statistiche, Università di Padova.

Chambers, J., Hastie, T.J. (1991) *Statistical Models in S,* Wadsworth/Cole & Brooks, Pacific Grove (Cal.).

Daboni L. (1993) *Lezioni di tecnica attuariale delle assicurazioni contro i danni,* II° ed., Edizioni LINT, Trieste.

Dong J., Simonoff J. S. (1994) The construction and properties of boundary kernels for smoothing sparse multinomials, *Journal of the American Statistical Society*, 3, 57-66.

Dong J., Simonoff J. S. (1995) A geometric combination estimator for *d*-dimensional ordinal sparse contingency tables, *The Annals of Statistics*, 23, 1143-1159.

Efron B., Tibshirani R. (1993) *An Introduction to the Bootstrap,* Chapman and Hall New York.

Grund B. (1993) Kernel estimators for cell probabilities, *Journal of Multivariate Analysis*, 46, 283-309.

Hall P., Di Ciccio T.J., Romano J.P. (1989) On smoothing and the bootstrap, *The Annals of Statistics*, 17, 2, 692-704.

Hall P., Titterington D.M. (1987) On smoothing sparse multinomials data, *Australian Journal of Statistics*, 39, 19-37.

Silverman B.W., Young G.A. (1987) The bootstrap: to smooth or not to smooth, *Biometrika*, 74, 3, 469-479.

Simonoff J.S. (1995) Smoothing categorical data, *Journal of Statistical Planning and Inference,* 47, 41-69.

Simonoff J.S. (1996) *Smoothing Methods in Statistics*, Springer, Berlin.

Reduction of Prediction Error by Bagging Projection Pursuit Regression

Simone Borra[(*)], Agostino Di Ciaccio[(**)]
[(*)]Univ. di Roma "Tor Vergata", Fac. di Economia, borra@economia.uniroma2.it
[(**)]Università di Urbino, Fac. di Economia, diciaccio@econ.uniurb.it

Abstract: In this paper we consider the application of Bagging to Projection Pursuit Regression and we study the impact of this technique on the reduction of prediction error. Using artificial and real-data sets, we investigate the predictive performance of this method with respect to the number of aggregated predictors, the number of functions in the single Projection Pursuit model and the signal-to-noise ratio of the sample data.

Keywords: Bagging, Projection Pursuit Regression, Prediction error.

1. Introduction

Recently new strategies have been explored to improve the prediction capability in nonparametric regression and classification, manipulating the explanatory or the dependent variables, randomizing the fitting procedure or forming new training sets by resampling from the original data set (sometimes weighing differently the units) (Dietterich, 1997). The last approach produced two relevant procedures: Bagging (Breiman, 1996, 1998, 1999) and Boosting (Freund and Schapire, 1997). In literature Bagging and Boosting had an empirical validation with classification and regression trees (Bauer & Kohavi, 1998) and in this contest they are used as an alternative to pruning methods. In this paper we show as Bagging can improve the prediction capability of the Projection Pursuit Regression (Friedman and Stuetzle, 1984). We show also that this improvement depends on both the complexity of the model and the characteristics of the data.

2. Bias/variance decomposition in nonparametric regression

In nonparametric regression we consider a dependent variable Y as function of a set of predictors $\mathbf{X}$:

$$Y = f(\mathbf{X}) + \varepsilon \qquad (1)$$

where $f(\mathbf{X})$ is an unknown smooth function and ε is a noise component. We use a sample $\mathbf{t} = \{y_i, \mathbf{x}_i\}_1^n$ (training data) drawn from the population to estimate $f(\mathbf{X})$ by

an approximating function $g(\mathbf{X} \mid \mathbf{t})$.
Given a training data set $\mathbf{t}$, we use as measure of prediction error the mean-squared generalization error:

$$PE(g/\mathbf{t}) = M_{y,\mathbf{x}}(y - g(\mathbf{x} \mid \mathbf{t}))^2 \tag{2}$$

where M indicates the average over all values (y, $\mathbf{x}$) of the population.
Considering the expectation of $PE(g/\mathbf{t})$ over all training samples of the same size drawn from the same population we obtain the average mean-squared generalization error $\overline{PE}(g)$. Using the decomposition proposed by Breiman (1996), we can write:

$$\overline{PE}(g) = E(\varepsilon^2) + E_{\mathbf{x}}(f(\mathbf{x}) - \bar{g}(\mathbf{x}))^2 + E_{\mathbf{x},\mathbf{t}}(g(\mathbf{x} \mid \mathbf{t}) - \bar{g}(\mathbf{x}))^2 \tag{3}$$

where $\bar{g}(\mathbf{x}) = E_{\mathbf{t}}(g(\mathbf{x} \mid \mathbf{t}))$ is the average of $g(\mathbf{x} \mid \mathbf{t})$ over all training samples of size n. The second and third term of (3) are, respectively, the square bias and the variance of the approximating function $g(\mathbf{X} \mid \mathbf{t})$. Decomposition (3) point up the fundamental trade-off between bias and variance showing the role of the complexity of the nonparametric method used.

3. Bagging projection pursuit regression

Instead of using a single $g(\mathbf{X}|\mathbf{t})$, accordingly to the Bagging approach, we can approximate $f(\mathbf{X})$ by a linear combination of approximating functions:

$$g_{bag}(\mathbf{X} \mid \mathbf{t}) = \sum_{k=1}^{K} \gamma_k g_k(\mathbf{X} \mid \mathbf{t}) \tag{4}$$

where γ_k is a weight, generally $1/k$, and any single $g_k(\mathbf{X}|\mathbf{t})$ is a member of a smoothed but otherwise unspecified class of functions estimated on the k-th bootstrapping of the sample $\mathbf{t}$. Using Projection Pursuit Regression (PPR) we define:

$$g_k(\mathbf{X} \mid \mathbf{t}) = \sum_{m=1}^{M} \beta_{m,k} h_{m,k}\left(\alpha_{m,k}^{\mathrm{T}} \mathbf{X} \mid \mathbf{t}\right) \tag{5}$$

where $h_{m,k}(\cdot)$ are smooth functions of different linear combinations of predictor variables. Substituting (5) in the expression (4) we obtain the Bagging PPR predictor (PPRbag):

$$g_{PPRbag}(\mathbf{X}\mid\mathbf{t})=\sum_{k=1}^{K}\gamma_k\sum_{m=1}^{M}\beta_{m,k}h_{m,k}\left(\alpha_{m,k}^{\mathrm{T}}\mathbf{X}\mid\mathbf{t}\right) \qquad (6)$$

We can note from (3) that $\overline{PE}(\bar{g})\leq\overline{PE}(g)$ as the variance term is zero. The $g_{PPRbag}(\mathbf{X})$ predictor has not the same nice property, as it is not based on independent training data sets but on K bootstrapping of a sample **t**, however it is likely to produce a reduced variance predictor.

Recently, Friedman and Hall (1999) studied the effect of Bagging in terms of decomposition of statistical estimators into linear and higher order parts. The authors pointed out that Bagging reduces the variability of the nonlinear component by replacing it with an estimate of its expected value, while leaving the linear part unaffected.

Our aim is to study when and how much the PPRbag predictor is better than PPR in term of $\overline{PE}$.

4. Simulation

We tried to point out by simulation studies the dependence of the PPRbag performance with respect to the number K of aggregated PPR, the complexity M of each single PPR, the noise importance measured as signal-to-noise ratio (the ratio of the standard deviation of the true function $f(\mathbf{X})$ over the standard deviation of the noise). We used two different simulation designs in correspondence to real and artificial data-sets. In the first case (see Fig.1), we split the real data-set in a training set and in a test set (depending on the size of data set we fixed the test set to 10% or 20%).

Figure 1: *Simulation schema for real data-set*

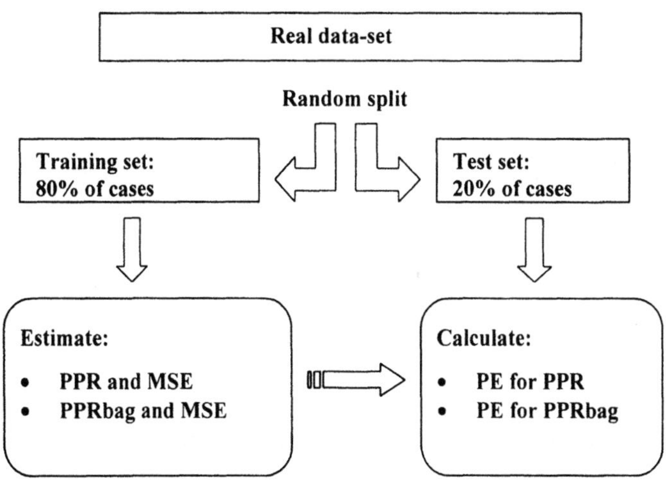

From the training data we estimated a single PPR, several PPRbag (for

K=1,2,...,30) and the corresponding values of MSE. Finally we calculated the mean-squared generalization error (PE) on the test set. This procedure was repeated 50 times, starting from different random splits of the data set. Then we estimated the average mean-squared generalization error $\overline{PE}(g_{PPR})$ and $\overline{PE}(g_{PPRbag})$ for PPR and PPRbag respectively.

For the artificial data-set we generated two independent data-sets (see Fig.2): the base data and the test data with 5000 cases for each. Then we calculated the average mean-squared generalization error in accordance with the following procedure:

1. we extract randomly a 200 cases sample from the base data and we use this sample to estimate the single PPR, the PPRbag (for K=1,2,...,30) and the corresponding values of MSE;
2. the mean-squared generalization error is calculated on the test sample for PPR and PPRbag;
3. steps 1 and 2 are repeated for 50 times and the average mean-squared generalization error is calculated.

Figure 2: *Simulation schema for artificial data-set*

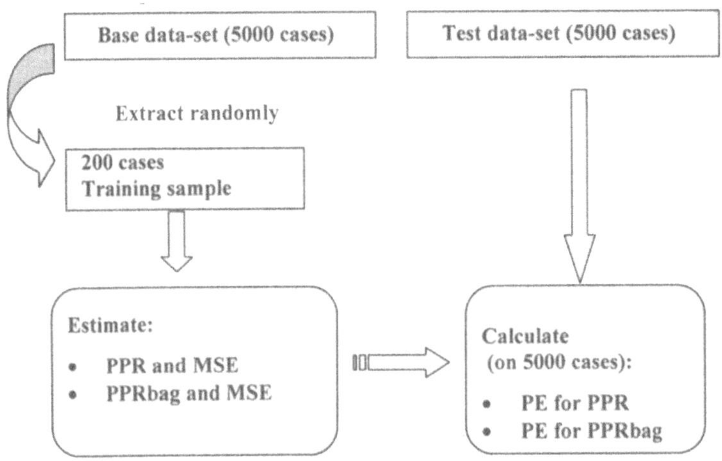

In the implementation of PPR we used the SMART program (Friedman, 1985). In this program, to employ a pruning strategy, the user must select the largest number of smooth functions, M_L, to use in the search as well as the final number of smooth functions, M_F. In our applications we fixed $M_L = M_F + 4$.

5. Empirical results

We considered three well known real-data sets: *Boston Housing* (506 cases, 12 explanatory variables), *Auto-MPG* (398 cases, 8 explanatory variables) and an artificial data set originated in Friedman (1991), named *Friedman 2*. The artificial data are originated by the model:

$$Y = \sqrt{X_1^2 + \left(X_2 X_3 - \left(1/X_2 X_4\right)\right)^2} + \varepsilon$$

where $\varepsilon \approx N\left(0,\sigma^2\right)$ and X_1, X_2, X_3, X_4 are uniformly distributed over the ranges: $0 \le X_1 \le 100$; $20 \le X_2/2\pi \le 280$; $0 \le X_3 \le 1$; $0 \le X_4 \le 11$.

Considering the number K of aggregated PPR, we note that in all our experiments, increasing the number K, $\overline{PE}(g_{PPRbag})$ decrease till it becomes stable (generally for K>15 or 20), therefore is not necessary to consider a large K to obtain a good predictor. For all real data-sets we have found that, keeping fixed the other factors, $\overline{PE}(g_{PPR}) - \overline{PE}(g_{PPRbag}) \geq 0$. For example, in Fig.3 we show the behavior of $\overline{PE}(g_{PPRbag})$ increasing the number K of aggregated PPR on Auto-MPG data, fixed M=8. We can observe that increasing K we obtain an improvement in the prediction of test set and fitting of training set.

Figure 3: *Comparison of performance of PPR and PPPbag on Auto-MPG by several values of K, fixed M=8*

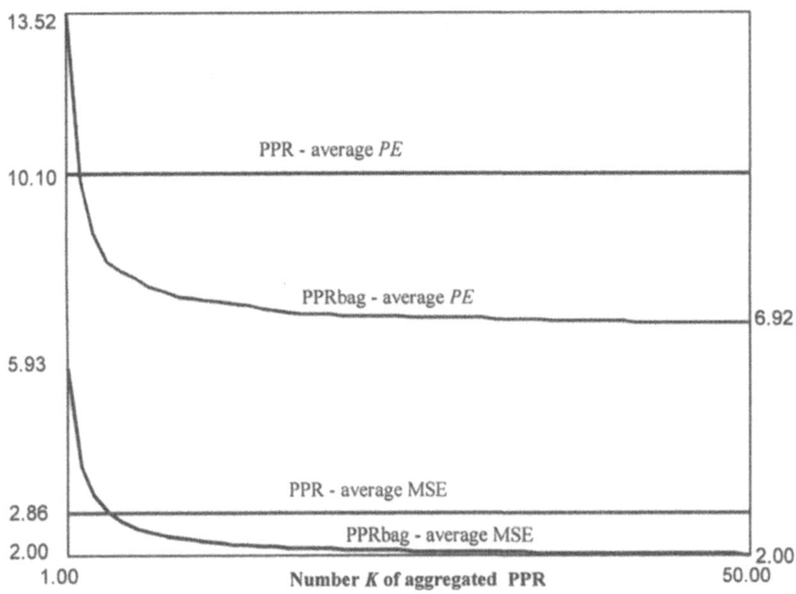

About the choice of M, we observe in all data-sets that, while for moderate values of M, PPR tends to overfit the test data-set and consequently increases the value of $\overline{PE}(g_{PPR})$, PPRbag keeps the value of $\overline{PE}(g_{PPRbag})$ stable. In Fig.4, for example, we can see the value of $\overline{PE}$, both for PPR and PPRbag, obtained on the

Boston Housing data with respect to different values of *M*, fixed *K*=50. In this case the smallest value of $\overline{\overline{PE}}$ for single PPR is given by *M*=2, increasing *M* leads PPR to overfit the training data and to get worse prediction on the test data. Conversely, PPRbag prediction improves increasing *M*. For this data set, PPRbag has a good performance, both with respect to the single PPR than to the application of other models (the value of $\overline{\overline{PE}}(g_{PPRbag})$ obtained for *M*=12 is lower than the values obtained by Breiman using Bagging regression tree and Bagging nearest neighbor predictors).

Figure 4: *Comparison of performance of PPR and PPPbag on Boston Housing by several values of M, fixed K=50*

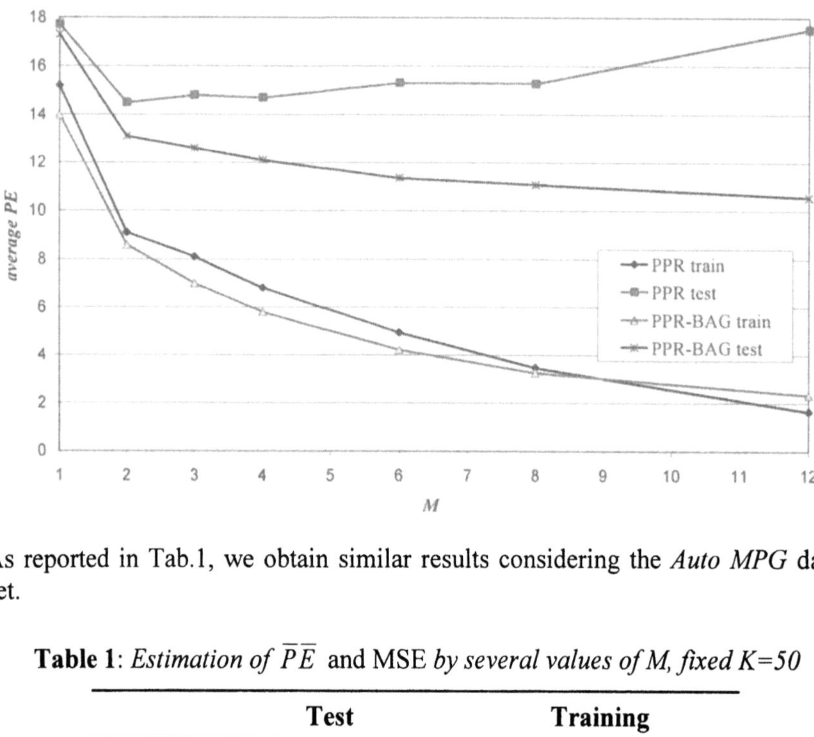

As reported in Tab.1, we obtain similar results considering the *Auto MPG* data-set.

Table 1: *Estimation of* $\overline{\overline{PE}}$ and MSE *by several values of M, fixed K=50*

	Test		Training	
M	*PPR*	*PPRbag*	*PPR*	*PPRbag*
2	8.45	7.44	6.15	4.93
4	8.68	7.11	4.37	3.37
6	9.53	7.03	3.32	2.53
8	10.10	6.92	2.86	2.00

Data set: *Auto MPG* - 10% cases test sample

Considering the artificial data set *Friedman2*, we found the same results obtained with the other two data sets. Using this artificial data set we studied the effect of complexity *M* through the bias/variance decomposition (3). From the Tab.2 we noted that, for both PPR and PPRbag, for large *M* the bias decreases and the variance increases, however the variance of PPR rises much more than PPRbag.

Table 2: *Bias/variance decomposition of $\overline{PE}$ by several values of M*

	PPR		PPRbag	
M	***bias***	***Variance***	***bias***	***Variance***
1	5007	1548	5069	1305
3	1277	5805	2277	1336
6	486	7893	709	1880
12	522	13558	384	2676

Data set: *Friedman 2* - 200 cases training sample

We investigated also the effect of noise measured as signal-to-noise ratio (SNR=$\sigma_f / \sigma_\varepsilon$) with respect to different values of *M.* We considered three cases: absence of noise, moderate noise (SNR=3/1) and high noise (SNR=3/1.58). The results of our simulations on *Friedman 2* are shown in Tab.3. We note in absence of noise the same performance between PPR and PPRbag, but increasing the component error in the sample the difference of performance is more evident. When *M* increases there is an evident advantage of PPRbag with respect to PPR in presence of moderate-high noise. This is due to the tendency of PPR to modeling the noise yet for M>2 when the SNR is low.

Table 3: *Effect of signal/noise ratio on $\overline{PE}$ by several values of M*

		Training		Test		
Number of smoothing functions	Component error	PPR	PPRbag	PPR	PPRbag	Difference
***M*=1**	Absent	4800	4367	5197	5275	-78
	Moderate	18903	17240	22250	22075	175
	High	42972	39344	49104	48853	251
***M*=3**	Absent	21	47	164	232	-68
	Moderate	13005	9660	22751	19303	3448
	High	30822	23234	54709	46851	7858
***M*=6**	Absent	9	26	175	229	-54
	Moderate	7364	5490	24088	18286	5802
	High	20911	14258	61565	46229	15336

Data set: *Friedman 2* - 200 cases training sample

6. Conclusion

In our simulation study we have shown as Bagging technique can improve the prediction capability of Projection Pursuit Regression. We have studied the effect of some factors on the performance of Bagging. Increasing *K*, the number of aggregated PPR calculated on the bootstrapped samples, the average of prediction error decreases until it becomes constant. When the number *M* of smoothing functions is high, the single PPR tends to overfit the training data, consequently the prediction capability tends to decrease. Conversely, using Bagging the overfitting is avoided and the prediction capability rises. In presence of moderate-high noise the effect of Bagging is more evident with respect to the single PPR. Future research could consider the use of Bagging for other nonparametric regression methods as MARS, DART etc., other resampling methods (for example, subseries block bootstrap or moving block bootstrap for time series), the use of boosting approach.

References

Bauer, E., Kohavi, R. (1998) An empirical comparison of voting classification algorithms: bagging, boosting, and variants. *Machine Learning*, vv, :1-38.

Breiman, L. (1996) Bagging predictors. *Machine Learning*, 26, N.2:123-140.

Breiman, L. (1998) Half and half bagging and hard boundary points. Technical Report n.534, Statistics Department, University of California.

Breiman, L. (1999) Using adaptive bagging to debias regressions. Technical Report, Dept. of Statistics, Stanford University, n. 547, February.

Dietterich T.G., (1997) Machine Learning Research: four current directions, *AI Magazine*, 18(4): 97-136.

Freunde, Y. and Schapire, R. (1997) A decision-theoretic generalization of on-line learning and an application to boosting. *Journal of Computer and System Sciences*, 55(1):119-139.

Friedman, J.H. (1984) SMART -User's Guide. Technical Report, Dept. of Statistics, Stanford University, n.1, October.

Friedman, J.H. (1985) Classification and multiple regression through projection pursuit. Technical Report, Dept. of Statistics, Stanford University, n.12, January.

Friedman, J.H., Stuetzle, W. (1984), Projection Pursuit Regression, *Journal of Americ. Statist. Assoc.*, 70: 817-823.

Friedman, J.H., Hall, P. (1999) On Bagging and nonlinear estimation, on Web-page: http://www.-stat.stanford.edu/~jhf/

Selecting Regression Tree Models: a Statistical Testing Procedure[1]

Carmela Cappelli
Department of Mathematics and Statistics, University of Naples Federico II
Monte S.Angelo, via Cinthia, 80126 Napoli, Italy
e-mail: cappelli@dms.unina.it

Francesco Mola
Department of Economics, University of Cagliari
Viale Fra Ignazio, 09127 Cagliari, Italy.
e-mail: mola@unica.it

Roberta Siciliano
Department of Mathematics and Statistics, University of Naples Federico II
Monte S.Angelo, via Cinthia, 80126 Napoli, Italy
e-mail: roberta@unina.it

Abstract: This paper provides a statistical testing approach to the validation of the pruning process in regression trees construction. In particular, the testing procedure, based on the F distribution, is applied to the CART sequence of pruned subtrees providing a single tree prediction rule which is statistically reliable and might not coincide with any tree in the sequence.

Keywords: Regression trees; pruning; validation.

1. Introduction

Regression tree procedures such as CART (Classification and Regression Trees of Breiman *et* al., 1984) and AID (Morgan and Sonquist, 1963) are a convenient way to produce a prediction rule from a set of examples described in terms of a vector of features and a response value. The aim is to define a general prediction rule which can be used to assign a response value to new cases solely on the basis of their features vectors. A regression tree procedure involves two main stages: 1) creating the totally expanded tree; 2) pruning the tree to give statistical reliability. Let $(Y, \mathbf{X})$ be a multivariate random variable where $\mathbf{X}$ is the vector of K predictors and Y is the numerical response variable. The totally expanded tree is obtained by recursively splitting a learning sample, namely $£=\{(y_n, \mathbf{x}_n), 1,...N\}$, taken from the distribution of $(Y, \mathbf{X})$, into two subsets according to a splitting criterion which allows to select at each node the best predictor to split the node. The final prediction rule results from pruning the totally expanded tree. Actually, this tree tends to be very large and complex and, above all, it overfits the learning data. In

[1] The present paper is financially supported by MURST funds.

other words it appears accurate respect to the learning cases but it is likely to perform poorly when it will be used to assess the response value of new cases. Pruning, by removing some of the branches, reduces the size of the tree improving in this way its understandability as well as its accuracy. So far, lot of attention has been paid to alternative pruning methods for classification trees (see for example Cappelli and Siciliano, 1998; Cappelli *et* al. 1998; Siciliano 1998). In this paper we focus the attention on pruning regression trees, providing in the framework of the CART regression tree procedure, a statistical testing procedure to decide whether a branch of the totally expanded tree should be pruned or not.

2. CART Regression Tree Procedure

CART regression tree construction consists in two main procedures: splitting and pruning.

CART splitting. The splitting criterion considers at each internal node the following quantity:

$$R(t) = \frac{1}{N} \sum_{x_n \in t} (y_n - \bar{y}(t))^2 \qquad (1)$$

where $\bar{y}(t)$ is the average of the response values y_n at node t. Note that the (1) can be understood as total sum of squares, denoted by $TSS_Y(t)$, divided by N. Any split s of node t induces a decrease in $R(t)$ defined as:

$$\Delta R(s,t) = R(t) - [R(t_l) + R(t_r)], \qquad (2)$$

where $R(t_l)$ and $R(t_r)$ are calculated similarly to (1) considering the averages $\bar{y}(t_l)$ and $\bar{y}(t_r)$ of the response values at the left decendant t_l and at the right descendant t_r of node t respectively. The split $s*$ which gives the highest decrease is said best and it is selected to split the node. It can be easily verified that the quantity $N[R(t_l) + R(t_r)]$ represents the within groups sum of squares, denoted by $WSS_{Y|s}(t)$. As a consequence, according to the decomposition of the total sum of squares it results:

$$N\,\Delta R(s,t) = TSS_Y(t) - WSS_{Y|s}(t) = BSS_{Y|s}(t), \qquad (3)$$

where $BSS_{Y|s}(t)$ represents the between groups sum of squares; thus, being N a costant, the best split of a node is the one which maximises the between groups sum of squares (Siciliano and Mola, 1996).

CART pruning. Concerning the pruning phase, the CART approach considers both the aspects of the accuracy (evaluated by the same measure $R(\cdot)$ used in the growing phase[2]) and the complexity (measured by the number of terminal nodes) of a tree by introducing the so called *error-complexity measure*. This is defined

[2] Note that when the measure R(.) is calculated over the whole tree, it corresponds to the mean squared error.

for any node t as:

$$R_\alpha(t) = R(t) + \alpha \tag{4}$$

and for the subtree T_t branching from t as:

$$R_\alpha(T_t) = R(T_t) + \alpha\left|\overline{T}_t\right| \tag{5}$$

where $R(T_t) = \sum_{h \in H_t} R(h)$, h is a terminal node of the subtree T_t in the set H_t whose cardinality is $\left|\overline{T}_t\right|$ and α is a sort of penalty for complexity. The idea is to prune the subtree T_t if its error-complexity measure is not lower than the error-complexity measure of its root t; actually, in such a case it is useless to retain the subtree because it does not decrease the error while increasing the complexity. In particular, the two error measures become equal for a critical value of α given by:

$$\alpha_t = \frac{R(t) - R(T_t)}{\left|\overline{T}_t\right| - 1} \tag{6}$$

in other words, for any internal node t, α, named *complexity parameter*, gives the reduction in error per terminal node.

The method is in two phases: first a sequence of nested pruned subtrees, $T_{\max} \supset T_{(1)} \supset \ldots \supset T_{(k)} \supset \ldots \supset \{t_1\}$, is created cutting at each step the subtree branching from the node (called weakest link) with the minimum value of α, then a single tree, i.e., a final prediction rule, is selected. The selection can be made on the basis of the accuracy evaluated either on a separate test set or by means of the cross-validation; then, that tree with the smallest error measure estimate (so called 0-SE rule) or the one which error measure estimate is within one standard error of the minimum (so called 1-SE rule) is selected. The main criticism to this method concerns just the selection strategy for it takes into account only the accuracy regardless of the statistical reliability of the trees in the created sequence. Empirical experience shows that in regression, the sequence of pruned trees tends to be larger than in classification because usually only two terminal nodes are cut off at a time. As a consequence, plotting on a graph the test sample estimates of the error measure against the number of terminal nodes of the trees in the sequence the resulting curve appears very flat and wide so that the selection of the particular tree which produces the smallest error measure estimate is arbitrary.

3. Statistical Testing in Pruning Process

The complexity parameter α can be expressed as follows:

$$\alpha_t = \frac{1}{|\bar{L}_t|}\sum_{l\in L_t}\Delta R(s^*,l) \tag{7}$$

where L_t is the set of internal nodes l of the subtree T_t having cardinality $|\bar{L}_t|$ (this relation holds for strictly binary trees where the number of internal nodes is equal to the number of leaves minus one). In order to prove the equivalence between the (6) and the (7) notice that for any node t it holds:

$$R(t) - R(T_t) = R(t) - \lfloor R(T_{2t}) + R(T_{2t+1}) \rfloor \tag{8}$$

where node $2t$ and $2t$+1 denote the left and the right descendant of node t respectively, so that the equivalence is verified by definition when $|\bar{T}_t| = 2$, being the (8) equal to $\Delta R(s^*,t)$. Now, let $|\bar{T}_t| = 3$ and $|\bar{T}_{2t}| = 2$ (i.e., node $2t$+1 is terminal), it can be easily verified that:

$$R(T_{2t}) = R(2t) - \Delta R(s^*,2t), \tag{9}$$

replacing the (9) in the (8) it results:

$$R(t) - R(T_t) = \Delta\, R(s^*,t) + \Delta R(s^*,2t) = \sum_{l\in L_t} \Delta R(s^*,l), \tag{10}$$

where again L_t includes the internal nodes of the subtree T_t, namely t and $2t$. By induction the (10) follows for any $|\bar{T}_t|$. The complexity parameter has the property to be related to the χ^2 distribution. The starting point of this result is the equivalence between the decrease in the measure $R(t)$ induced by the split of node t and the between groups sum of squares shown in equation (3); in the case of the parameter α, we have:

$$\alpha_t = \frac{1}{|L_t|\,N}\sum_{l\in L_t} BSS_{Y|s^*}(l), \tag{11}$$

so that, for any internal node t the complexity parameter can be viewed as the mean of the between groups sums of squares arising from splitting node t (and its non terminal descendants), multiplied by a constant factor. Since under the usual probability assumptions each $BSS_{Y|s^*}(l)/\sigma^2$ is distributed as a χ^2 with one degree of freedom, we propose testing the significance of each pruning operation by applying the analysis of variance testing procedure which compares for any internal node t and its branch, the variance between the groups resulting by splitting t (and eventually its non terminal descendants) with the variance within the groups. Moreover, since the effects of a predictor depend on prior predictors selected in the tree, in order to avoid the problem of dependence the proposed

testing procedure considers the cases of a separate test set opposed to the pruning process based on the same learning set employed to grow the tree.
The employed statistic is then defined as follows:

$$F_e^{ts} = \frac{\sum_{l \in L_t} BSS^{ts}_{Y\,|\,s}(l)}{WSS^{ts}_{Y\,|\,s}(t)} \times \frac{n^{ts}(t) - |\bar{T}_t|}{|\bar{L}_t|}, \tag{12}$$

where the subscription *ts* indicates that the corresponding quantities are computed on the test set and $n^{ts}(t)$ is the number of test set cases falling into node *t*. The above statistic has, under the null hypothesis of independence of the response variable from the splitting variable, a Snedecor-Fisher distribution with $|\bar{L}_t|$ and $n^{ts}(t) - \left[|\bar{L}_t| + 1\right] = n^{ts}(t) - |\bar{T}_t|$ degrees of freedom. In other words, when datasets are large (so that the underlying assumption of multinormality may be supposed to be satisfied) by fixing a significance level, it is possible to verify at each step whether the branch to be pruned produces a significant increase in the variance between the groups (it should be kept) or not (it should be pruned).

4. How the Procedure Works

In order to show how the proposed testing procedure works, we have applied it to a well known real data set: the Boston Housing, collected by Harrison and Rubinfeld (1978) and extensively used in regression analysis literature (see for example the CART book itself). The data consists of 506 census tracts in the Boston metropolitan area; the response variable is the median value of homes in thousands of dollars. The predictors are described in Table 1.

Table 1: *Description of the predictors*

Labels	Predictors
X_1	crime rate
X_2	percent land zoned for lots
X_3	percent nonretail business
X_4	1 if on Charles River, 2 =otherwise
X_5	nitric oxides concentration
X_6	average number of rooms
X_7	percent built before 1940
X_8	weighted distances employment centers
X_9	accessibility to radial highways
X_{10}	tax rate
X_{11}	pupil/teacher ratio
X_{12}	percent of blacks
X_{13}	percent lower status population

The totally expanded tree, grown using the 70% of the cases presents 46 terminal nodes; this tree has been pruned according to the CART pruning procedure, producing a sequence of trees which is reported in Table 2.

Table 2: *Sequence of pruned subtrees with the values of the F statistic associated to the corresponding weakest link*

Subtrees	Number of terminal nodes	α_{min}	Weakest linked node	F_e^{ts}	R^{ts}(T)
T_1	46	0.00000	6058	2.2769	109.1095
⋮	⋮	⋮	⋮	⋮	⋮
T_{15}	31	0.00028	765	5.4000	109.1294
T_{16}	30	0.00029	762	5.9583	109.1294
T_{17}	29	0.00030	32	5.4029	109.1294
T_{18}	28	0.00035	378	0.5078	109.1294
T_{19}	26	0.00049	382	7.1039	109.1494
T_{20}	25	0.00056	188	3.7950	109.1494
T_{21}	24	0.00058	189	0.0428	109.1494
T_{22}	23	0.00060	190	2.0909	109.1494
T_{23}	21	0.00065	94	0.4181	109.1494
T_{24}	20	0.00150	191	**15.2857**	109.1394
T_{25}	19	0.00190	126	1.5676	109.1394
T_{26}	18	0.00320	95	**17.4400**	109.2989
T_{27}	17	0.01150	47	0.6225	109.2989
T_{28}	16	0.19970	23	0.1632	109.3388
T_{29}	15	0.91480	16	0.6655	109.7776
T_{30}	14	0.95930	11	**21.2700**	111.5226
T_{31}	13	0.96480	63	0.7589	113.6565
T_{32}	12	1.02340	5	**10.8800**	114.5839
T_{33}	11	1.42920	8	0.4082	110.7448
T_{34}	10	2.34880	124	0.0272	104.1336
T_{35}	9	2.42205	2	**13.9784**	102,1493
T_{36}	7	2.59495	31	1.7762	107.9528
T_{37}	5	2.68583	7	0.6278	96.18625
T_{38}	3	5.14220	3	0.0091	86.20465
T_{39}	2	13.60080	1	**15.9768**	79.22452

For each tree in the sequence are indicated: the number of terminal nodes, the value of the smallest complexity parameter α, the corresponding weakest linked node with the associated test set empirical *F* value (in bold are those significant at 0.01) and the test sample error measure estimate. The tree circled in the first column with only 2 terminal nodes, corresponds to both the CART best choices 0-SE rule and 1-SE rule. Moreover, as expected, the sequence is large (for sake of brevity we omit trees from T_2 to T_{14} characterized by a value of the complexity parameter equal to 0.0) and the test sample estimates for the error measure R^{ts}(T) do not differ enough to justify the selection rules adopted by the CART methodology exclusively based on such estimates.

A different selection, which assure statistical reliability, can be made by means of the proposed F testing procedure. This testing follows the CART pruning process either cutting off or keeping those branches that induce a significant reduction of the variance within the groups on the basis of the F statistic. The testing takes at each step account of the previous outcomes, for example since node 191 is characterised by a significant F value, when testing node 95 also the branch descending from node 191 itself is considered; in other words, since the testing outcomes sometimes disagree with the CART choices, *the resulting tree does not necessarily coincide with any subtree in the sequence.*

Figure 1 shows the final F testing tree; within each node are indicated its label and the number of cases falling in it while the numbers in bold underneath terminal nodes are the corresponding values of the average of the response variable for that terminal node. This tree presents 6 terminal nodes and moreover, it is characterized by a test sample error measure estimate $R^{ts}(T)$ equal to 68.14, lower than any tree in the CART sequence.

Figure 1. *The final tree resulting from the statistical testing procedure*

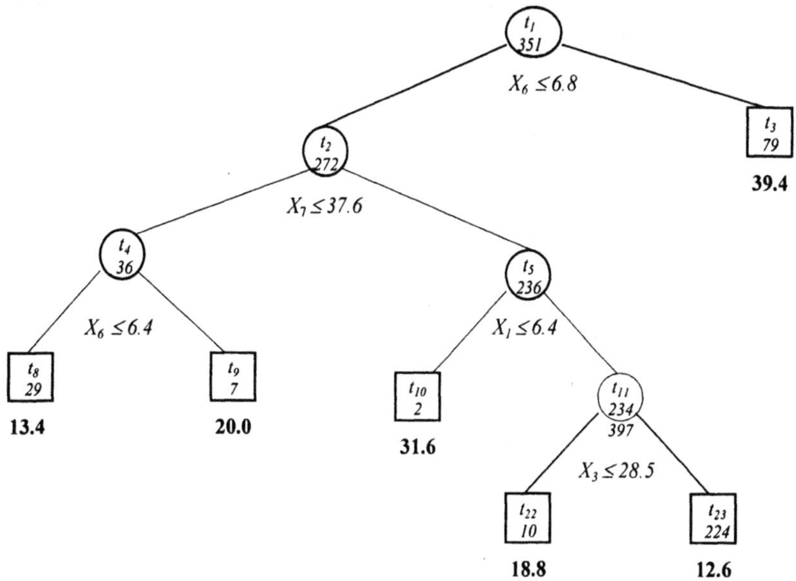

5. Concluding Remarks and Perspectives

In this paper a statistical testing procedure for pruning regression trees has been proposed. Actually, this topic is quite neglected by literature where the attention focuses on classification trees. In this framework, statistical testing has been proposed by Mingers (1987; 1989) if a probability measure has been used in creating the totally expanded tree and Zhang (1999).

Concerning regression, the F statistic has also been employed by Lanubile and

Malerba (1998) as a top down stopping rule to stop growing *one depth* branch. Opposed to them, we have used it retrospectively to validate the CART sequence of pruned subtrees i.e., to assess whether to cut or not *any depth* branches. Moreover, since the use of a testing procedure as stopping rule does violate the independence hypothesis, to tackle this problem we have considered an independent test set.

Future work will address the problem of instability of tree based predictors by bagging (Breiman, 1996). To this purpose we will consider multiple versions of a tree predictor selected by means of our testing procedure. These predictors will be weighted taking into account their overall significance as well as the relative frequence the same predictors reappears by bootstraping the original sample.

References

Breiman L., Friedman J.H., Olshen R.A., Stone C.J. (1984) *Classification and Regression Trees,* Wadsworth, Belmont CA.

Breiman L. (1996) Bagging Predictors, *Machine Learning*, **24**, 123-140.

Cappelli, C., Siciliano, R., (1998), Strategies for Choosing the best Decision Tree, in: *Analyse Multidimensionnelle des Donnèes, IV Congrès International N'GUS97* CERESTA-CISIA ed.

Cappelli C., Mola F., Siciliano R. (1998) An alternative pruning method based on the impurity-complexity measure, in: *Proceedings in Computational Statistics*, R. Payne and P. Green eds, 221-226, Physica-Verlag.

Harrison D., Rubinfeld D.L. (1978) Hedonic prices and the demand for clean air, *Journal of Environmental Economics and Management*, **5**, 81-102.

Lanubile A., Malerba D. (1998) Induction of Regression Trees with Regtree, *Classification and Data Analysis: Book of Short Papers,* 253-256, Meeting of the Italian Group of Classification, Pescara.

Mingers J.(1987) Expert System- Rule Induction with Statistical Data, *Journal of the Operational Research Society*, **38**, 39-47.

Mingers J.(1989) An Empirical Comparison of Pruning Methods for Decision Tree Induction, Machine Learning, **4**, 227-243.

Morgan, J.N., Sonquist, J.A. (1963). Problems in the analysis of survey data and a proposal, *Journal of American Statistical Association*, **58**, 415-434.

Siciliano, R., Mola F. (1996) A Fast Regression Tree Procedure, in: *Statistical Modelling: Proceedings of the 11th International Workshop on Statistical Modelling*, A. Forcina *et* al. eds, 332-340, Perugia: Graphos.

Siciliano, R. (1998) Exploratory versus Decison Trees, in *Proceedings in Computational Statistics*, invited paper, R. Payne and P. Green eds, 113-124, Physica-Verlag.

Zhang H., Singer B. (1999). *Recursive Partitioning in the Health Sciences*, New York: Springer Verlag.

Linear Fuzzy Regression Analysis with Asymmetric Spreads

Pierpaolo D'Urso, Tommaso Gastaldi
Dipartimento di Statistica, Probabilità e Statistiche Applicate,
Università degli Studi di Roma "*La Sapienza*", P.le A. Moro, 5 - 00185 Roma.
E-mail: durso@pow2.sta.uniroma1.it, gastaldi@pow2.sta.uniroma1.it

Abstract: We discuss a regression model for the study of asymmetrical fuzzy data and provide a method for numerical estimation of the relevant regression parameters. The proposed model is based on a new approach and has the capability to take into account the possible relationships between the size of the spreads and the magnitude of the centers of the fuzzy observations. Two illustrative examples are also presented.

Keywords: Fuzzy Data, Asymmetric and Symmetric Spreads, Fuzzy Linear Regression, Adaptive Model, Least Squares Estimators.

1. Introduction

There are observations which cannot be known or quantified exactly and their description is, hence, approximate. For instance, such data arise when, in order to describe a perception of a phenomenon, we provide an interval of values. This kind of data is sometimes referred to as *vague* or *fuzzy* and finds several application in fields, such as psychometry, technology, marketing, etc. An example is quantification of semantic scales used for opinion surveys of consumers about commercial products.

In the analysis of fuzzy data, it is particularly important the study of the relationships between a set of crisp independent variables and one or more dependent fuzzy variables. Several relevant contributions have been given on such a topic. We shall mention that fuzzy linear regression was introduced by Tanaka *et al.* (1980, 1982) and several works followed (Bardossy, 1990; Celimins, 1987; Chang *et al.*, 1996a, 1996b; Diamond, 1988; Kacprzyk *et al.*, 1992; Klir *et al.*, 1995; Redden *et al.*, 1996; Savic *et al.*, 1991; Tanaka, 1987; Xizhao *et al.*, 1992).

The common approach to fuzzy regression analysis is based on linear optimization techniques. In this paper, we discuss a new approach based on the introduction of an adaptive model fitted to the data through the least squares method. The model is called "adaptive" because of its special structure which is made essentially of three distinct models: one for the centers and the other two for the left spreads and the right spreads, respectively. The two models for the spreads are "built upon" the first one in order to allow the spreads to possibly depend on the magnitude of the centers.

The main contribution of our approach is the possibility to introduce in the model a form of dependence of the spreads upon the centers. These possible relationships are not taken into consideration in the classic fuzzy regression analysis. However, several real world situations call for this kind of formalization, where spreads depend on the intensity of the phenomenon under study. For instance, in surveys about personal income, it is natural to think that the vagueness associated with higher incomes may be greater. Besides, in this example, it is also natural to think that right spreads are significantly larger than left spreads, as often people tend to claim lower incomes.
The structure of this paper is as follows. In Section 2 we will introduce the concept of fuzzy data. In Section 3 we present our model for linear regression analysis with fuzzy data. An example is shown in Section 4. Section 5 contains final remarks.

2. Fuzzy data

A fuzzy number may be defined as $\mathrm{F} = (c, p, q)_{\mathrm{LR}}$, where c denotes the *center* (or *mode*), p and q are the *left spread* and *right spread*, respectively, L and R denote the *left* and *right shape functions*. It is common to define a fuzzy number by using a *triangular membership function* (Diamond *et al.*, 1994; Klir *et al.*, 1995; Zedeh, 1965; Zimmermann, 1991):

$$\mu_F(u) = \begin{cases} 1 - \dfrac{(c-u)}{p}, \text{ if} & c - p \le u \le c \\ 1 - \dfrac{(u-c)}{q}, \text{ if} & c \le u \le c + q \\ 0 & \text{otherwise.} \end{cases}$$

A graphical representation of the above membership function is shown in the picture below (Figure 1).

Figure 1: *Triangular membership function*

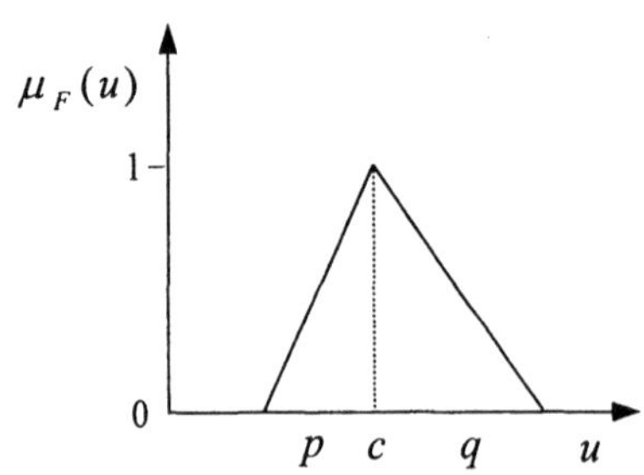

When $p = q$, a fuzzy number is referred to as *symmetrical*.
In the following section, we will study a general method for regression analysis with *asymmetric* fuzzy data.
In literature several type of membership function have been used, e.g., trapezoidal (Cherkassky, *et al.*, 1998), however, for most application is useful the triangular form. We will focus on the case of triangular membership functions. Our method is however general and can be extended to all cases where the membership function depends upon three parameters. Further extensions to a greater number of parameters (as, for instance, for the trapezoidal membership function, where the parameters are four) is straightforward.

3. Linear regression analysis for fuzzy data

We study the relationships between an *asymmetrical* fuzzy variable $Y \equiv (c, p, q)$ and a set of crisp independent variables $X_1, \ldots, X_k$. For instance, $Y \equiv (c, p, q)$ could be the score assigned by a commission to a candidate after an examination, while $X_1, \ldots, X_k$ might be some pieces of evidence provided by the candidate to document his/her competence.
The study of the relationships between Y and $X_1, \ldots, X_k$ is carried out through the definition of a regression model for the components (center and spreads) of the fuzzy variable and the independent crisp variables.
For simplicity, we adopt a descriptive perspective. Accordingly, we do not make distributional assumptions.
The fuzzy regression model we propose, and which will be referred to as *general doubly linear adaptive fuzzy regression model*, is defined as follows:

$$\begin{cases} \mathbf{c} = \mathbf{c}^* + \mathbf{e}_c, & \text{where } \mathbf{c}^* = \mathbf{X}\mathbf{a} \\ \mathbf{p} = \mathbf{p}^* + \mathbf{e}_p, & \text{where } \mathbf{p}^* = \mathbf{c}^* b + \mathbf{1} d \\ \mathbf{q} = \mathbf{q}^* + \mathbf{e}_q, & \text{where } \mathbf{q}^* = \mathbf{c}^* g + \mathbf{1} h \end{cases}$$

where:

$\mathbf{X}$: $n \times (k+1)$-matrix containing the vector $\mathbf{1}$ concatenated to k input variables;
$\mathbf{a}$: $(k+1) \times 1$-vector of regression parameters for the *core regression model*;
b, d, g, h : regression parameters for the *spread regression models*;
$\mathbf{c}, \mathbf{c}^*$: $n \times 1$-vectors of the *observed centers* and *interpolated centers*, respectively;
$\mathbf{p}, \mathbf{p}^*$: $n \times 1$-vectors of *observed left spreads* and *interpolated right spreads*;
$\mathbf{q}, \mathbf{q}^*$: $n \times 1$-vectors of *observed right spreads* and *interpolated left spreads*;
$\mathbf{1}$: $n \times 1$-vector of all 1's;
$\mathbf{e}_c, \mathbf{e}_p, \mathbf{e}_q$: $n \times 1$-vectors of residuals.

As it is evident, such a model is capable to incorporate the possible influence of the size of the centers on the spreads.
We consider the following distance between asymmetrical fuzzy numbers:

$$\delta_i \equiv \delta(Y_i, Y_i^*) = \sqrt{(c_i - c_i^*)^2 w_c + (p_i - p_i^*)^2 w_p + (q_i - q_i^*)^2 w_q}$$

where w_c, w_p, w_q are arbitrary positive weights, $Y_i \equiv (c_i, p_i, q_i)$, $i \in \{1,\ldots, n\}$, are the observations, and $Y_i^* \equiv (c_i^*, p_i^*, q_i^*)$ the theoretical (or interpolated) observations of the dependent (fuzzy) variable. These weights allow the user to give more importance to the fit of the centers or to the fit of the (left and right) spreads.
We provide a numerical solution to the following problem:

$$\begin{aligned}
&\textit{minimize}: \psi(\mathbf{a}, b, d, g, h) \equiv \sum_{i=1}^{n} \delta_i^2 \\
&= (\mathbf{c}-\mathbf{c}^*)'(\mathbf{c}-\mathbf{c}^*) w_c + (\mathbf{p}-\mathbf{p}^*)'(\mathbf{p}-\mathbf{p}^*) w_p + (\mathbf{q}-\mathbf{q}^*)'(\mathbf{q}-\mathbf{q}^*) w_q \\
&= (\mathbf{c}'\mathbf{c} - 2\mathbf{c}'\mathbf{c}^* + \mathbf{c}^{*\prime}\mathbf{c}^*) w_c + (\mathbf{p}'\mathbf{p} - 2\mathbf{p}'\mathbf{p}^* + \mathbf{p}^{*\prime}\mathbf{p}^*) w_p + (\mathbf{q}'\mathbf{q} - 2\mathbf{q}'\mathbf{q}^* + \mathbf{q}^{*\prime}\mathbf{q}^*) w_q \\
&= (\mathbf{c}'\mathbf{c} - 2\mathbf{c}'\mathbf{X}\mathbf{a} + \mathbf{a}'\mathbf{X}'\mathbf{X}\mathbf{a}) w_c + (\mathbf{p}'\mathbf{p} - 2\mathbf{p}'(\mathbf{X}\mathbf{a}b + \mathbf{1}d) + \mathbf{a}'\mathbf{X}'\mathbf{X}\mathbf{a}b^2 + 2\mathbf{a}'\mathbf{X}'\mathbf{1}bd + nd^2) w_p \\
&\quad + (\mathbf{q}'\mathbf{q} - 2\mathbf{q}'(\mathbf{X}\mathbf{a}g + \mathbf{1}h) + \mathbf{a}'\mathbf{X}'\mathbf{X}\mathbf{a}g^2 + 2\mathbf{a}'\mathbf{X}'\mathbf{1}gh + nh^2) w_q.
\end{aligned}$$

A recursive solution to the problem of the weighted least squares estimation with fuzzy data is found through *alternating least squares*:

$$\begin{cases}
\mathbf{a} = \dfrac{1}{(w_c + b^2 w_p + g^2 w_q)} ((\mathbf{X}'\mathbf{X})^{-1} \mathbf{X}'(\mathbf{c} w_c + (\mathbf{p} - \mathbf{1}d) b w_p + (\mathbf{q} - \mathbf{1}h) g w_q)) \\
b = (\mathbf{a}'\mathbf{X}'\mathbf{X}\mathbf{a})^{-1} (\mathbf{p}'\mathbf{X}\mathbf{a} - \mathbf{a}'\mathbf{X}'\mathbf{1}d) \\
d = \dfrac{1}{n} (\mathbf{p}'\mathbf{1} - \mathbf{a}'\mathbf{X}'\mathbf{1}b) \\
g = (\mathbf{a}'\mathbf{X}'\mathbf{X}\mathbf{a})^{-1} (\mathbf{q}'\mathbf{X}\mathbf{a} - \mathbf{a}'\mathbf{X}'\mathbf{1}h) \\
h = \dfrac{1}{n} (\mathbf{q}'\mathbf{1} - \mathbf{a}'\mathbf{X}'\mathbf{1}g).
\end{cases}$$

Notice that our method is based on fitting a model for fuzzy data by minimizing a squared distance function. Under this point of view, it presents a new approach w.r.t. the common techniques, based on Tanaka work, which are instead based on linear programming.
It is straightforward to prove the following identities (the sums and, hence, the averages, of all kind of errors are zero, cfr. D'Urso and Gastaldi, 2000):

$$\mathbf{1}' (\mathbf{c} - \mathbf{c}^*) = 0 \text{ (i.e., } E(\mathbf{c}) = E(\mathbf{c}^*)),$$

$$\mathbf{1}'(\mathbf{p}-\mathbf{p}^*)=0 \text{ (i.e., } E(\mathbf{p})=E(\mathbf{p}^*)),$$
$$\mathbf{1}'(\mathbf{q}-\mathbf{q}^*)=0 \text{ (i.e., } E(\mathbf{q})=E(\mathbf{q}^*)).$$

Also, it can be shown that, in the case we have only one independent variable, the proposed method yields the same solutions as we would carry out two (independent) regression on the centers and the spreads (cfr. D'Urso and Gastaldi, 2000).

4. Examples

We will now apply our method to two real world problems.
For the two examples, all weights in the objective function have been set equal to $1: w_c = w_p = w_q = 1$. We first show a univariate case with the corresponding graphical representation and, then, an example with two independent variables.
All the computing were done by using SAS/IML language.

Example 1

We have a collection of average scores given by a group of experts about the Italian wine "Chianti" for the years 1972-1983. Scores are fuzzy numbers, each one defined by a center, a left spread and a right spread.
We wish to study the trend of the wine quality during these years, as reflected by the experts' opinions.
In Table 1, we show our dataset.

Table 1: *Fuzzy scores of experts about the Italian wine "Chianti" for the years 1972-1983*

Dataset			
x	**c**	**p**	**q**
1972	4	0.5	1.5
1973	4	0.5	1.5
1974	6	0.5	1.0
1975	8	1.5	1.0
1976	4	0.5	1.5
1977	8	1.5	1.0
1978	10	2.0	1.0
1979	8	1.5	1.0
1980	8	1.5	1.0
1981	6	0.5	1.0
1982	8	1.5	1.0
1983	8	1.5	1.0

In Table 2, the model parameters we found are summarized.

Table 2: *Estimated parameters*

Parameter Estimates
a = (-643.1142, 0.33)'
b = 0.25
d = -0.58
g = -0.12
h = 1.96

In Figure 2 our model and the scatterplot of the input data are represented.

Figure 2: *Scatterplot and linear fuzzy trend*

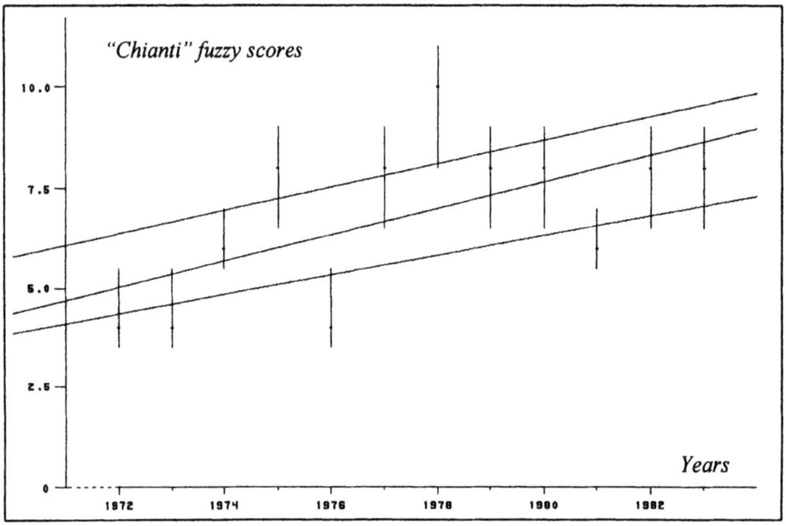

Example 2

As second example, we apply our technique to a dataset obtained through an experiment conducted on 20 computer users, who were asked to express their opinions about a popular e-mail client program.
The purpose of the experiment was to investigate if the users' evaluation of the software (Y) were dependent on their experience (X_1), measured in months, and the speed of the processor (X_2), measured in Mhz, in the system which hosts the e-mail client (cfr. Table 3).

Table 3: *Dataset and estimated centers, left and right spreads*

x_1	x_2	c	p	q	c*	p*	q*
12	400	8.0	2.0	1.0	8.2081174	1.0962467	0.9776145
1	600	9.5	1.0	0.5	8.3798583	1.0921699	0.8989921
18	500	6.0	1.0	2.0	5.7517403	1.1545555	2.1021346
12	366	7.0	1.5	1.0	8.5923130	1.0871267	0.8017312
6	400	10.0	0.5	0.0	9.5345074	1.0647612	0.3703981
24	200	9.0	2.0	1.0	7.8153115	1.1055710	1.1574395
16	300	8.0	1.0	1.0	8.4538444	1.0904137	0.8651215
4	600	7.5	0.0	0.5	7.7166633	1.1079127	1.2026003
10	400	9.0	1.0	1.5	8.6502474	1.0857515	0.7752090
8	550	6.5	1.0	2.0	7.3973968	1.1154913	1.3487593

(Notice that all the properties pointed out in Section 3 hold)

The following Table 4 reports the values obtained for the estimates of the regression parameters in our fuzzy regression model.

Table 4: *Estimated parameters*

Parameter Estimates
a = (15.380846, -0.221065, -0.0113)'
b = -0.023738
d = 1.2910887
g = -0.457796
h = 4.7352595

In our example, the results point out that the e-mail client is more appreciated by new users. Less certain is the correlation with the processor speed, which does not seem too important for the users of this e-mail client.

As we have seen our procedure has made possible to process the fuzzy data as they are, without substituting a set of values indicated by the user with a single crisp value. In general, when the observed data is not obtained exactly, this allows to introduce in the analysis the original fuzziness, thus avoiding manipulations of the data which may be distortive or arbitrary.

5. Conclusions

The model we have studied can be applied to cases when we have fuzzy data with asymmetrical spreads. This is often the case in experimental studies (see, for instance, Chang *et al.*, 1996b) where *asymmetric* shapes of the membership function are often observed. As a subcase, the model also applies to the more common *symmetrical* fuzzy data, which is also often encountered in practical applications (e.g., Laviolette *et al.*, 1995).

References

Bardossy, A. (1990) Note on fuzzy regression, *Fuzzy Sets and Systems*, 37, 65-75.

Celimins, A. (1987) Multidimensional least-squares fitting of fuzzy models, *Mathematical Modelling*, 9, 669-690.

Chang, P. T., Lee, E. S. (1996a) A generalized fuzzy weighted least-squares regression, *Fuzzy Sets and Systems*, 82, 289-298.

Chang, P.T., Lee, E.S., Konz, S.A. (1996b) Applying fuzzy linear regression to VDT legibility, *Fuzzy Sets and Systems*, 80, 197-204.

Cherkassky, V., Mulier, F. (1998) *Learning from Data. Concepts, Theory, and Methods*, John Wiley and Sons, New York.

D'Urso, P., Gastaldi, T. (2000) A Least Squares Approach to Fuzzy Linear Regression Analysis, *Computational Statistics & Data Analysis*, to appear.

Diamond, P., Kloeden, P. (1994) *Metric Spaces of Fuzzy Sets, Theory and Application.* World Scientific, Singapore.

Diamond, P. (1988) Fuzzy least squares, *Information Science*, 46, 141-157.

Dubois, D., Prade, H. (1980) *Fuzzy Sets and Systems: Theory and Applications.* Academic Press, New York.

Kacprzyk, J., Fedrizzi, M. (Eds.) (1992) Fuzzy Regression Analysis, Omnitech and Heidelberg: Physica-Verlag, Warsaw.

Kim, K. J., Moskowitz H., Koksalan M. (1996) Fuzzy versus statistical linear regression, *European Journal of Operational Research*, 92, 417-434.

Klir, G. J., Yuan, B. (1995) *Fuzzy Sets and Fuzzy Logic. Theory and Application*, Prentice Hall, New Jersey.

Laviolette, M., Seaman, J.W., Barrett, J.D., Woodall, W.H. (1995) A probabilistic and statistical view of fuzzy methods, with discussion , *Technometrics*, 37, 3, 249-292.

Redden, D. T., Woodall W. H. (1996) Further examination of fuzzy linear regression, *Fuzzy Sets and Systems*, 79 203-211.

Savic, D. A., Pedrycz, W. (1991) Evaluation of fuzzy linear regression models, *Fuzzy Sets and Systems*, 39, 51-63.

Tanaka, H. (1987) Fuzzy data analysis by possibilistic linear models, *Fuzzy Sets and Systems*, 24, 363-375.

Tanaka, H., Uejima, S., Asai, K. (1980) Fuzzy linear regression model, *IEEE Trans. Systems Man Cybernet.*, 10, 2933-2938.

Tanaka, H., Uejima, S., Asai K. (1982) Linear regression analysis with fuzzy model, *IEEE Trans. Systems Man Cybernet.*, 12, 903-907.

Tanaka, H., Watada J. (1988) Possibilistic linear systems and their application to the linear regression model, *Fuzzy Sets and Systems*, 27, 145-160.

Xizhao, W., Minghu, H. (1992) Fuzzy linear regression analysis, *Fuzzy Sets and Systems*, 51, 179-188.

Zadeh, L. A. (1965) Fuzzy Sets, *Informat. Control*, 8, 338-353.

Zimmermann, H. J. (1991) *Fuzzy Set Theory and its Application*, Kluwer Academic Press, Dordrecht.

PART III

SPATIAL AND TIME SERIES DATA ANALYSIS

Forecasting Non-Linear Time Series: Empirical Evidences on Financial Data [(1)]

Alessandra Amendola*

Francesco Giordano*

Cira Perna*
*Dipartimento di Scienze Economiche, Università di Salerno
alamendola@unisa.it; giordano@unisa.it; perna@unisa.it;

Abstract: This paper presents a preliminary comparison of forecasting performance for alternative non-linear methods using daily returns from the Italian Stock Market. In particular, some non-linear models and non-parametric techniques are considered. The accuracy of the forecast is evaluated using the sign prediction criteria, the mean square error and the mean absolute error.

Keywords: Forecasting, GARCH, TAR, Kernel, Neural Networks

1. Introduction

A primary topic in time series analysis is to provide accurate predictions for an observed time series. However, empirical evidence on financial time series indicates that returns on financial assets seem unforecastable at short horizons.
The development of a wide variety of non-linear models and the growing attention in the application of non-parametric techniques in time series analysis, seem to create a renewed interest in forecasting financial data.
In the recent literature many authors investigate how more elaborate structure performe better than simpler linear systems for forecasting complex data. Among the others Tiao and Tsay (1994) compare the post-sample forecasts of a TAR model versus a linear AR model, Franses and Van Dijk (1996) analyse the performances of a GARCH model in forecasting stock market volatility, Brooks (1997) evaluates the accuracy of exchange rate forecasts using linear and non-linear models.
The aim of the paper is a preliminary comparison of the forecasting performance of some alternative methods which can be used to forecast stock market returns.
In particular we forecast daily Italian stock market returns using some classes of non-linear models largely applied in finance, such as heteroscedastic and threshold models, and some non-parametric methods, such as kernel and neural

[1]The paper is partially supported by MURST98, "Modelli Statistici per l'Analisi delle serie temporali".

networks, whose flexible functional form makes them particularly suited to financial applications.
The outline of the paper is as follows: the next section gives a preliminary analysis of the data; section 3 briefly describes the parametric and non-parametric methods used for forecasting and presents the estimation results; section 4 compares the forecasting performances and gives some comments.

2. The Data

The data we analyse in this paper are daily returns, expressed in Euro, of three equities quoted on the Italian Stock Market: Fiat, Generali and Stet. They exhibit all the features which are usually present in financial time series commonly arising in practical applications. The sample covers the period from January 1997 to December 1998, involving 500 observations for each series. The analysis is based on the first 485 observations with the remainder 15 being left for post-sample forecast evaluation. Returns are calculated as logarithmic first differences, $R_t = \nabla(\log X_t)$.
Some summary statistics and the results of non-linearity tests are given in table 1 and table 2.

Table 1: *Summary Statistics*

	FIAT	GENERALI	STET
Mean	0.000524	0.001867	0.001588
Median	0.001021	0.000882	0.001878
St.Dev	0.023119	0.019070	0.021975
Skewness	-0.072798	-0.069356	-0.308570
Kurtosis	4.047934	4.002357	3.647958

Table 2: *Non-linearity Tests*

	FIAT	GENERALI	STET
Jarque-Bera	23.27345	21.28986	16.64815
Lin-Mudholkar	2.208012	1.6590*	2.4821
Ljung-Box	37.627	24.9587*	31.602
McLeod-Li	61.580	97.1659	153.4620
LM (ARCH)	6.125	43.982	33.509
LR (TAR)	24.580	24.65	43.49

Note: all the tests are significant at 5% level except for those labelled (*) which are significant at 10% level; the Ljung-Box and McLeod-Li tests are computed on up to 20 lags and are asymptotically distributed as a χ^2_g with respectively g=19, g=17, g=17 for each series; the asymptotic distribution of the LR leads to critical values of 22.65, 19.16, 24.32.

It is evident from table 1 that all the series show leptokurtosis. Hence the excess kurtosis leads the Jarque-Bera normality test, distributed under the null as a χ^2_2, to reject normality. The same result is obtained by the Lin-Mudholkar (1980)

normality test, which under the null hypothesis has a standardised normal distribution. The Ljung-Box test for autocorrelations, and the McLeod and Li (1983) test for linearity, carried out on the residuals of linear AR(p) models (identified in tab.3), confirm the presence of non-linear structure.
The Lagrange Multiplier (LM) test of Engle (1982) supports the evidence of ARCH components. In order to verify the validity of a threshold specification we applied the Likelihood Ratio (LR) test for linearity of Tong (1990) where the null hypothesis of linearity is tested against a specified threshold model.

3. Methods and models for forecasting

In order to evaluate the forecasting performance on financial asset returns we compare the predictive ability of linear time series models, non-linear autoregressive models and some non parametric methods.
In the parametric setting we have restricted our attention to those models which classically allow to capture the main features of financial time series. In particular we have considered the class of ARCH models, where the non-linearity stems from the conditional variance specification, and the class of TAR models where non-linear effects are introduced in the conditional mean structure.
In the non-parametric context we have considered kernel estimators and neural networks. This choice is motivated by the fact that the first estimators has been widely applied in time series due to their statistical properties (Robinson, 1983); while neural networks are particularly successful in point prediction of financial variables.

3.1 Parametric Models

We start from the linear autoregressive models, (Box, and Jenkins, 1970), that still remain the most popular set of models for economic applications and can be considered as a benchmark for forecasting comparisons. An AR(k) for the returns series R_t can be defined as:

$$R_t = a_0 + \sum_{i=1}^{k} a_i \, R_{t-i} + \varepsilon_t \qquad (1)$$

Following the classical three-stage procedure (Box and Jenkins,1970) we identified and estimated an AR(1) for Fiat and an AR(3) for Generali and Stet (Tab.3).
In the analysis of financial time series the classical assumption of homoscedasticity is often not satisfied. In order to introduce the effect of a changing conditional variance we consider the GARCH (Generalized Conditional Heteroscedastic, Bollerslev 1986) component and estimate an AR-GARCH given by:

$$R_t = a_0 + \sum_{i=1}^{k} a_i \, R_{t-i} + \varepsilon_t \qquad (2)$$

where $\varepsilon_t \sim N(0,\sigma^2)$ and $\sigma_t^2 = \alpha_0 + \sum_{i=1}^{q} \alpha \varepsilon_{t-1_i}^2 + \sum_{j=1}^{p} \beta_j \sigma_{t-1}^2$.

With a GARCH structure the estimated parameters in the mean equation change significantly. However it is not necessary to forecast the variance in order to be able to forecast the mean. To allow the conditional variance to affect the mean forecast we use the GARCH-M model (Engle *et al.* 1987) in which the mean equation is defined as:

$$R_t = a_0 + \sum_{i=1}^{k} a_i R_{t-i} + \delta \sigma_t^2 + \varepsilon_t \tag{3}$$

The fitted models are an AR(1)-GARCH(1,1) for Fiat, and an AR(3)-GARCH(1,1) for Generali and Stet with the correspondent GARCH-M formulation. The estimated parameters and the corresponding standard errors are shown in table 2.

The response of stock market prices exhibits a pronounced asymmetric cyclical behaviour due to the different reaction to negative and positive shocks. For modelling asymmetry in the conditional mean we consider a piecewise linear model, the SETAR (Self Exiting Threshold, Tong 1983) model given by:

$$R_t = a_0^{(j)} + \sum_{i=1}^{k} a_i^{(j)} R_{t-i} + h^{(j)} \varepsilon_t \text{ with } r_{j-1} \le R_{t-d} < r_j \tag{4}$$

for j=1,...,l, where ε_t is i.i.d., the *threshold variable* R_{t-d} is given by the lagged returns, and the *threshold values*, $(r_0, r_1, r_2,..,r_l)$, are such that $r_0<r_1<...<r_l$, $r_0 = -\infty$ and $r_l = +\infty$ with $\Re_j = (r_{j-1}, r_j]$.

Following the recursive procedure of identification and estimation proposed by Tong (1990), we identified a SETAR(2;4,1) for Fiat with delay d=2 and threshold value r=0.0093; a SETAR(2;4,2) for Generali with d=1 and r=0.0026; and a SETAR(2;3,1) for Stet with d=2 and r=0.0014. The least squares estimates of the autoregressive parameters in each regime are given in tab.3.

3.2 Non-parametric Methods

In the non-parametric context we refer to the autoregressive models:

$$X_t = g(X_{t-1},\ldots,X_{t-p}) + \varepsilon_t \tag{5}$$

where ε_t are i.i.d random variables with zero mean and unit variance and g is an unknown smooth function.

As an estimator of the function g we consider the Nadaraya-Watson kernel estimator, with one smoothing parameter:

$$\hat{g}(\mathbf{x}) = \frac{\sum_{t=1}^{n} K_H(\mathbf{x} - Z_{t,p}) X_t}{\sum_{t=1}^{n} K_H(\mathbf{x} - Z_{t,p})} \tag{6}$$

where $H=h^2I$ is the bandwidth matrix, $K_H(\mathbf{x}) = |H|^{-1/2} K(H^{-1/2}\mathbf{x})$, K is a p-variate kernel function satisfying $\int_{\Re^p} K(\mathbf{x})d\mathbf{x} = 1$, and $Z_{t,p} = (X_{t-1},\dots,X_{t-p})$.

In order to use this estimator it is necessary to identify the parameter p and estimate h. The latter is determined through a cross-validation criterion while for the specification of p we have used a step procedure based on the AIC.
For the three series considered the results are shown in table 4.

Table 3: *Estimate parameters*

	k	p	q	a_0	a_1	a_2	a_3	a_4	δ	α_0	α_1	β_1
FIAT												
AR	1			0.0015	0.1558							
				(0.003)	(0.045)							
AR-Garch	1	1	1		0.1653					0.0001	0.0727	0.8983
					(0.048)					(0.000)	(0.024)	(0.033)
Garch-M	1	1	1		0.1641				0.0485	0.0001	0.0914	0.8641
					(0.044)				(0.052)	(0.000)	(0.028)	(0.039)
SETAR I	4				0.1498			0.1925				
					(0.081)			(0.074)				
II	1				0.1838							
					(0.082)							
GENERALI												
AR	3			0.0410	0.0972		0.1372					
				(0.020)	(0.045)		(0.045)					
AR-Garch	3	1	1	0.0017	0.1220					0.0003	0.1186	0.7680
				(0.001)	(0.046)					(0.000)	(0.036)	(0.071)
Garch-M	3	1	1		0.1234		-0.1161		-0.0393	0.0003	0.1082	0.8029
					(0.044)		(0.050)		(0.025)	(0.000)	(0.035)	(0.070)
SETAR I	4						-0.2548	0.1432				
							(0.061)	(0.062)				
II	2				0.1899	-0.0997						
					(0.106)	(0.046)						
STET												
AR	3			0.0077	0.1475		-0.1677					
				(0.005)	(0.044)		(0.045)					
AR-Garch	3	1	1	0.0021	0.1276		-0.1005			0.0001	0.1133	0.8597
				(0.001)	(0.044)		(0.047)			(0.00)	(0.032)	(0.065)
Garch-M	3	1	1	0.0050	0.1322		-0.0977			0.0001	0.1110	0.8630
				(0.003)	(0.045)		(0.047)			(0.00)	(0.031)	(0.057)
SETAR I	3				0.1626		-0.2093					
					(0.064)		(0.065)					
II	1				0.1205							
					(0.063)							

Table 4: *Autoregressive order (p) and Smoothing parameter (h)*

	FIAT	GENERALI	STET
p	5	8	3
h	0.1594629	0.1088889	0.1464347

A different approach for the estimation of the function $g(\cdot)$ is based on the use of artificial neural networks (Cheng and Titterungton, 1994). In particular we use a *single hidden layer feedforward network* with the following structure:

$$\hat{g}_{NN} = \sum_{j=1}^{s} \beta_j \phi\left(\sum_{i=1}^{p} \alpha_{ij} X_{t-i} \right) \tag{7}$$

where β_j is the weight of the link between the *j-th* neuron in the hidden layer and the output; α_{ij} is the weight of the connection between input X_j and the *j-th* neuron in the hidden level, s is the number of hidden units. We suppose that the bias terms are both zero, the activation function of the input is the standard Normal cumulative function and the parameters d and m are respectively 1 and 5.

4. Forecasting Comparisons

In order to compare the different forecast performances we consider, for each of the different methods, the m-step-ahead forecast that can be written as $f_{t,m} = E(x_{t+m} | I_t)$, where I_t is the set of all information available at time t, $I_t = (x_1, \ldots, x_t)$. To obtain $f_{t,m}$ we use an iterative procedure starting from the one-step-ahead forecast $f_{t,1}$. The forecasts are generated by a recursive method in which at any step the forecasting origin is rolled forward one period and the parameters of the model are recalculated.

To assess the accuracy of the forecasting methods we use the Percentage of Correct Sign (PCS) to predict direction changes, the Mean Square Error (MSE) and the Mean Absolute Percentage error (MAPE) to evaluate their magnitude. These criteria are defined as:

$$PCS = \frac{1}{N-n} \sum_{t=n}^{N} z_{t+n}$$
$$MSE = \frac{1}{N-n} \sum_{t=n}^{N} (X_{t+m} - f_{t,m})^2 \tag{8}$$
$$MAPE = \frac{1}{N-n} \sum_{t=n}^{N} \left| \frac{X_{t+m} - f_{t,m}}{X_{t+m}} \right|$$

where N is the total number of observations, n is the number of observations used for in-sample modelling, with z_{t+n}=1 if $(x_{t+m} f_{t,m})>0$, z_{t+n}=0 otherwise.

The values of the three different indexes calculated for each of the different forecasting methods and for each of the considered data sets are given in table 5.

Table 5: *Forecasting comparison*

SERIES	METHODS	PCS	MSE x10^{-3}	MAPE
FIAT	**AR**	0.2857143	0.2681289	1.549657
	AR-GARCH	0.3571429	0.2650209	1.827892
	GARCH(M)	0.4285714	0.2652143	5.119120
	SETAR	0.5000000	0.2543417	0.985343
	KERNEL	0.5714286	0.2363878	1.558336
	N. NET	0.5000000	0.2345576	1.217304
GENERALI	**AR**	0.5714286	0.1438333	4.011517
	AR-GARCH	0.6428571	0.1459086	3.983359
	GARCH(M)	0.7142857	0.134693	5.264807
	SETAR	0.5000000	0.1824852	3.158735
	KERNEL	0.571428	0.1451929	2.597753
	N. NET	0.6428571	0.1608869	2.631148
STET	**AR**	0.5714286	0.1505698	1.278614
	AR-GARCH	0.7142857	0.1430961	0.993827
	GARCH(M)	0.5714286	0.1580384	1.207781
	SETAR	0.5714286	0.1528627	0.9175703
	KERNEL	0.8571429	0.1689807	0.8488313
	N. NET	0.7142857	0.2019829	0.891114

The interpretation of the PCS and MSE results is quite easy due to their regularity; the MAPE index, instead, could be seriously increased by small values. The indexes for the first series, Fiat, seem to confirm that the use of more complex structures improves the forecasting performance. For the second series, the gain over the benchmark linear model is not significant, especially in terms of MSE. For the last series, the performance of linear and non-linear parametric models is almost the same However an improvement is observed for the non-parametric methods.

The different performances of the estimated models are due to the particular structures characterising the three series analysed. In the case of Generali and Stet series, the values of the MSE obtained for Kernel and Neural Networks methods, respectively, are substantially different. It can be explained taking into account the relevant change in the series structure registered for the final period (heteroscedasticity) and observing that neural networks have less "degrees of freedom" because of the higher number of parameters. Therefore kernel methods, with only one smoothing parameter, can fit the observed values faster than neural networks. In general, non-linear models have a better performance in fitting the data (Amendola *et al.* 1998), with respect to the non-parametric techniques, but their superiority from a forecasting point of view is less convincing.

The computational complexity of the procedures used has led us to make some choices that necessarily affect the results of the analysis. Hence the achievements obtained in this paper have to be considered as a starting point for future research.

Acknowledgements

The authors would like to thank an anonymous referee for his useful comments which helped to improve the paper.

References

Amendola A., Giordano F., Perna C., (1998) Parametric and Non-parametric Methods in Non-linear Time Series Analysis: A Critical Evaluation, *Pre-Proceedings of "New Techniques & Technologies for Statistics"*, 2, 25-30.

Bollerslev T., (1986) Generalized Autoregressive Conditional Heteroskedasticity, *Journal of Econometrics*, 31, 307-327.

Box G.E.P., Jenkins G.M., (1970) *Time Series Analysis Forecasting and Control*, Holden-Day, San Francisco.

Brooks C., (1997) Linear and Non-linear (Non-)Forecastability of High-frequency Exchange Rates, Journal of Forecasting, 16, 125-145.

Cheng, B., Titterington D.M., (1994) Neural Networks: a review from a statistical prospective, Statistical Science, 9,1, 2-54.

Engle R.F., (1982) Autoregressive conditional heteroschedasticity with estimates of the variance of united kingdom inflation, *Econometrica*, 50, 4, 987-1007.

Engle R.F., Lilien D., Robins R., (1987) Estimating time varying risk premia in the return structure: The ARCH-M model, *Econometrica*, 55, 391-407.

Franses P.H., Van Dijk D., (1996) Forecasting Stock Market Volatility Using (Non-linear) Garch Models, Journal of Forecasting, 15, 229-235.

Hardle, W., Lütkepohl, H. & Chen, R. (1997) A Review of Non-parametric Time Series Analysis, *International Statistical Review*, 65, 49-72.

Lin C.C., Mudholkar G.S., (1980), A Simpler test for normality against asymmetric alternatives, Biometrika, 67,455-61

McLeod A.I., Li W.K., (1983) Diagnosting checking ARMA time series using squared residual autocorrelations, *Journal of Time Series Analysis*, 4.

Robinson, P.M. (1983) Non-parametric Estimators for Time Series, Journal of Time Series Analysis, 4, 185-207.

Tiao G.C., Tsay R.S., (1994) Some advances in nonlinear and adaptive modeling in time series, Journal of Forecasting, 13, 109-31.

Tong H. (1983) *Lecture Notes in statistics: Threshold models in nonlinear time series analysis*, Springer–Verlag.

Tong H., (1990) *Non-Linear Time Series: A Dynamical System Approach*, Oxford Publications, Oxford University Press.

Dynamics and Comovements of Regional Exports in Italy[1]

Gianluca Cubadda
Università del Molise, Dipartimento di Scienze Economiche Gestionali e Sociali
e-mail: cubadda@uniroma1.it

Pierluigi Daddi
Università degli Studi di Perugia, Dipartimento di Scienze Statistiche
e-mail: daddi@stat.unipg.it

Abstract: This paper investigates the dynamic behaviour of regional exports in Italy. After aggregating the Italian regions in six macro-areas, we use a recently developed multivariate technique in order to analyse the dynamics and comovements in the relative exports quarterly time-series. Our empirical findings indicate that the north-east exports move separately in the long run from the remaining eastern regions and the western regions are linked by a single long-run relation. Moreover, there is evidence of strong similarities between propagation mechanisms of shocks within central and northern exports respectively.

Keywords: Regional comovements, Seasonal cointegration, Codependence.

1. Motivations

A major theme in recent economic research is the analysis of the factors determining different growth rates and convergence across various geographical areas (see e.g. Barro and Sala-I-Martin, 1992). This paper fits into the framework of literature on the quantification of regional development in Italy and is addressed at a stylised analysis of the dynamics of regional exports, which, as is well-known, represent a generally important component in the definition of the regional product (GDP). Its main aim is to identify the not unimportant aspects of local growth processes which have influenced, at times in a decisive though differentiated way, the development of Italian regions in recent decades.
The research carried out, which considers six macro-areas at the level of territorial disaggregation,[2] has been restricted by the limited availability of the quarterly

---

[1] The authors wish to thank the Italian Ministry of the University and Scientific Research (MURST) for financial support.
[2] In particular, these macro-areas are the North-West, including Valle d'Aosta, Piemonte, Lombardia and Liguria, the North-East, including Trentino, Veneto and Friuli, the Centre-West, including Lazio and Toscana, the Centre-East, including Emilia-Romagna, Umbria and Marche, the South-West, including Campania, Calabria, Basilicata, Sicilia and Sardegna, and the South-East, including Abruzzo, Molise and Puglia.

time series of exports for the years subsequent to 1984. Quarterly time series of logs of regional exports, provided by *Istat*, to cover the period 1984:q1-1997:q4, were used for the empirical analysis on data adjusted for inflation by the same national price index of exports, as price indices of exports are not available on regional level.

In particular, the main contribution is given by an empirical analysis of the characteristics inherent in the dynamics of each area with reference to both long run and short run dynamics, as well as investigating leading and lagging comovements across export time series of the Italian economy. As regards geographical division, the limited number of observations has likewise necessitated a relatively low number of macro-areas. In light of the results of previous studies, in which the hypothesis of dividing the country into centre-north and south has been repeatedly shown to be misleading and simplistic, it has been adopted a classification which also takes into account potential east-west dissimilarities. In this way, we are able to control for the main geographical factors which generate disparities in regional exports dynamics.

2. Methodological issues and empirical results

In order to analyse the dynamics and comovements of macro-regional exports, we relied on a recently developed statistical method (Cubadda, 1999a) which allows for the extension of the common cycles analysis (Vahid and Engle, 1993) to quarterly time-series possessing unit roots both at the zero and seasonal frequencies (see Hylleberg, Engle, Granger and Yoo, henceforth HEGY, 1990). This methodology is briefly summarised below.

Consider the Wold representation of a n-vector of quarterly time series x_t which are integrated of the first order [henceforth, I(1)] at the zero and seasonal frequencies:

$$\Delta_4 x_t = C(L)\varepsilon_t$$

where $\Delta_4 = (1 - L^4)$, L is the usual lag operator, and $C(L)$ is a polynomial matrix such that $C_0 = I_n$ and $\sum_{j=0}^{\infty} j\left|C_j\right| < \infty$.

By proper expansion of $C(L)$, we get the HEGY (1990) decomposition:

$$x_t = \sum_{j=0}^{\infty} \Psi_1 \varepsilon_{t-j} + \sum_{j=0}^{\infty} \Psi_2 \cos(j\pi)\varepsilon_{t-j} +$$

$$\sum_{j=0}^{\infty} \left[\Psi_3 \cos(j\pi/2) + \Psi_4 \sin(j\pi/2)\right]\varepsilon_{t-j} + C^{*}(L)\varepsilon_t \qquad (1)$$

where $\Psi_1 = C(1)/4$, $\Psi_2 = C(-1)/4$, $\Psi_3 = \mathrm{Re}[C(i)]/2$, $\Psi_4 = \mathrm{Im}[C(i)]/2$, and $C_j^* = \sum_{i=1}^{\infty} C_{j+4i}$ for $j \geq 0$.

Equation (1) decomposes the vector x_t to a random walk component or trend, a biannual component which is I(1) at frequency π, an annual component which is I(1) at frequency π/2, and a stationary remainder or cycle.[3]

Let us now focus on the cyclical component. Specifically, consider the following definitions:

Definition 1: Elements of $\Delta_4 x_t$ are codependent of order q, denoted by $\Delta_4 x_t \sim$ CD(q), if there exists a $n \times k$ $(k < n)$ full-rank matrix α such that

$$\alpha' C_j \begin{cases} \neq 0 & \text{if } j = q \\ = 0 & \text{if } j > q \end{cases}$$

The subspace spanned by the matrix α is the codependence space and its dimension, k, is the codependence rank (Gourieroux and Paucelle, 1989).

Definition 2: Elements of x_t have k^* $(k^* < n)$ common cycles in their HEGY decomposition if there exists a $n \times (n - k^*)$ full-rank matrix α^* such that $\alpha^{*\prime} C^*(L) = 0$.

Following Cubadda (1999a), we know that elements of x_t have $(n-k)$ common cycles in their HEGY decomposition if, and only if, the series $\Delta_4 x_t \sim$ CD(q) with $q \leq 3$ and α^* is a codependence matrix of rank k.

If we further assume that the cointegration ranks at frequencies 0, π, and π/2 are respectively r_1, r_2, and r_3 (see, e.g., HEGY 1990; Lee 1992),[4] we can rewrite equation (1) in the following common trend-common seasonals-common cycle representation:

$$x_t = \gamma_1 \tau_t + \gamma_2 b_t + \gamma_3 a_t + \phi c_t \tag{2}$$

where $\tau_t = \sum_{j=0}^{\infty} \delta_1' \varepsilon_{t-j}$, $b_t = \sum_{j=0}^{\infty} \delta_2' \cos(j\pi) \varepsilon_{t-j}$, $a_t = \sum_{j=0}^{\infty} \delta_3' \cos(j\pi/2) \varepsilon_{t-j}$, $c_t = \tilde{c}(L)' \varepsilon_t$, γ_j and δ_j are $n \times r_j$ full-rank matrices for $j = 1, \ldots 3$, ϕ is a full-rank $n \times (n-k)$ matrix such that $\alpha^{*\prime} \phi = 0$, and $\tilde{c}(L)$ is an $n \times (n-k)$ polynomial matrix.

---

[3] Note that term "cycle" has here a different meaning from spectral analysis, see Cubadda (1999b).

[4] Following Lee (1992), we assume that annual cointegration is synchronous.

In order to test for the common components model in model (2), we need to assume that series x_t have the VAR(p) representation $A(L)x_t = \varepsilon_t$. In this case, the presence of cointegration at the various frequencies allows to represent series x_t by the seasonal error-correction model [henceforth, SECM]:

$$x_{4,t} = \beta_1 z_{1,t-1} + \beta_2 z_{2,t-1} + \beta_3 z_{3,t-2} + A^*(L)x_{4,t-1} + \varepsilon_t \tag{3}$$

where $x_{4,t} = \Delta_4 x_t$, $z_{1,t} = \alpha_1'(1+L+L^2+L^3)x_t$, $z_{2,t} = \alpha_2'(1-L+L^2-L^3)x_t$, $z_{3,t} = \alpha_3'(1-L^2)x_t$, $\beta_1\alpha_1' = -A(1)/4$, $\beta_2\alpha_2' = A(-1)/4$, $\beta_3\alpha_3' = A(i)/2$, α_j and β_j are $n \times r_j$ full-rank matrices for $j = 1,\ldots,3$, and $A^*(L)$ is a polynomial matrix of order $(p-4)$.

Building on Johansen (1988), Lee (1992) developed maximum likelihood [henceforth, ML] analysis of model (3). Interestingly, the ML estimates of the α_j's are T-consistent and hence they can be substituted to the true cointegration vectors without affecting the asymptotic distribution of the estimators of other SECM parameters.

Unfortunately, codependence constraints on the SECM coefficients are generally non-linear (Cubadda, 1999a). These constraints become linear only in the special case that series $x_{4,t}$ are CD(0). Notwithstanding this complication, statistical inference on common HEGY cycles can be based on the scalar component model proposed by Tiao and Tsay (1989). Particularly, a test for the hypothesis of k codependence relations of order, at most, q, is a test for k zero canonical correlations between series $x_{4,t}$ and the instruments w_{t-q-1}:

$$\mathrm{C}(q,k) = -(T-p-q)\sum_{j=1}^{k}\ln\left(1 - \frac{v_j(q)}{\sigma_j(q)}\right)$$

where $v_j(q)$ is the $j-$th smallest squared canonical correlation between $x_{4,t}$ and w_{t-q}, w_t are the right-hand side variables in equation (3) where the ML estimates of the α_j's are inserted in place of the unknown parameter values, and $\sigma_j(q)$ is defined is defined as

$$\sigma_j(q) = 1 + 2\sum_{i=l}^{q}\hat{\rho}_i(\hat{\alpha}_x' x_{4,t})\hat{\rho}_i(\hat{\beta}_\omega' \omega_{t-q})$$

where $\hat{\rho}_i(\cdot)$ is the lag $i-$th sample autocorrelation of the series in argument and $\hat{\alpha}_x' x_{4,t}$ and $\hat{\beta}_\omega' \omega_{t-q}$ are the sample canonical variates associated to $v_j(q)$.

Vahid and Engle (1997) noted that an eigenvector $\hat{\alpha}_x$ corresponding to an insignificant $v_j(q)$ is a general method of moments [henceforth, GMM] estimator (Hansen, 1982) of a codependence vector. However, this GMM estimator turns out to be optimal only when series $x_{4,t}$ are CD(0). In case of codependence of order higher than zero, the above eigenvectors can be used as initial estimates in order to compute the optimal GMM estimators, see Cubadda (1999a) for details.

For our proposals, the practical advantage of using the above approach is twofold. First, there is both theoretical reasoning and empirical evidence indicating that the presence of common cycles is not invariant to seasonal adjustment (Hecq, 1988; Cubadda, 1999a). Secondly, the strong seasonal character of macro-regional exports suggests that seasonal shifts in preference and technology may play an important role in the generation of dynamics and comovements of these variables. However, a note of caution is needed in interpreting our statistical results due to some limitations in the data set, such as the limited time span.

Keeping in mind the above *caveat*, our empirical analysis starts with the estimation of a VAR(4) model, including a time trend and seasonal dummies, for the logs of the macro-regional exports.[5] These series are graphed in Figure 1. The VAR order was selected according to the longest significant lag rule. This model was also checked for residual autocorrelation (with the LM test), conditional heteroskedasticity (with the ARCH test), and normality (with the Jarque-Bera test); all these diagnostics gave no indication of misspecification.[6]

Figure 1: *Logs of regional exports at constant prices*

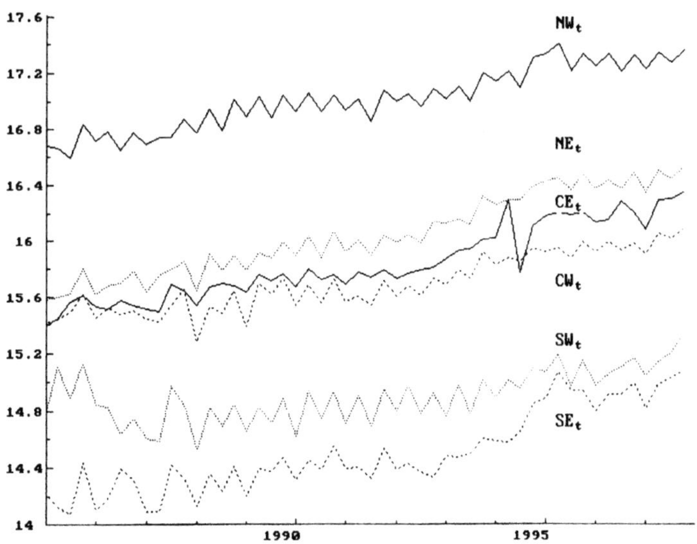

[5] These variables are hereafter indicated, with obvious notation, NW_t, NE_t, CW_t, CE_t, SW_t, and SE_t.

[6] The tests results, not reported in order to save space, are available from the authors upon request.

The results of the Lee's tests for seasonal cointegration suggest the existence of a cointegration rank equal to two at frequency zero, to one at frequency π, and to three at frequency $\pi/2$.[7] It is of some interest to verify whether there are macro-regional exports affected solely by idiosyncratic shocks in the long-run. This can be done by testing for zero-restrictions in the zero-frequency cointegration space, see Johansen (1991). In Table 1, we report the restricted zero-frequency cointegration vectors and the associated error correction terms are graphed in Figure 2. Note that a test for the implied restrictions produced a largely insignificant $\chi^2(5)$ statistic such as 7.71.

Table 1: *Zero-frequency cointegration vectors*

Variable	NW_t	NE_t	CW_t	CE_t	SW_t	SE_t
First vector	0.00	0.00	0.00	1.00	0.00	–0.75
Second vector	–0.74	0.00	1.00	0.00	–0.84	0.00

Figure 2: *Zero-frequency error correction terms*

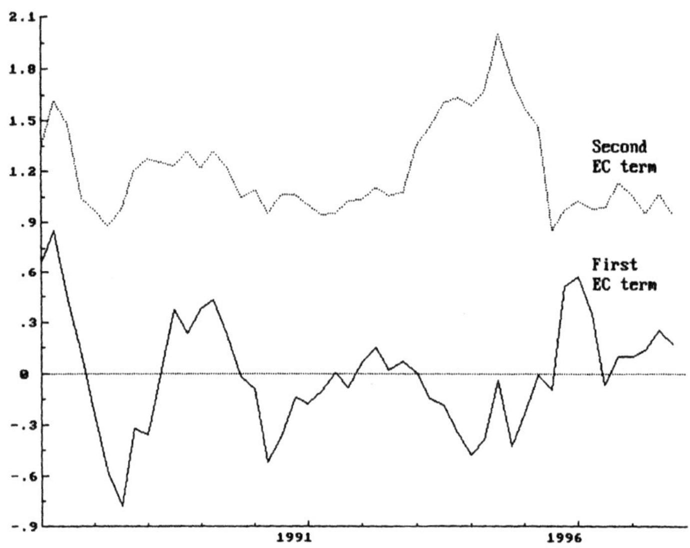

Interestingly enough, the long-run behaviour of NE_t has apparently an idiosyncratic nature since this macro-area has null coefficients in both the cointegration vectors.[8] Moreover, there is evidence of a long-run relation among

[7] Again, the tests results are available upon request.

[8] Indeed, a vector with the only non-null element corresponding to NE_t must belong to the space spanned by the common trends coefficient matrix γ_1 since we know from the Granger representation theorem (Engle and Granger, 1987) that $\alpha_1'\gamma_1 = 0$.

exports of the western regions only, which can be interpreted as a tendency of these regions to keep stable their relative comparative advantages.
Regarding common cycle analysis, application of the test based on canonical correlations indicate that there are six codependence relations of order three. This evidence indicates that the dynamics of Italian regional exports appears to be strongly characterised by the absence of relevant transitory components.
Another interesting result is the finding of two codependence relations of order zero among annual growth rates of regional exports. An estimate of the elements of the codependence vectors, including their standard errors, can be obtained by the GMM. In Table 2, we report these estimates with the imposition of some zero-restrictions as resulting from a general to specific testing procedure. We see that one codependence relation involves solely the northern regions and the other regards the central regions only. Hence, there is evidence of strong similarities between propagation mechanisms of shocks among regional exports within these areas.

Table 2: *Zero-order codependence vectors*

Variable	$\Delta_4 NW_t$	$\Delta_4 NE_t$	$\Delta_4 CW_t$	$\Delta_4 CE_t$	$\Delta_4 SW_t$	$\Delta_4 SE_t$
First vector	–0.72	1.00	0.00	0.00	0.00	0.00
Second vector	0.00	0.00	1.00	–0.94	0.00	0.00

3. Concluding remarks

The main intent of the paper has been a stylised analysis of the regional economy in Italy, through statistical measures of dynamics and comovements across macro-regional exports from a time series perspective. Specifically, we used a recently developed multivariate time-series technique which enabled us to analyse the presence of common cycles when unit roots both at the zero and seasonal frequencies are considered. The main empirical results are that macro-regional export series show some degree of commonality at the zero and seasonal frequencies whereas no significant transitory components have been detected.
Although a 14-years time span is a relative short one to detect common dynamic patterns with a good degree of statistical resolution, some interesting conclusions may still be drawn. Indeed, the idiosyncratic nature of the long-run behaviour of the north-east area, the finding of an equilibrium relation involving exports of western regions only, the uneven shock propagation mechanisms of southern regions, are all further elements of evidence that the dynamic analysis of regional exports indicate a clear non-uniform pattern of development among the Italian macro-areas, even if the economic convergence has been one of the most emphasised and continuos pursued purpose of economic policies for the last decades.

References

Barro R.J., and X. Sala-I-Martin (1992), Convergence, *Journal of Political Economy*, 100, 223-251.

Cubadda G. (1999a), Common Cycles in Seasonal Non-stationary Time Series, *Journal of Applied Econometrics*, 14, 273-291.

Cubadda G. (1999b) Common Serial Correlation and Common Business Cycles: a Cautious Note, *Empirical Economics*, 24, 529-535.

Engle R.F., and Granger C.W.J (1987), Co-integration and Error Correction: Representation, Estimation and Testing, *Econometrica*, 55, 271-276.

Gourieroux, C., and I. Paucelle (1989), Detecting a Long Run Relationship (with an application to the P.P.P. hypothesis), *CEPREMAP Working Paper*, 8902.

Hansen, L. P. (1982), Large Sample Properties of Generalized Method of Moments Estimator, *Econometrica*, 50, 1029-1054.

Hecq, A. (1998), Does Seasonal Adjustment Induce Common Cycles?, *Economics Letters*, 59, 289-297.

Hylleberg S., Engle R.F., Granger C.W.J., and B.S. Yoo (1990), Seasonal Integration and Cointegration, *Journal of Econometrics*, 44, 215-238.

Johansen, S. (1988), Statistical Analysis of Cointegration Vectors, *Journal of Economic Dynamics and Control*, 12, 231-254.

Johansen, S. (1991), Estimation and Hypothesis Testing of Cointegration Vectors in Gaussian Vector Autoregressive Models', *Econometrica*, 59, 1551-1580.

Lee, H. S. (1992), Maximum Likelihood Inference on Cointegration and Seasonal Cointegration, *Journal of Econometrics*, 54, 1-47.

Tiao, G. C., and R. S. Tsay (1989), Model Specification in Multivariate Time Series (with discussion), *Journal of the Royal Statistical Society*, B, 51, 157-213.

Vahid F., and R.F. Engle (1993), Common Trends and Common Cycles, *Journal of Applied Econometrics*, 8, 341-360.

Vahid, F., and R. F. Engle (1997), Codependent Cycles, *Journal of Econometrics*, 80, 199-211.

Large-sample Properties of Neural Estimators in a Regression Model with φ-mixing Errors[1]

Francesco Giordano
Dipartimento di Scienze Economiche, Università degli Studi di Salerno
Sichelgaita S.p.A. Società di Studi Economici e Sociali

Cira Perna
Dipartimento di Scienze Economiche, Università degli Studi di Salerno
Centro di Specializzazione e Ricerche, Portici (Na)

Abstract: In this paper the large sample properties of neural networks estimators in a regression model with φ-mixing errors are investigated. In particular, using the theory of M-estimators, it is proved that the minimum squared error estimators of the connection weights and of the fitted values are consistent and asymptotically Normal

Keywords: Feed-forward Networks, M-estimators, φ-mixing

1. Introduction

Let $\{Y_t\}$, t=1, ..., T be a time series generated according to the model:

$$Y_t=f(\mathbf{X}_t)+e_t \qquad t=1, ..., T \tag{1}$$

where f is a non linear continuous function, $\mathbf{X}_t = (X_{1t}, X_{2t}, ...,X_{dt})$ is d-dimensional vector of non stochastic explanatory variables defined on a compact $\aleph\subset\Re^d$, and $\{e_t\}$ are zero mean random variables with constant variance.
In order to estimate the function f, neural networks have become increasingly popular (Barron, 1993). Of course, to consider them a statistical technique it is necessary for the parameter estimators to verify the properties of convergence in probability and in distribution. In a previous paper (Giordano and Perna, 1998), using the theory of M estimators (Huber, 1981) we have derived the large sample properties of neural estimators in a regression model with i.i.d. errors.
The aim of this paper is to extend the previous results to the case of a regression model with fourth order stationary and φ-mixing errors (Billingsley, 1968).
The paper is organised as follows. In section 2 the use of *feed-forward* neural networks in the non parametric regression model is illustrated. In section 3 the consistency of the neural networks estimators is proved while in section 4 their

[1] The paper is partially supported by MURST '98 "Modelli statistici per l'analisi delle serie temporali"

asymptotic distribution is obtained. In section 5 some concluding remarks are presented. Finally, to make clear the generalisation to the case of φ-mixing errors, the outlines of the proofs for the case of i.i.d. errors are reported in Appendix.

2. Neural networks

In this paper we have considered the following architecture:

$$Y_t^* = \sum_{k=1}^{m} c_k \phi\left(\sum_{j=1}^{d} a_{kj} X_{jt} \right) \tag{2}$$

where c_k is the weight of the link between the *k-th* neuron in the hidden layer and the output; a_{kj} is the weight of the connection between input X_j and the *k-th* neuron in the hidden level. In the formulation (2) the bias term of the hidden layer and that of the output are both zero. Moreover, the activation function ϕ of the input layer and of the hidden layer are supposed to be respectively the standard Normal cumulative function and the identity function.
An estimate of the parameters $\boldsymbol{\theta} = \left(c_1,\dots,c_m,\mathbf{a}_1',\dots,\mathbf{a}_m'\right)$, where $\boldsymbol{a}_k = (a_{k1}, \dots, a_{kd})'$, can be obtained minimising a least squares criterion:

$$\hat{\boldsymbol{\theta}}_T = \underset{\boldsymbol{\theta}\in\Theta}{\arg\min} \frac{1}{T} \sum_{t=1}^{T} \frac{1}{2} \left(Y_t - Y_t^*(\boldsymbol{\theta})\right)^2 \tag{3}$$

where $\Theta = \Re^{m(d+1)}$.
An estimate of Y_t* is thus:

$$\hat{Y}_t = Y^*{}_t(\hat{\boldsymbol{\theta}}) = \sum_{k=1}^{m} \hat{c}_k \phi\left(\sum_{j=1}^{d} \hat{a}_{kj} X_{jt} \right) \tag{4}$$

White (1989), using stochastic approximations, has derived some asymptotic properties of the recursive procedure used to estimate (3) and (4).
In a previous paper (Giordano e Perna, 1998), using the theory of M estimators (Huber, 1981), we have verified the consistency of the estimators $\hat{\boldsymbol{\theta}}$ and $\hat{Y}_t$ and derived their asymptotic distributions in the case of model (1) when the errors are i.i.d. The outlines of the proofs for this case are reported in the appendix.
In the next section we extend these results to the case of a regression model with φ-mixing errors. In order to achieve this result, it is necessary to make some assumptions on the parameter m; throughout this paper we suppose that:

$$m \equiv m(T);\ m(T)\to\infty \text{ per } T\to\infty;\ \ ((m(T))^2/T \to 0 \text{ per } T\to\infty \tag{5}$$

3. The consistency of the estimators $\hat{\theta}$ and $\hat{Y}_t$

Following Huber (1981) for the notation about M estimators, we indicate with $\psi(Y_t;h)$ the derivative of the least squares criterion defined in (3) with respect to h. So it can be shown that for c_r $\forall$r $\in$ {1,...,m}, it is

$$\psi(Y_t;c_r) = \left(Y_t - Y_t^*(\boldsymbol{\theta})\right)\phi\left(\sum_{j=1}^{d} a_{rj} X_{jt}\right) \tag{6}$$

while, for the a_{rs} parameters $\forall$r $\in$ {1,...,m} e $\forall$s $\in$ {1,...,d}, it is

$$\psi(Y_t;a_{rs}) = \frac{1}{\sqrt{2\pi}}\left(Y_t - Y_t^*(\boldsymbol{\theta})\right)c_r \exp\left(-\frac{1}{2}\left(\sum_{j=1}^{d} a_{rj} X_{jt}\right)^2\right) X_{st}\,. \tag{7}$$

Under the hypothesis (5) the following theorems hold.

Theorem 1. Consider model (1) and suppose that the errors e_t are fourth order stationary and φ-mixing. Moreover suppose that $\sum_{n=1}^{\infty}(\varphi(n))^{1/2} < \infty$. Under these hypotheses, the vector estimator $\hat{\boldsymbol{\theta}}$ converges in probability to the parameter θ.

Proof. Because of the hypothesis of stationarity, it is:

$$\mathrm{E}(e_t)=\mu,\ \ \mathrm{Var}(e_t)=\sigma^2 \text{ e } \mathrm{Cov}(e_t, e_s) = \gamma(|t-s|).$$

Without losing in generality we can assume $\mu=0$.
Given that $|\mathrm{E}(e_1 e_k)| \le 2\varphi(k-1)^{1/2}\mathrm{E}(e_1^2)$ and for the hypotheses on the series $\varphi(n)$, we have:

$$\sum_{k=2}^{\infty} |\mathrm{E}(e_1 e_k)| < +\infty$$

From lemma 20.3 of Billingsley (1968), putting $S_n = \sum_{i=1}^{n} e_i$, it is:

$$\frac{1}{n}\mathrm{E}(S_n^2) \to \upsilon^2$$

where:

$$\upsilon^2 = \sigma^2 + 2\sum_{k=1}^{\infty}\gamma(k).$$

From the theorem (1) of Giordano and Perna (1998), (see theorem A1 of the Appendix for an outline of the proof), it is, for a θ_0:

$$\mathrm{Var}\left(\frac{1}{T}\sum_{t=1}^{T}\psi(Y_t;\theta_0)\right) \le \frac{\upsilon^2}{T} + \frac{c_f}{mT}$$

Because of the hypotheses (5) on the parameter m, the last quantity converges to zero for $T \to \infty$ □

Corollary 1. Under the hypotheses of the theorem 1, the estimator $\hat{Y}_t$ converges in probability to $f(\mathbf{X}_t)$.

Proof. The proof follows from theorem 1. Since the random variables $\hat{a}_{kj}$ converge in probability and $\phi(.)$ is a continuous function, then

$$\hat{J}_k = \phi\left(\sum_{j=1}^{d}\hat{a}_{kj}X_{jt}\right)$$

converges in probability to J_k.
The estimator $\hat{Y}_t$ is a linear combination of random variables that converge in probability . It follows, from Barron's result (1993), that $\hat{Y}_t$ converges in probability to $f(\mathbf{X}_t)$. □

4. The asymptotic distributions of the estimators $\hat{\theta}$ e $\hat{Y}_t$

In this section the asymptotic distribution of $\hat{Y}_t$ and the element of vector estimator $\hat{\boldsymbol{\theta}}$, given all other, are derived. In particular, it is proved that both the estimators are asymptotically Normal and unbiased

Theorem 2. Consider the model (1) and suppose that the errors e_t are fourth order stationary and φ-mixing. Moreover suppose that $\sum_{n=1}^{\infty}(\varphi(n))^{1/2} < \infty$ and that

$\int \frac{\partial}{\partial t}\psi(x;t)\Big|_{t=\theta} dF(x) \neq 0$. θ is an element of parameter vector $\boldsymbol{\theta}$. Under these hypotheses the component of estimator (3), given all other, converges in distribution to a Normal distribution with zero mean and variance equal to:

$$\frac{1}{T}\frac{\int \psi^2(x;\theta)dF(x)}{\left[\int \frac{\partial}{\partial t}\psi(x;t)\Big|_{t=\theta} dF(x)\right]^2}$$

Proof. From the theorem (2) of Giordano and Perna (1998) (see theorem A2 of the Appendix for an outline of the proof), it follows that the asymptotic distribution of $\sqrt{T}\left(\hat{\theta}-\theta\right)$is equal to that of $T^{-1/2}\sum_{t=1}^{T}\psi(Y_t,\theta)$.

The random variables $\psi(Y_t,\theta)$ are functions of the random variables e_t and these are stationary and φ-mixing. Because the function φ is continuous, the random variables $\psi(Y_t,\theta)$ are φ'-mixing, where φ' is a suitable transformation of φ. It can be proved that the transformation $\varphi'$is such that $\sum_{n=1}^{\infty}\left(\varphi'(n)\right)^{1/2} < \infty$.

The theorem is proved observing that the random variables can be transformed in zero mean random variables and using the theorem 20.1 in Billingsley (1968). □

Since the random variable $\hat{a}_{ij}$ converges in probability to a_{ij}, in order to derive the asymptotic distribution of $\hat{Y}_t$ it is sufficient to consider the asymptotic distribution of

$$\widetilde{Y}_t = \sum_{k=1}^{m}\hat{c}_k\phi\left(\sum_{j=1}^{d}a_{kj}X_{jt}\right)$$

The random variables $\widetilde{Y}_t$ are linear combinations of Normal random variables and therefore they are asymptotically Normal with

$$\mathrm{E}(\widetilde{Y}_t) = Y_t^*; \tag{8}$$

$$\mathrm{Var}(\widetilde{Y}_t) = \sum_{k=1}^{m}\phi^2\left(\sum_{j=1}^{d}a_{kj}X_{jt}\right)\mathrm{Var}(\hat{c}_k) + \sum_{k\neq h}\phi\left(\sum_{j=1}^{d}a_{kj}X_{jt}\right)\phi\left(\sum_{j=1}^{d}a_{hj}X_{jt}\right)\mathrm{Cov}(\hat{c}_k,\hat{c}_h)$$

Using theorem 2 and the approximation in Barron (1993), it is possible to obtain the following upper bounds:

$$\operatorname{Var}(\hat{c}_k) \le \frac{1}{T} \frac{v^2 + \frac{c_f}{m}}{\left[\int \phi^2 \left(\sum_{j=1}^{d} a_{kj} X_j\right) d\mathbf{X}\right]^2} \tag{9}$$

$$\operatorname{Cov}(\hat{c}_k, \hat{c}_h) \le \frac{1}{T} \frac{v^2 + \frac{c_f}{m}}{\left[\int \phi^2 \left(\sum_{j=1}^{d} a_{kj} X_j\right) d\mathbf{X}\right]\left[\int \phi^2 \left(\sum_{j=1}^{d} a_{hj} X_j\right) d\mathbf{X}\right]} \tag{10}$$

and, from these, an upper bound for $\operatorname{Var}(\tilde{Y}_t)$

5. Concluding remarks

From the previous results it follows that the neural networks estimators are consistent and asymptotically Normal even if the errors of the regression model are φ-mixing and therefore not independent.
With respect to the case of i.i.d. errors, it is necessary to formulate an additional hypothesis concerning the series $\varphi(i)$ i=1, ... ; this condition is necessary to assure the rate of convergence to zero of the function $\varphi(i)$.
From an applicative point of view, the proved properties allow to use neural networks in time series analysis. For example, it is well known that the AR(p) processes are φ-mixing and verify all the conditions of the theorems.

Appendix

Theorem A1. Consider the model (1) and suppose that the errors e_t are i.i.d. random variables with $E(e_t)=0$ and $\operatorname{Var}(e_t)=\sigma^2$. Under these hypotheses the vector estimator $\hat{\theta}_T = \arg\min_{\theta \in \Theta} \frac{1}{T}\sum_{t=1}^{T}\frac{1}{2}\left(Y_t - Y_t^*(\theta)\right)^2$ converges in probability to the parameter θ.

Proof. Following theorem 1 of Giordano and Perna (1998), it is only necessary to prove that

$$\frac{1}{T}\sum_{t=1}^{T}\psi(Y_t;\theta) \tag{A.1}$$

converges in probability to $E\left(\psi(Y_t,\theta)\right)$.

The random variables in (A.1) are independent but not identically distributed.
To prove the convergence we have demonstrated the existence of the second moment of (A.1). From this result it follows the mean square convergence and thus also the probability one.
It results, for a θ_0,

$$\mathrm{Var}\left(\frac{1}{T}\sum_{t=1}^{T}\psi(Y_t;\theta_0)\right) \leq \frac{\sigma^2}{T} + \frac{c_f}{mT} \tag{A.2}$$

in which $c_f=(2rC)^2$
Here r is the radius of the compact $\aleph$ where the vector $\boldsymbol{X}=(X_1, \ldots X_d)$ is defined; $C = \int_{\Re^d} |w| |\tilde{f}(w)| dw$ and $\tilde{f}(w)$ is the Fourier transform of the function f.
The second term in (A.2), under the hypotheses (5) on the parameter m, converges to zero for $T \to \infty$.
Using lemma 7.2.1B in Serfling (1980) it follows that the estimator $\hat{c}_r$ converges in probability to c_r.
Even if the function ψ has a different structural form it is easy to prove that $\hat{a}_{ij}$ converges in probability to a_{ij}. □

Remark 1. In the previous theorem we refer to the existence of an optimal solution to the problem of determining of the parameters of the neural network and not to the algorithm for their determination.

Theorem A2 Consider the model (1) and suppose that the errors e_t are are i.i.d. random variables with $\mathrm{E}(e_t)=0$ and $\mathrm{Var}(e_t)=\sigma^2$ and φ-mixing. Moreover suppose that $\sum_{n=1}^{\infty}(\varphi(n))^{1/2} < \infty$ and that $\int \frac{\partial}{\partial t}\psi(x;t)\Big|_{t=\theta} dF(x) \neq 0$, where θ is an element of parameter vector $\boldsymbol{\theta}$. Under these hypotheses the component of vector estimator $\hat{\theta}_T = \arg\min_{\theta \in \Theta} \frac{1}{T}\sum_{t=1}^{T}\frac{1}{2}\left(Y_t - Y_t^*(\theta)\right)^2$, given all other, converges in distribution to a Normal distribution with zero mean and variance equal to:

$$\frac{1}{T}\frac{\int \psi^2(x;\theta)dF(x)}{\left[\int \frac{\partial}{\partial t}\psi(x;t)\Big|_{t=\theta} dF(x)\right]^2}$$

Proof. Under the hypotheses and using the theorem 7.2.2B in Serfling (1980) it follows that:

$$\sqrt{T}\left(\hat{\theta}-\theta\right)=-\frac{A_T}{B_T}$$

where $A_T = T^{-1/2}\sum_{t=1}^{T}\psi(Y_t,\theta)$ and $B_T = T^{-1}\sum_{t=1}^{T}\frac{\partial}{\partial t}\psi(Y_t,c)\Big|_{c=\tilde{\theta}}$ with $\left|\theta-\tilde{\theta}\right|\le\left|\theta-\hat{\theta}\right|$.

Put $B_{T'} = T^{-1}\sum_{t=1}^{T}\frac{\partial}{\partial t}\psi(Y_t,c)\Big|_{c=\theta}$.

The asymptotic distribution of $\sqrt{T}\left(\hat{\theta}-\theta\right)$ is the same as the distribution of the ratio A_T/B_T.

It is easy to show that B_T converges to $B_{T'}$ so that, in order to derive the distribution, it can be considered as a constant. It follows that to derive the asymptotic distribution of $\sqrt{T}\left(\hat{\theta}-\theta\right)$ it is necessary to derive the asymptotic distribution of $T^{-1/2}\sum_{t=1}^{T}\psi(Y_t,\theta)$. Under the hypotheses of the theorem the random variables $\psi(Y_t,\theta)$ are independent but not identically distributed. Using the Liapunov theorem, the conclusion holds □

Remark 2. In the previous theorem, the distribution function *F(x)* is that of a Uniform random variable *x* being a deterministic variable.

References

Barron, A.R. (1993) Universal Approximation Bounds for Superpositions of a Sigmoidal Function, *IEEE Transactions on Information Theory*, 39, No. 3, 930-945.

Billingsley, P. (1968) *Convergence of Probability Measures*, J. Wiley & Sons, New-York

Giordano, F. Perna, C. (1998) Proprietà asintotiche degli stimatori neurali nella regressione non parametrica, Atti della XXXIX Riunione Scientifica S.I.S.

Huber, P. (1981) *Robust Statistics*, J. Wiley & Sons, New-York.

Serfling, R. (1980) *Approximation theorems of mathematical statistics*, J. Wiley & Sons, New-York.

White, A. (1989) Some Asymptotic Results for Learning in Single Hidden-Layer Feedforward Network Models, *Journal of the American Statistical Association*, 84, No. 408, 1003-1013.

Subseries Length in MBB Procedure for α-mixing Processes[(1)]

Michele La Rocca, Cosimo Vitale
DISE, Università di Salerno, e-mail: larocca@dise.unisa.it

Abstract: In this paper we propose a new procedure to determine the length of the subseries in the MBB bootstrap which takes into account the structure of the model and the strength of dependence in the observed series. It can be easily implemented and easily extended to much more complex structures: multivariate ARMA processes; non linear models; STARMA processes.

Keywords: Bootstrap, Time series, α-mixing processes, Subseries.

1. Introduction

In their classical form, as first proposed by Efron (1979), bootstrap methods are designed for application to samples of independent data. Extensions of the bootstrap to dependent data are not straightforward and modifications of the original procedures are needed in order to preserve the dependence structure of the original data in the bootstrap samples. In the context of stationary time series a number of nonparametric, purely model free bootstrap schemes have been proposed. In such procedures blocks of length k of consecutive observations are resampled randomly, with replacement, from the original time series and assembled by joining the blocks together in random order, to obtain a simulated version of the original process. These approaches, known as moving block bootstrap, generally work satisfactory and enjoy the properties of being robust against misspecified models. The only assumptions needed are suitable mixing and moment conditions for the observations (Politis and Romano, 1992 *inter alia)*. However, these procedures depend on the block-length k, to be chosen by the user, which has to increase with the sample size N. Künsch (1989) showed that, for $N \to \infty$ and $k \to \infty$, it should be

$$k / N \to 0. \tag{1}$$

These conditions, however, are too general to be used from an operational point of view to fix k. Scherman and Carlstein (1996) and Scherman (1997), in the MBB context, showed that $k = \left[cN^{\gamma}\right]$ where $c > 0$ and $\gamma \in (0,1)$ with $[x]$ denoting the closest integer to x. The constant c is a function of the statistics of

(1) The present paper is financially supported by MURST98 "Modelli Statistici per l'Analisi delle Serie Temporali"

interest and of the dependence among the observations (Davison and Hinkley, 1997). Common proposals, on some empirical base, are to choose $c = 1$ or $c = 2$. More refined schemes are based on quite complex iterative algorithms. A data driven choice of k has been proposed by Hall *et al.* (1995). They consider the performance of different block lengths for subsamples of length $m < N$ and then extrapolate the value of k for the given sample. In any case the tuning constant m has to be fixed somehow. More recently Buhlmann and Kunsch (1999) proposed an alternative approach based on the equivalence of the blockwise bootstrap variance to a lag weight estimator of a spectral density at the origin. In this equivalence the block length is the inverse of the bandwidth. Their approach is quite general but requires the knowledge of the influence function of the statistics considered.

In this paper we focus on the selection of a value for k which explicitly reflects the dependence among the observations and propose a procedure to choose k in the MBB context. The new procedure can be easily generalised to general α-mixing stochastic random fields including, as particular cases, non linear temporal processes and spatial processes. In section 2 we emphasize the importance of the strength of dependence on the choice of the length k of the subseries in the MBB. Then, we propose an alternative methodology, which explicitly takes into account the maximum degree of dependence existing among the random variables composing the given stochastic process, and discuss some of its properties. In section 3 some results of a simulation study are presented and discussed. Some concluding remarks and extensions will close the paper.

2. Strength of dependence and length of subseries

Given the stationary process $\{X_t, t = \pm 1, \pm 2, \ldots\}$, the strength of dependence is measured by the α-mixing coefficients defined as

$$\alpha(k) = \sup_k \{|\Pr(A \cap B) - \Pr(A)\Pr(B)|\}$$

where the events subspaces A and B belong, respectively to the minimal σ-algebra generated by $\{\ldots, X_{-1}, X_0\}$ and $\{X_k, X_{k+1}, \ldots\}$. If $\alpha(k) \to 0$ for $k \to \infty$, $\{X_t\}$ is said strong α-mixing. If $\{X_t\}$ is stationary and strong α-mixing, subseries of proper chosen length will contain all the information on the temporal dependence of the process.

Let $(x_1, x_2, \ldots, x_N)$ be a time series generated by the stationary, strong α-mixing, process $\{X_t\}$. The problem is to choose k so that the subseries of length k contain all the dependence of $\{X_t\}$. Hence, these subseries can be considered valid copies of $\{X_t\}$ and used to make inference in the moving block bootstrap (MBB) framework. A value for k is usually chosen without considering explicitly the strength of dependence among the random variables of the process as previously seen. Scherman and Carlstein (1996) admit the presence of this relationship but, using an approach based on Edgeworth expansions up to the second order term, they argue, from the results of some illustrative examples, that the value of k is basically not influenced by the strength of dependence in $\{X_t\}$, when k is chosen to minimise the MSE of the parameter estimates.

On the contrary this dependence is present, cannot be neglected and can be easily exploited to improve blockwise bootstrap performance. Let $\{X_t\}$ be a gaussian, stationary and invertible process with $ARMA(p,q)$ structure, $\phi(B)X_t = \theta(B)a_t$ with $a_t \sim WN(0,\sigma^2)$. In this case only linear dependence, measured by the autocorrelation function, is present in the process. By simply looking at the model structure it should be $k \geq p+q+1$. Indeed, if $p=q=0$ we have $X_t \equiv a_t$ and the bootstrap procedure can be directly used on the original single observations. If $p+q>0$ each subseries should contain at least $p+q+1$ terms in order to estimate its $p+q+1$ parameters.

As regards to the strength of dependence, it is well known that a Gaussian MA(q) process is q-dependent, hence $\alpha(k)=0$ for $k>q$. On the other hand, a stationary AR(p) is equivalent to an MA(∞) process, but it is possible to find $q>p$ so that AR(p)≈MA(q), up to a given level of approximation and again it is $k>q$. Accuracy is a function of the strength of dependence in the AR(p) process: the bigger is the strength the higher is q. If $X_t \sim ARMA(p,q)$, it is dominated by the AR structure of $\{X_t\}$. Let $\{B_i, i=1,2,\ldots,p\}$, be the p roots of the characteristic polynomial $\phi(B)=0$. The model can be equivalently written as

$$\prod_{i=1}^{p}(1-w_iB)X_t = \theta(B)a_t$$

and its maximum characteristic root, $\phi = \max|1/B_i| = \max|w_i| < 1$, will dominate the strength of dependence.

Hence, we propose to model the maximum strength dependence of $\{X_t\}$ by $(1-\phi B)X_t = e_t$ which is AR(1). For such a process we have that $\hat{\rho}(k)=\hat{\phi}^k$ where $\rho(k)$ is the autocorrelation function at lag k. In the particular case that $X_t = a_t$, $\mathrm{var}(\hat{\rho}(k)) = \mathrm{var}(\hat{\phi}^k) \approx 1/N$. So, the dependence is negligible at the level α, taking into account the ARMA structure of $\{X_t\}$, if $\hat{\phi}^{k-(p+q+1)} \leq z_{\alpha/2}/\sqrt{N}$ with $z_{\alpha/2}$ being the percentile of the standard normal and so

$$k \geq -\frac{\log\sqrt{N} - \log(z_{\alpha/2})}{\log\phi} + (p+q+1). \qquad (2)$$

If k respects the equality in the condition (2), condition (1) is always verified. Moreover, if $\phi \to 0$ then $k \to q+1$, which is consistent with a q-dependent process. If $\phi \to 0$ and $q = 0$, then $k \to 1$, as expected. On the other hand, if $\phi \to 1$ (the process is non-stationary and not strong α-mixing) then $k \to \infty$ and the MBB procedure cannot be used to make inference, as shown in Lahiri (1993).

If a_t is *iid* but not gaussian, in the process we find linear and non linear dependence. But, to have an ergodic and strong stationary process, linear dependence should dominate the other (see Tjøstheim, 1994) and the above considerations still apply.

If $\{X_t\}$ is strong stationary and ergodic with $X_t = f(X_{t-1},\ldots,X_{t-d}) + a_t$, where $f(\cdot)$ is some nonlinear function and a_t is *iid* and gaussian, the maximum Lyapunov exponent, λ, can be computed (which is, along with its estimate, invariant to initial conditions). In the linear case we have $\lambda = \log(\phi) < 0$. This condition, with analogous considerations of the non-gaussian case, should still be valid in a non-linear context in order to have stationary, ergodic and non-chaotic processes. In this framework we can consider the relation

$$k \geq -\frac{\log\sqrt{N} - \log(z_{\alpha/2})}{\lambda} + (d+1)$$

All together, we state that an operational procedure to select *k* is the following: (*i*) use a unit root test to operate on the level series X_t or on the differenced series ∇X_t; (*ii*) fit a linear (or a non linear) model to X_t or to ∇X_t; (*iii*) compute ϕ (or λ) and choose *k*; (*iv*) use *k*, resulting from the equality, as the subseries length in the MBB for the bootstrap nonparametric inference. This latter is performed as follows. Assume we want to estimate the standard error of a real valued statistics $T_N = T(X_1,\ldots,X_N)$. Let $b = [N/k]$. Resample blocks of length k from the original data, i.e. form the bootstrap sample $X^*_{(j-1)k+t} = X_{S_j+t}$ with $1 \leq j \leq b$ and $1 \leq t \leq k$ where $S_1,\ldots,S_b$ are *iid* uniform random variables on $\{1,\ldots,N-k\}$. Then, consider the bootstrap counterpart of the statistics T_N, $T^*_N = T(X^*_1,\ldots,X^*_N)$ and $\hat{\sigma}_* = \sqrt{\text{var}_*(T^*_N)}$ with $\text{var}_*(\cdot)$ denoting the variance of T^*_N conditional on the observed series.

Observe that here the model is used only to estimate the strength of dependence in the series and the procedure remains truly non-parametric in its spirit.

3. Simulations

To assess the performance of the proposed procedure a simulation study was conducted. The data were generated according to an AR(1) model, $X_t = \phi X_{t-1} + \varepsilon_t$ with ε_t *iid* $N(0,\sigma^2)$. We also considered non-normal errors generated according to a Student-t distribution with 6 degrees of freedom. The parameter ϕ varies in the range $[0.1, 0.9]$, in order to cover different strengths of dependence, and the scaling is such that $\mathrm{var}(X_t)=1$. In addition to our method (NM in the following) we also considered the basic, simple choice $k=\lfloor N^{1/3}\rfloor$ (Basic in the following). We investigate the procedure's performance for the case of the mean and for the median (a non-linear statistics). We choose the sample sizes $N=128$ and $N=512$. All the results are based on 1000 Monte Carlo runs and 500 bootstrap replicates. The true standard error $\sigma_N=\sqrt{N\,\mathrm{var}(T_N)}$ was estimated with 1000 simulation runs. We estimated the MSE as $E\left[\left(\sqrt{N}\,\hat{\sigma}_* - \sigma_N\right)^2\right]$ and in Table 1 we reported the efficiency (computed as MSE(Basic)/MSE(NM)) for the different models and for different sample sizes.

Table 1: *Efficiency of the proposed method with respect to the basic choice*

	Mean				Median			
	Normal		*Student-t*		*Normal*		*Student-t*	
ϕ	*N*=128	*N*=512	*N*=128	*N*=512	*N*=128	*N*=512	*N*=128	*N*=512
0.1	1.463	1.419	1.538	1.467	1.130	1.166	1.221	1.190
0.2	1.112	1.081	1.249	1.128	1.097	1.160	1.036	1.169
0.3	0.997	0.859	1.153	0.991	1.135	0.950	1.025	1.211
0.4	0.894	0.688	1.027	0.807	1.057	0.911	0.984	1.047
0.5	0.952	0.760	1.043	0.779	0.993	1.057	1.051	0.970
0.6	1.000	0.817	1.000	0.890	1.000	0.923	1.000	0.969
0.7	1.115	1.178	1.168	1.204	1.035	1.123	1.08	1.093
0.8	1.333	1.726	1.397	1.668	1.253	1.395	1.215	1.466
0.9	1.662	3.088	1.648	2.894	1.495	2.463	1.452	2.447

Table 2: *Comparison of block length choices to estimate $\sigma_N=\sqrt{N\,\mathrm{var}(T_N)}$ for the sampling mean.*

Model	N	Method	σ_N	Bias	SD	RMSE
AR(1)	128	NM	3.0716	-0.9938	0.4612	0.1272 (0.0083)
		OPT	-	-0.8701	0.5882	0.1169 (0.0093)
		Basic	-	-1.3067	0.3073	0.1910 (0.0074)
	512	NM	2.9406	-0.5418	0.2975	0.0442 (0.0017)
		OPT	-	-0.3435	0.3935	0.0316 (0.0016)
		Basic	-	-0.7836	0.2148	0.0763 (0.0017)
AR(2)	128	NM	1.4549	-0.0550	0.2508	0.0312 (0.0035)
		OPT	-	0.0133	0.2412	0.0276 (0.0035)
		Basic	-	0.1735	0.2194	0.0370 (0.0044)
	512	NM	1.3791	0.0531	0.1385	0.0116 (0.0008)
		OPT	-	0.0625	0.1453	0.0132 (0.0009)
		Basic	-	0.1803	0.1189	0.0245 (0.0011)

The new method outperforms the basic choice in most cases, especially for weak or strong dependence, for non-normal errors and for non-linear statistics. The basic choice works quite well for intermediate strength of dependence, in presence of normal errors and for linear statistics.

Table 3: *Comparison of block length choices to estimate* $\sigma_N = \sqrt{N \,\mathrm{var}(T_N)}$ *for the median*

Model	N	Method	σ_N	Bias	SD	RMSE
AR(1)	128	NM	3.2027	-0.8560	0.6700	0.1152 (0.0093)
		OPT	-	-0.7048	0.7948	0.1100 (0.0100)
		Basic		-1.1645	0.5341	0.1600 (0.0098)
	512	NM	3.0923	-0.4900	0.4620	0.0474 (0.0021)
		OPT	-	-0.2992	0.5256	0.0383 (0.0021)
		Basic	-	-0.7054	0.3666	0.0661 (0.0024)
AR(2)	128	NM	1.6977	-0.0288	0.4447	0.0689 (0.0085)
		OPT	-	0.0417	0.4417	0.0683 (0.0097)
		Basic	-	0.1882	0.4152	0.0721 (0.0108)
	512	NM	1.6776	0.0280	0.2919	0.0306 (0.0020)
		OPT	-	0.0307	0.2870	0.0296 (0.0020)
		N^1/3	-	0.1494	0.2921	0.0382 (0.0028)

We also considered as point reference for the block length the asymptotic optimal value (OPT in the following) determined by the procedure suggested by Buhlman and Kunsch (1999). Of course this value is not available for real data, where the true model is unknown, but it is interesting to evaluate how far our results are from those optimal ones. The data were generated according to an AR(1) model, $X_t = 0.8\,X_{t-1} + \varepsilon_t$ with ε_t *iid* $N(0, 0.36)$, and to an AR(2) model, $X_t = 1.372\,X_{t-1} - 0.677\,X_{t-2} + \varepsilon_t$ with ε_t *iid* $N(0, 0.179)$. The two models were scaled in order to get $\mathrm{var}(X_t) = 1$. In Table 2 and Table 3 we reported the bias, i.e. $E\left(\sqrt{N}\,\hat{\sigma}_*\right) - \sigma_N$, the standard deviation, i.e. $SD = \sqrt{\mathrm{var}\left(\sqrt{N}\,\hat{\sigma}_*\right)}$ and the relative mean-square errors (RMSE), i.e. $E\left[\left(\sqrt{N}\,\hat{\sigma}_* - \sigma_N\right)^2\right] / \sigma_N^2$ for the two different models, for different sample sizes and for the three different choices of the block length. In figures 1 and 2 we reported box-plots for $\sqrt{N}\,\hat{\sigma}_* / \sigma_N$ as well. In all the cases considered the results obtained with the NM approach are somewhat between the basic and the optimal choice. In some cases, the AR(2) model, they are surprisingly near the optimal ones.

Hence, in both models the bootstrap standard errors estimates obtained using MBB with a block-length k chosen according to our procedure are competitive with, and often better than, the basic choice and quite similar to the optimal procedure considered here (unavailable for real data sets).

Figure 1: *Box-plots of* $\sqrt{N}\,\hat{\sigma}_* / \sigma_N$ *for the mean for the AR(1) and AR(2) model with sample sizes N=128 and N=512*

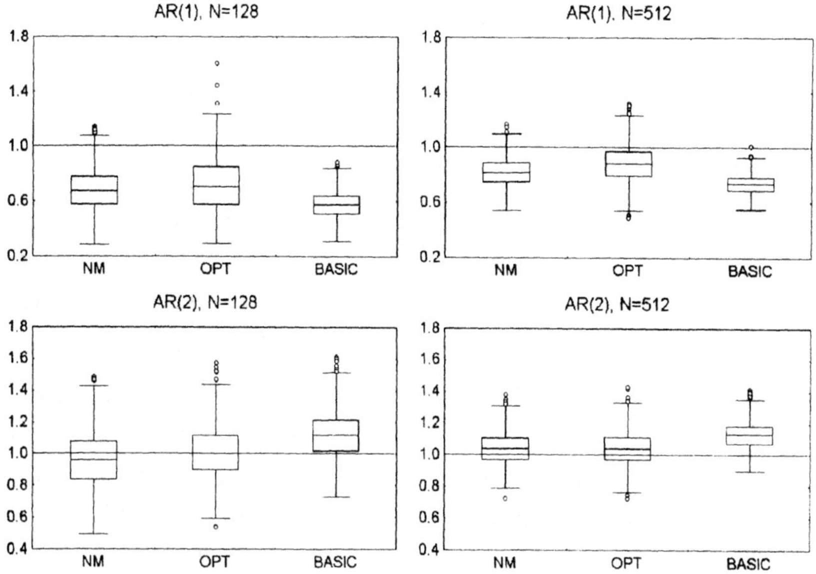

Figure 2: *Box-plots of* $\sqrt{N}\,\hat{\sigma}_* / \sigma_N$ *for the median for the AR(1) and AR(2) model with sample sizes N=128 and N=512*

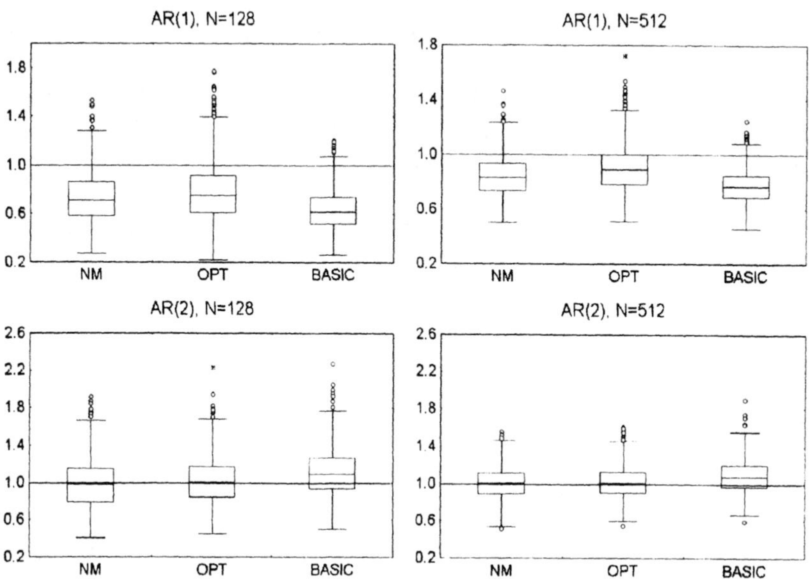

4. Concluding remarks and extensions

In this paper a new procedure to determine the length of the subseries in the MBB bootstrap has been proposed. This procedure takes into account explicitly the strength of dependence which is present in the observed series, it can be easily implemented (no numerical iterative procedure is required) and it can be easily extended to much more complex structures than the ones considered here. Similar arguments can be used to extend the proposed approach to multivariate ARMA processes and to univariate and multivariate STARMA processes.

The procedure used to choose k in the MBB can be easily extended to the subseries block-bootstrap case (here k is the number of observations to be deleted when forming the blocks), as well. Let k_n be a real number such that $\phi^{k_n} \leq z_{\alpha/2}/\sqrt{N}$ in the whole observed time series $(x_1, ..., x_N)$ and let k_l be a real number such that $\phi^{k_l} \leq z_{\alpha/2}/\sqrt{l}$ in the subseries of length l $(x_{h+1}, ..., x_{h+l})$. We immediately have $\phi^{2(k_n - k_l)} \leq l/N \leq 1$ and $N\phi^{2(k_n - k_l)} \leq l \leq N$ If we choose $k_n - k_l = 1$ corresponding to the minimum loss of dependence in the subseries with respect the original whole series of length N, we have $N\phi^2 \leq l \leq N$. Let's observe that if $\phi = 0$ the process is *iid* and, consistently, we obtain $l \geq 0$. In this case, it is reasonable to choose $l = 1$, obtaining the classical bootstrap approach for the *iid* case. If $\phi = 1$ the process is non stationary and so it is not α-mixing, consistently we obtain $l = N$, and no resampling procedure is feasible.

References

Efron, B. (1979) Bootstrap methods: another look at the jackknife, *The Annals of Statistics*, vol. 7, 1-26.

Buhlman, P. and Kunsch, H. R. (1999) Block length selection in the bootstrap for time series, *Computational Statistics and Data Analysis*, 31, 295-310.

Davison A. C., Hinkley D. V. (1997) *Bootstrap Methods and their Application*, Cambridge University Press.

Hall P., Horowitz J.L., Jing B. (1995) On blocking rules for the block bootstrap with dependent data, *Biometrika*, 82, 561-574.

Künsch H.R. (1989) The jackknife and the bootstrap for general stationary observations, *The Annals of Statistics*, 17, 1217-1241.

Lahiri S.N. (1993) On the moving block bootstrap under long range dependence, *Statistics & Probability Letters*, 18, 405-413.

Sherman M. (1997) Subseries methods in regression, *Journal of the American Statistical Association*, 92, 1041-1048.

Sherman M., Carlstein E. (1996) Replicate histogram, *Journal of the American Statistical Association*, 91, 566-576.

Tjøstheim D. (1994) Non-linear time series: a selective review, *Scandinavian Journal of Statistics*, 21, 97-130.

Modelling a Change of Classification by a Structural Time Series Approach

Filippo Moauro
ISTAT, Dipartimento di Contabilità Nazionale e Analisi Economica

Abstract: The change of classification problem for economic sectoral time series data is examined by a conversion matrix approach. A state space form for data reconstruction by structural time series models is proposed. The Doran (1992) methodology of constraining the Kalman filter to satisfy time varying restrictions is applied to increase efficiency of the estimates. Results of an application on Italian Quarterly Accounts are discussed.

Keywords: Sectoral Time Series Data, Conversion Matrix, State Space Models, Time Varying Constraints, Kalman Filter, Smoothed Estimates.

1. Introduction

In this paper a framework for a conversion of sectoral time series from old to new classification standards is provided. This is based on a conversion matrix to represent time-varying compositional effects among different sectoral standards.
This is a typical problem of data production which Statistical Agencies should face when new classification standards are introduced. An example is given by the European System of National Accounts, which has been recently revised from ESA79 to ESA95. National Accounts has been totally refounded and retrapolating techniques are broadly applied to gain long and coherent estimates.
A powerful instrument to handle missing observations and data reconstruction is given by the Kalman filter (KF hereinafter): see, for example, Jones (1980), Harvey (1990), Gomez and Maravall (1994), Dagum et al. (1998). In this respect the Doran (1992) proposal of constraining the KF to obey time varying restrictions could increase efficiency of smoothed estimates.
Here a new approach of backward calculation is suggested. A state space form is set up considering the new sectoral standards figures to be reconstructed as unobserved and the few available observations as time-varying restrictions. In this terms the dependent variables are given by the fully observed old sectoral standards figures, supposed to be generated by structural models with components which include in additive form the new standard ones to be reconstructed.
The plan of the paper is as follows: in section 2 the problem of the change of classification is introduced by a formal definition of the conversion matrix; section 3 provides the state space form to handle data reconstruction; section 4 shows the results of an application to perform a backward calculation of energy accounts. Finally, conclusions are briefly discussed in section 5.

2. The Conversion Matrix

Let x_t^* denote a *m-vector* measuring an elementary economic variable at time t. It is assumed that x_t^* is evenly sampled at a given frequency and it is available for the time period $t = 1, 2, ..., T$. Contemporaneous aggregation of x_t^* into n $(n<m)$ sectors is obtained defining a binary *0/1 (n×m) aggregation matrix A* such that $y_t = Ax_t^*$. y_t represents an *n-vector* of aggregated data in n sectors. A is a full row rank matrix where every column sums to unity and it is unique for each classification standard. A scalar sum Y_t of the elements of y_t is also defined.
In these terms, a different standard means a new $(p \times m)$ aggregation matrix B $(p<m)$, such that $z_t = Bx_t^*$, where z_t represents the new *p-vector* of aggregated data in p sectors. As Y_t, a scalar Z_t as sum of the elements of z_t can be defined. Then, a *(p×n) conversion matrix* C_t is defined through the following equations:

$$z_t = a_t + C_t y_t, \tag{1}$$

$$C_t = Bx_t^{*d} A' \left(Ax_t^{*d} A'\right)^{-1}, \tag{2}$$

where x_t^{*d} is an *m-diagonal matrix* with $x_{1t}^*, x_{2t}^*, ..., x_{mt}^*$ as diagonal terms.
The *p-vector* a_t summarizes the effects on the system given by situations in which classification changes are accompanied by inclusion of new products and definition changes of relevant variables. In this form the sum over the p elements of a_t represents the difference between Y_t and Z_t.
C_t is a full column rank matrix and, as for A and B, every column sums to 1. Each element of C_t, c_t^{ij} $(i=1,2,..,p$ and $j=1,2,..,n)$ is bounded between zero and one.
Note that each element c_t^{ij} of C_t gives a *transition weight* among sectors of different classifications. If the definition of the *i-th* new sector (in terms of joined economic activities) is exactly the same as the *j-th* sector of the old classification, then $c_t^{ij}=1$; if there is a *split* from the *j-th* old sector into more than one new sector, then $0< c_t^{ij} <1$; finally, if there are no linkages $c_t^{ij}=0$. Thus, for every i, j such that $c_t^{ij}=0$ or $c_t^{ij}=1$, this holds for all t, $(t=1,2,...T)$.
Such an analytical framework provides a deterministic representation of compositional effects that a change of sectoral classification involves. It seems reliable for cases in which detailed data sets are observed and it represents a good reference when a data reconstruction has to be performed.

3. The State Space Form

Let the vector y_t be observed S times each year for T periods $(t = 1, 2, ..., T)$, whereas the vector z_t be available with the same frequency but only for the last l periods: an estimation of the first T-l observations of z_{it} $(i=1,2,..,p)$ can be obtained by applying the KF to the following state space model:

$$y_t^i = G_i\alpha_t^i + \varepsilon_t^i, \tag{3}$$

$$\alpha_{t+1}^i = T_i\alpha_t^i + v_{t+1}^i, \tag{4}$$

where y_t^i is a k_i-*vector* representing the most convenient selection from y_t in terms of conversion towards z_{it}; G_i and T_i are, respectively, $(k_i \times r)$ and $(r \times r)$ matrices of parameters and α_t^i is the $(r \times 1)$ state vector. Assumptions are that y_{jt}^i ($j = 1, .., k_i$) and z_{it} are generated by structural time series models such that

$$y_{jt}^i = \left(\mu_t^j + \mu_t^z\right) + \left(\gamma_t^j + \gamma_t^z\right) + \left(\varepsilon_t^j + \varepsilon_t^z\right), \tag{5}$$

$$z_{it} = \mu_t^z + \gamma_t^z + \varepsilon_t^z, \tag{6}$$

where $(\mu_t^j + \mu_t^z)$, $(\gamma_t^j + \gamma_t^z)$ and $(\varepsilon_t^j + \varepsilon_t^z)$ represent, respectively, the trend, the seasonal and the error component of y_{jt}^i. Such representation is extremely convenient to obtain a reconstruction of z_{it} because its structural components, i.e. μ_t^z, γ_t^z and ε_t^z, are modelled as additive terms of the y_{jt}^i's. A choice of models is available for μ_t^k, γ_t^k and ε_t^k $(k=j, z)$ of equations (5) and (6). In this paper we use a basic one, i.e.

$$\mu_t^k = \mu_{t-1}^k + \beta_{t-1}^k + \xi_t^k, \qquad \xi_t^k \sim \left(0, \sigma_{\xi^k}^2\right) \tag{7}$$

$$\beta_t^k = \beta_{t-1}^k + \eta_t^k, \qquad \eta_t^k \sim \left(0, \sigma_{\eta^k}^2\right) \tag{8}$$

$$\gamma_t^k = -\sum_{s=1}^{S-1} \gamma_{t-s}^k + \omega_t^k, \qquad \omega_t^k \sim \left(0, \sigma_{\omega^k}^2\right) \tag{9}$$

$$\varepsilon_t^k \sim \left(0, \sigma_{\varepsilon^k}^2\right) \text{ for } k=j, z$$

This form correspond to the Harvey's basic structural model with dummy variable seasonality. In terms of *SARIMA* reduced form, it implies a *(0, 1, 1)(0, 1, 1)* model which is a quite general formulation for many economic time series. Alternatively, fixed slope and/or trigonometric seasonality can be considered by modifying equations (7)-(9) following Harvey (1990)[1].

In terms of the state space form of equations (3)-(4), the assumptions above involve the following results: the dimension of the state vector is $r=(k_i+1)(S+1)+1$ and α_t^i, ε_t^i, v_t^i, G_i and T_i are given, respectively, by:

$$\alpha_t^i = \left(\mu_t^1 .. \mu_t^{k_i} \mu_t^z \beta_t^1 .. \beta_t^{k_i} \beta_t^z \gamma_t^1 .. \gamma_{t-S+2}^1 .. \gamma_t^{k_i} .. \gamma_{t-S+2}^{k_i} \gamma_t^z .. \gamma_{t-S+2}^z \; \varepsilon_t^z\right)', \tag{11}$$

$$\varepsilon_t^i = \left(\varepsilon_t^1 \varepsilon_t^2 .. \varepsilon_t^{k_i}\right)', \tag{12}$$

---

[1] Common trend and seasonal components may even be included into this framework by an appropriate reduction of the dimension of the state vector of equation (11) and considering the factor loading coefficients in order to modify the matrix G_i.

$$v_t^i = \left(\xi_t^1 .. \xi_t^{k_i} \xi_t^z \eta_t^1 .. \eta_t^{k_i} \eta_t^z \omega_t^1 ..0.. \omega_t^{k_i} ..0\, \omega_t^z ..0\, \varepsilon_t^z\right)', \tag{13}$$

$$G_i = \left[I_{k_i}, 1_{k_i,1}, 0_{k_i,(k_i+1)}, I_{k_i} \otimes e_{S-1}, 1_{k_i} \otimes e_{S-1}, 1_{k_i,1}\right], \tag{14}$$

$$T_i = diag\left(\begin{bmatrix} I_{k_i+1} & I_{k_i+1} \\ 0_{(k_i+1),(k_i+1)} & I_{k_i+1} \end{bmatrix}, I_{k_i+1} \otimes \begin{bmatrix} \text{-1 -1} \dots \text{-1} \\ I_{S-2}, 0_{(S-2),1} \end{bmatrix}, 0\right), \tag{15}$$

where I_g denotes the *g-identity* matrix, $0_{g,f}$ and $1_{g,f}$ $(g{\times}f)$ matrices of *zeros* or *ones* respectively, $\otimes$ the *Kronecker product* and e_g the first row of the *g-identity* matrix. Following Doran (1992, 1996), the last l observed values of z_{it} are incorporated into the model as linear time-varying restrictions on α_t^i in the form $R\alpha_t^i = z_{it}$, where R is the appropriate selection matrix to pick up from the state vector α_t^i the structural components of z_{it}. Applying the method to this case consists in augmenting at time $t = T\text{-}l\text{+}1, \dots, T$ the measurement equation (3) by $z_{it} = R\alpha_t^i$. The result is that the smoothed and unsmoothed KF estimates also obey the specified linear constraints.

4. An Application of Data Reconstruction

The application refers to the Italian quarterly production of energy accounts at 1995 prices. Seasonal unadjusted figures in the sample *80.1-98.4* (T=*76*) are considered to form the vector y_t. Following the notation of sections 2 and 3, the dimension of y_t is n=*4*, corresponding to the ESA79 sectors of *Coal and Lignite* (E79_1), *Coke* (E79_2), *Natural Gas Crude and Refined Petroleum* (E79_3), *Electricity and Water Supply* (E79_4).

z_t is the p=*3* vector of seasonal unadjusted energy production in terms of the ESA95 sectoral classification, which is, *Mining of Energy Products* (E95_1), *Coke and Refined petroleum products* (E95_2), *Electricity Gas and Water Supply* (E95_3). Their graphs are shown in Figure 1 together with the ESA79 groupings; E79_T and E95_T represent the totals in terms of both standards.

All the figures are obtained grouping the production of *37* elementary energy products according to either ESA79 or ESA95, by using a unique weighting system of 1995 prices. Note that differences in the totals are due to the introduction of semi-manufactured oil products in ESA95 and a different computation of natural gas among schemes; in fact, with respect to the old standards, ESA95 splits up miming from distribution of natural gas.

In this analysis only the last *28* observations of z_t are supposed to be available (i.e. the sample *92.1-98.4*), although z_t can be computed over all the sample period. Then, the application represents an exercise to verify the effectiveness of the structural time series model proposed in section 3 to data reconstruction.

Reconstruction of each term of z_t over the sample *80.1-91.4* (i.e. the first *48* observations) has been carried out by the KF estimating, separately, parameters of three state space models as in equations (3) and (4). To allow the additive property of the structural components of equation (5) no transformation of data has been considered; results are then referred to the level of y_t. Restricted estimates of z_t for $t = 1, .., 48$ have been obtained by the KF following Doran (1992).

Figure 1: *Energy Production by sector (billion of Italian lira at 1995 prices)*

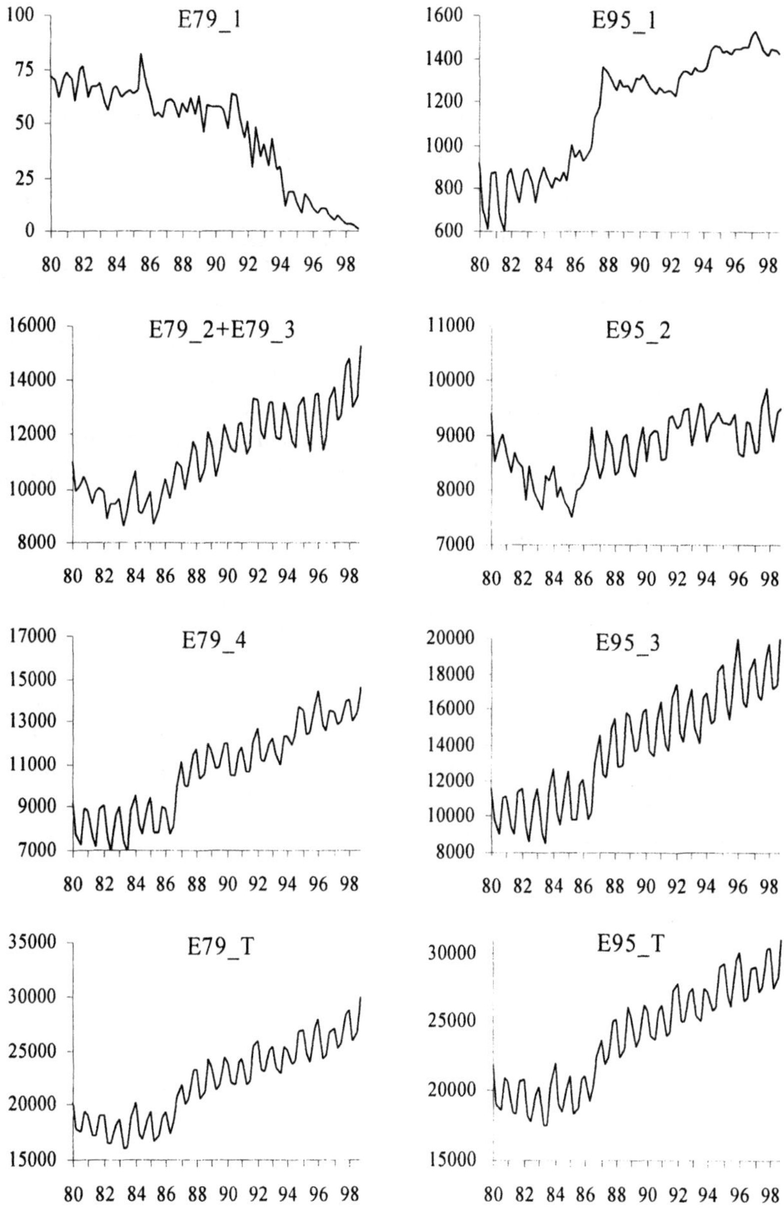

The three models have involved the reconstruction of E95_2, E95_3 and E95_T. Dependent variables in terms of fully observed ESA79 sectors have been set up as follows: a bivariate vector given by E79_3 and E79_2+E79_3 to gain a backcast of E95_2, E79_3 to E95_3 and E79_T to E95_T; E95_1 has been then obtained subtracting the estimates reached for E95_2 and E95_3 from E95_T.

Very poor results of a direct estimation of E95_1 suggested such a reparameterisation: in fact, these observations are completely dominated by the larger sectors, which implies that the likelihood function contains very little information on the parameters relating to E95_1.

Estimates of model parameters are shown in Table 1 and some relevant diagnostics on filtered residuals in Table 2. For models in which E95_2, E95_3 and E95_T are the figures to be reconstructed, $\hat{L}_i$ is the maximum of the sample log likelihood reached by the Newton-Raphson optimisation routine; $\hat{\sigma}_\xi$, $\hat{\sigma}_\eta$, $\hat{\sigma}_\omega$ and $\hat{\sigma}_\varepsilon$ are, respectively, the estimates of the standard errors as in equations from (7) to (10). In brackets are the standard errors of the estimates.

Relevant points of these estimates are as follows: a) the KF estimates of z_t and the actual values are quite similar; b) constraints given by the last *28* observations of z_t are always satisfied; c) over the reconstruction sample (i.e. the first *48* observations) the KF estimates always perform standard errors, enabling precision of estimation to be assessed d) Table 2 reveal as only in one case (*LR heteroschedasticity test* of E79_3 for the E95_2 model) diagnostics are significant, with a general confidence in normality of filtered residuals for all the estimation performed.

For reconstruction purposes it is useful to revise the KF estimates by applying a smoothing algorithm. It offers the advantage to compute the estimates by moving backwards through time from $t=T$ to $t=1$, so that all the available information is used. As a result, the smoothed estimates are more efficient than the unsmoothed and, in a historical perspective, optimal relative to a given information set.

Following Harvey (1990, pp.154-5) the 'fixed-interval' smoothing is applied: figure 2 provides a graph of actual values and smoothed KF estimates of E95_T.

It should be pointed out that: a) comparability among actual and estimated values is always increased by the smoothing algorithm; b) the constraints are also satisfied; c) in every case the standard errors of the smoothed estimates are smaller than the unsmoothed.

Table 1: *Estimates of model parameters*

ESA95 Sector	$\hat{L}_i$	$\hat{\sigma}_\xi$	$\hat{\sigma}_\eta$	$\hat{\sigma}_\omega$	$\hat{\sigma}_\varepsilon$
E95_2	*-1183.13*	*168.41*	*9.61*	*76.63*	*0.05*
		(14.45)	*(3.67)*	*(8.82)*	*(0.21)*
E95_3	*-736.24*	*216.16*	*15.12*	*99.92*	*208.34*
		(33.51)	*(6.43)*	*(14.71)*	*(30.13)*
E95_T	*-730.94*	*303.60*	*34.01*	*42.97*	*131.80*
		(57.84)	*(11.77)*	*(13.15)*	*(73.18)*

Table 2: *Diagnostics on filtered residuals*

Residuals	*Durbin Watson*	*Jarque Bera*	*Sum of sq. Res.*	*Ljung Box (15)*	*ARCH*	*Chow*	*LR het.*	*Shapiro Wilks*
				Model for E95_2				
E79_3	*1.704*	*2.277*	*2476*	*25.07*	*1.148*	*0.641*	*11.304**	*0.973*
E79_2 + E79_3	*1.867*	*1.463*	*1807*	*19.29*	*0.173*	*0.013*	*4.263*	*0.985*
				Model for E95_3				
E79_3	*1.650*	*1.122*	*223300*	*15.05*	*9.283*	*1.632*	*2.833*	*0.989*
				Model for E95_T				
E79_T	*1.748*	*1.810*	*36970*	*13.00*	*2.198*	*1.064*	*7.189*	*0.989*

Results in terms of smoothed KF estimates can be evaluated in several ways. In this application two criteria are considered: the first one concerns the correlation between actual observations and smoothed estimates; the second consists in performing backwards statistics of mean absolute percentage error (MAPE), where errors are given by the difference between actual values and smoothed estimates.

Figure 2: *Actual values of E95_T and smoothed Kalman filter estimates*

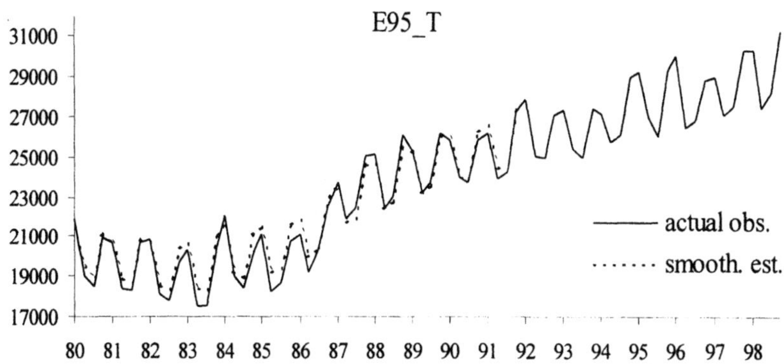

Table 3 summarises results of these measures: for each sector, the correlation refers to the overall reconstruction sample *80.1-91.4*; in contrast, the mean absolute percentage error has been computed over four different backwards periods, i.e. *89.1-91.4, 86.1-91.4, 83.1-91.4* and *80.1-91.4*.

Table 3: *Correlation between actual observations and smoothed estimates and average absolute percentage errors*

ESA95 Sector	*E95_1*	*E95_2*	*E95_3*	*E95_T*
Correlation	*0.803*	*0.964*	*0.992*	*0.996*
MAPE(89-91)	*10.74%*	*2.24%*	*1.20%*	*0.86%*
MAPE(86-91)	*15.59%*	*2.29%*	*2.52%*	*1.32%*
MAPE(83-91)	*19.23%*	*2.43%*	*3.40%*	*1.98%*
MAPE(80-91)	*20.37%*	*2.06%*	*3.45%*	*1.97%*

Note how large is the difference in terms of both measures between results involving the backcast of E95_1 and the other groupings. As stressed before, E95_1 is the smaller sector, dominated by the others and less affected by seasonality (see Figure 1). In any case its estimation results very complicated when information on the energy mining is hidden.
Both correlation and mean absolute percentage errors of the other sectors show reliable results. In particular, for E95_T MAPE never reaches 2%; whereas E95_2 and E95_3 are always below 2.5% and 3.5%, respectively.

5. Conclusions

A deterministic representation of a change of classification of sectoral data has been provided. The structural time series approach has been then considered to the related problem of data reconstruction. In this respect a state space form which is able to incorporate exogenous information has been set up.
An application to Italian energy accounts has revealed the ability of the presented method to gain reliable smoothed KF estimates. A general agreement of the estimates with the actual values has been shown, apart from a case in which hidden information was indeed necessary.
This framework provides different advantages: it is transparent as based on an explicit model of data reconstruction; subject to the model which formalises the relationship with the available old standards figures, the estimates are statistically optimal in the minimum mean square error sense; by following the Doran (1992) results, the smoothed estimates always satisfy the inherent constraints.

References

Dagum E.B., Cholette A., Chen Z. (1998) A Unified View of Signal Extraction, Benchnarking, Interpolation and Extrapolation of Time Series, *International Statistical Review*, 66, 3, 245-269.

Doran H.E. (1992) Constraining Kalman Filter and Smoothing Estimates to Satisfy Time Varying Restrictions, *The Review of Economics and Statistics*, 74, 568-572.

Doran H.E. (1996) Estimation under Exact Linear Time Varying Constraints, with an Application to Population Projections, *Journal of Forecasting*, 15, 527-541.

Gomez V. and Maravall A. (1994) Estimation, Prediction and Interpolation for Non-Stationary Series with the Kalman Filter, *Journal of the American Statistical Association*, 89, 611-624.

Harvey, A.C. (1990) *Forecasting, Structural Time Series Analysis and the Kalman Filter*, Cambridge University Press, Cambridge.

Jones R. (1980) Maximum Likelihood Fitting of ARMA Models to Time Series with Missing Observations, *Technometrics*, 22, 389-395.

Spatial Discriminant Analysis Using Covariates Information

Marco Alfò
Ministero della Sanità

Paolo Postiglione
Ministero dei Trasporti e della Navigazione

Abstract: The analysis of spatially distributed observations implies a number of theoretical problems due to the multidirectional dependence among nearest sites. The presence of such a dependence often causes the standard statistical method, instead based on independence assumptions, to provide inefficient estimates or, even, to fail badly. This paper concerns the problem of discrimination and classification of spatial polytomous data. It extends the approach discussed by Alfò and Postiglione (1999) for binary observations to polytomous data, presents a discrimination function based on markovian automodels and suggests a natural solution to the resulting allocation problem through a Gibbs sampler based procedure.
The proposed approach is contrasted with standard logistic discrimination and applied to a real data set consisting of a remote sensed image from Nebrodi mountains (Italy).

Keywords: Spatial discrimination, polytomous observations, Strauss automodel, Gibbs sampler.

1. Introduction

Spatial observations often arise in empirical applications, from archaeological studies to pattern recognition, from epidemiological spreading to survey analysis (see Haining, 1990).
The hypothesis of independence among observations assumed by traditional statistical techniques is a very convenient one and often allows statistical theory to be more tractable. This conjecture is obviously violated in the present context, given that in all geographical and territorial studies "(...) *everything is related to everything else, but near things are more related than distant things*", as stated in Tobler's first law of geography (Tobler, 1970).
To take into account this multidirectional dependence among nearest sites a general class of well-established models has been introduced in the statistical literature. For earlier developments see, for example, the seminal papers by Besag (1974) and Strauss (1977); for a review of this topic see Cressie (1991).

In this paper, we concentrate on the definition of a discrimination function for spatial observations, incorporating the notion of contextual information.
We proceed as in Alfò and Postiglione (1999), extending the approach therein proposed for binary observations to the more general case of spatial polytomous observations with g categories ($g \geq 2$). Generally speaking, we can recognise in discriminant analysis two different phases: discrimination and classification. The discrimination phase concerns the process of deriving classification (or allocation) rules from samples of labelled units, while the classification phase refers to the application of these rules to identify new units group membership (for a comprehensive review of this topic see Krzanowski, Marriott, 1995).
The paper is organised as follows: section 2 is devoted to briefly recall logistic discriminant analysis in the case of independent observations with g categories and to discuss the limits of this approach when applied to spatially distributed observations. Section 3 focuses on the generalization of the previous approach to spatial observations. We introduce a suitable discrimination function based on an extension of the Strauss automodel, discuss some instances related to parameter estimation using maximum pseudolikelihood and describe a relevant solution, through a Gibbs sampler (Geman and Geman, 1984) based algorithm, to the outlined allocation problem. Section 4 presents an empirical application: the aim is to discriminate between pixels from a remote sensed image of Nebrodi Mountains using a set of soil characteristics; in this context, the proposed method is contrasted with standard logistic discrimination. Finally, the last section is devoted to outline some concluding remarks and possible future research lines.

2. The logistic approach

Earliest ideas in logistic discrimination date back to the 60's, but the essential features of the method were developed by Anderson (1982). In the case of g groups, the posterior probabilities of group membership for the i-th unit with covariates vector $\mathbf{x}_i=(x_{i1},\ldots,x_{iq})^{\mathrm{T}}$ are given by (see McLachlan, 1992):

$$\Pr(\mathrm{Y}_i = l \mid \mathbf{x}_i) = \frac{\exp\left(a_l + \mathbf{b}_l^{\mathrm{T}} \mathbf{x}_i\right)}{1 + \sum_{h=1}^{g-1}\left[\exp\left(a_h + \mathbf{b}_h^{\mathrm{T}} \mathbf{x}_i\right)\right]} \qquad 1 \leq l < g \quad (1)$$

$$\Pr(\mathrm{Y}_i = g \mid \mathbf{x}_i) = \frac{1}{1 + \sum_{h=1}^{g-1}\left[\exp\left(a_h + \mathbf{b}_h^{\mathrm{T}} \mathbf{x}_i\right)\right]} \qquad l = g \quad (2)$$

where $\mathrm{Y}_i=l$ if the i-th unit belongs to the l-th group (i.e. is labelled with the l-th category), while a_l and, respectively, $\mathbf{b}^{\mathrm{T}}_l=(b_{l1},\ldots,b_{lq})$, $l=1,\ldots,g$-1, represent the group-specific intercepts and vectors of parameters. According to previous expressions, we can write the log-odds as:

$$\log\left[\frac{\Pr(Y_i = l \mid \mathbf{x}_i)}{\Pr(Y_i = g \mid \mathbf{x}_i)}\right] = a_l + \mathbf{b}_l^{\mathsf{T}}\mathbf{x}_i \qquad 1 \le l < g \quad (3)$$

Now, we can proceed as follows. Let us assume we are able to exactly assess the group membership for a subset of sample units: in the following, we will refer to this set as the training set. To apply the method, we can estimate the parameters a_l and $\mathbf{b}^{\mathsf{T}}_l=(b_{l1},...,b_{lq})$, $l=1,\dots,g-1$, from the training set (*discrimination phase*) and successively substitute these estimates in expressions (1) and (2) allocating units to the group having the highest posterior probability (*classification phase*).
But the use of a logistic approach to analyse spatial data is statistically incorrect, in consequence of their mutual dependence: closeness to other units belonging to a given group modifies our expectation of units group membership. Hence, it seems reasonable to formulate a model that links information contained in the covariates vector $\mathbf{x}_i$ together with the notion of spatial dependence (contextual information).

3. The contextual approach to logistic discriminant analysis

Several *ad hoc* approaches have been introduced in statistical literature to take into account spatial information in discriminant analysis. See, for example, Switzer (1980) and Mardia (1984).
To introduce our proposal of spatial discrimination function, let us start assuming that we are analysing a study area which can be discretized into M contiguous quadrate cells and that N of these cells ($N<M$) represent the training set.
This is only a working simplification which can be relaxed, when analysing irregular lattices, imposing a more general connectivity structure between sample units (see Besag et al., 1991).
Each unit of the training set can be labelled via a spatially located polytomous dependent variable with g categories $Y_i=1,\dots,g$, $i=1,\dots,N$. For $g=3$, this situation can be described using the regular lattice displayed in Figure 1.

Figure 1: *An example of a regular lattice grid*

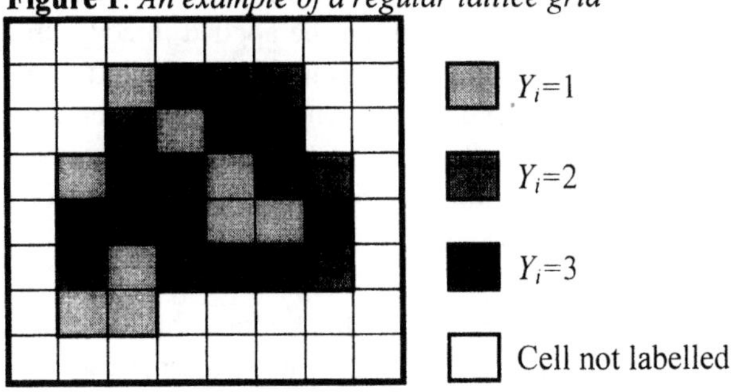

Our aim is to label the (M-N) remaining cells using the information and, hence, the parameter estimates derived from the training set.

To define an optimal approach to spatial discrimination in the case of g groups, we propose to extend the model introduced by Strauss (1977) by considering the information stored in the covariates vector, obtaining the following posterior probabilities of group membership:

$$\Pr(\mathrm{Y}_i = l \mid y_j, j \in \mathrm{N}_1(i); \mathbf{x}_i) = \frac{\exp\left(a_l + ru_i(l) + \mathbf{b}_l^{\mathrm{T}} \mathbf{x}_i\right)}{1 + \sum_{h=1}^{g-1} \left[\exp\left(a_h + ru_i(h) + \mathbf{b}_h^{\mathrm{T}} \mathbf{x}_i\right)\right]} \qquad 1 \leq l < g \quad (4)$$

$$\Pr(\mathrm{Y}_i = l \mid y_j, j \in \mathrm{N}_1(i); \mathbf{x}_i) = \frac{1}{1 + \sum_{h=1}^{g-1} \left[\exp\left(a_h + ru_i(h) + \mathbf{b}_h^{\mathrm{T}} \mathbf{x}_i\right)\right]} \qquad l = g \quad (5)$$

Here $\mathrm{N}_1(i)$, $u_i(l)$ and r represent, respectively, the first order neighbourhood set of the i-th site, the number of first order neighbours of this site having $\mathrm{Y}_j=l$ and a constant parameter describing spatial interaction. In the following, we will refer to the previous model as the Extended Strauss Model (ESM).

With g=2, the ESM become the LAM defined by Alfò and Postiglione (1999).

The ESM approach, as the Strauss automodel, is particularly suited to model units belonging to g unordered groups, where these groups are finite and exchangeable. This hypothesis is often satisfied in the study of remote sensed images and, in particular, when thematic maps of soil use are examined. This situation is presented with a real example in the next section.

Previous statements lead to the following expression for the generic log-odds:

$$\log\left[\frac{\Pr(\mathrm{Y}_i = l \mid y_j, j \in \mathrm{N}_1(i); \mathbf{x}_i)}{\Pr(\mathrm{Y}_i = g \mid y_j, j \in \mathrm{N}_1(i); \mathbf{x}_i)}\right] = a_l + ru_i(l) + \mathbf{b}_l^{\mathrm{T}} \mathbf{x}_i \qquad 1 \leq l < g \quad (6)$$

As stated in Arbia and Espa (1996) and discussed in Alfò and Postiglione (1999), there are some statistical problems connected with the previous model which are still unsolved: the spatial discrimination function can not be applied directly. In fact in the spatial model Y_i is present both as a dependent variable (on the left side of expressions (4) and (5)) and as an independent one (on the right side of the same expressions). Hence, in the classification step, to allocate units not belonging to the training set, we should predict Y_i using the covariates vector $\mathbf{x}_i$ (known for all the analysed units) and Y_j ($j \in \mathrm{N}_1(i)$), which are not known for units not belonging to the training set.

To solve this problem, we propose to extend the approach introduced by Alfò and Postiglione (1999) in the case of spatial binary observations and suggest to use the following two-step procedure:

Step 1 (*Discrimination phase*): We estimate the model parameters from the training set via the maximum pseudolikelihood (MPL) procedure (Besag,

1975), regressing Y_i on the q+1 covariates ($x_{i1},...,x_{iq}$, $u_i(l)$). The MPL estimator is consistent (Geman, Graffigne, 1987) and gives good results in practical applications (Ripley, 1990).
Furthermore, the pseudolikelihood estimates can be carried out using standard statistical packages (Strauss, Ikeda, 1990).

Step 2 (*Classification phase*): We consider these parameter estimates as representatives for the whole lattice grid. Hence, we allocate the remaining cells via a Gibbs sampler (Geman, Geman, 1984) procedure, simulating a Strauss automodel with parameters equal to those estimated on the units of the training set. From a randomly chosen starting configuration, we allocate cells outside of the training set to the g groups using posterior probabilities given by expressions (4) to (5).

Hence, in the classification phase we proceed on defining a sequence of configurations until convergence is achieved according to a given stop criterion (no units group memberships or only a relatively small proportion of them can be modified) and a final configuration is obtained (Gilks et al., 1996; Besag et al., 1995). In this way, all units not belonging to the training set are allocated to one of the g groups.

4. A real example

The proposed method was applied to a data set consisting of a remote sensed image from Landsat TM (Thematic Mapper) regarding Nebrodi Mountains (Italy). The image covers a surface of 2,016 Km^2 and is composed by 2.240 (40×56, vertical×horizontal) pixels, each one representing a ground surface of 30m×30m.
Our aim was to discriminate between arboreus farming (Y_i=1), sowing and pastures (Y_i=2), spontaneous vegetation, woods and other surfaces (Y_i=3) on the basis of a number of related soil characteristics. To give more details, the first group is composed by lavic arboreus, citrus fruit and other arboreus farming (in particular olives and almonds).
The related soil characteristics are: soil density (measured in grammes per cm^3), permeation capacity in saturation conditions (in millimetres per day), horizon thickness (in centimetres) and DTM (digital terrain model, in metres).
To check the sensitivity of the proposed approach to different specifications of the training set, we repeated the analysis adopting 4 different choices for this set (each one characterised by a different dimension and shape).
The dimension of the training sets was fixed in the range from 150 to 400 pixels; the analysis was then repeated adopting 4 different choices for the training set with, respectively, 210, 392, 182 and 278 pixels. To choose the training sets, we divided the lattice in four quadrants and sampled a pixel from each quadrant. The training sets were defined as regular geometric figures having the chosen pixels as one of their corners.

The proposed algorithm was iterated 100 times for each of the 4 training sets, leading to 400 different allocation scenarios. A unique solution for each training set was calculated by applying a consensus rule to corresponding allocation scenarios. With respect to technical issues, the algorithm was implemented in Matlab code; it presented low computational costs (the iterations were performed using a PII based architecture) and was quite stable in all empirical situations we faced, always requiring a reasonable number of iterations (<100 in 90% of the performed simulations) to converge to a final configuration.
In all the 4 analysed cases we found out values for the interaction parameter ranging from 0.22 to 0.29, showing a good interaction among neighbouring sites. Figure 2 shows results obtained using our spatial approach to discriminant analysis and a standard logistic discriminant analysis.

Figure 2: *Misclassification rates*

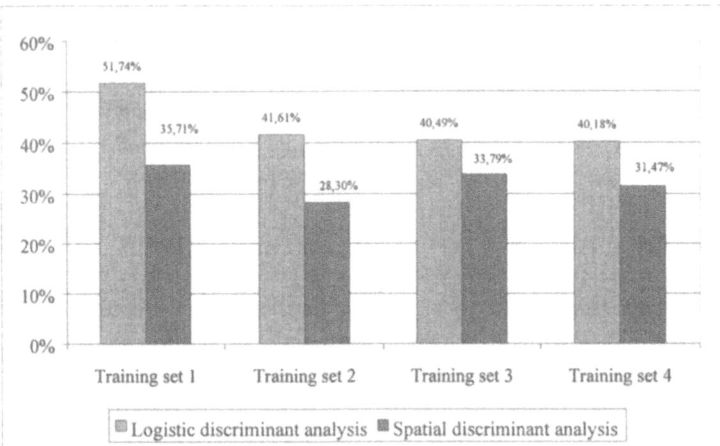

Note that results derived using the proposed approach and a median consensus rule are significantly better, in terms of misclassification rates, than those obtained using standard logistic discrimination.
Furthermore, observing Figure 2, it seems reasonable to assert that different dimensions and shapes of the used training sets do lead to low differences in terms of rightly classified pixels.

5. Concluding remarks

In this paper we outlined the limitations arising when logistic discriminant analysis is applied to spatial polytomous observations.
We contrasted this approach with a more congruent one, based on an extended version of the Strauss automodel, which takes into account the mutual dependence of geographical sites together with auxiliary information available in the form of

covariates measurements on the cells of the analysed lattice.
In particular, to provide an optimal solution to the corresponding allocation problem, we proposed a Gibbs sampler algorithm which extends the method discussed by Alfò and Postiglione (1999) to polytomous response variables. The algorithm has been applied to a real data set consisting in a portion of a remote sensed image from Nebrodi Mountains (Italy).
In this context, results obtained using the extended Strauss automodel and an adequate consensus rule have been contrasted with those obtained via a standard logistic discriminant function, and seem to confirm that, when working with spatial data, there is no reason to leave out contextual information from model specification.
To confirm these conclusions in the general case, more research is however needed, especially in the form of a large-scale simulation study characterised by different values of interaction parameters and by different sizes of the whole lattice and of the used training sets. At the same time, additional research is also necessary in order to compare our proposal with other spatial approaches to discriminant analysis.
The ESM can be easily extended to the case of g ordered groups, considering instead of the Strauss automodel the binomial automodel introduced by Besag (1986). All these topics will be subjects for future research.

Acknowledgements

The authors would like to thank Prof. G. Arbia and Prof. M. Vichi for helpful comments on earlier versions of this paper. Finally we gratefully acknowledge M.L. Paracchini of Space Application Institute of the Joint Research Centre of the European Commission (Ispra, Varese) for kindly providing the remote sensed image of section 4.

References

Alfò, M., Postiglione, P. (1999). Discriminant Analysis using Markovian Automodels, in: *Classification and Data Analysis. Theory and Application*, Vichi M. and Opitz, O. (eds.), Springer-Verlag, Heidelberg.

Anderson, J.A. (1982). Logistic discrimination, in: *Handbook of Statistics vol.2*, Krishnaiah P.R. & Kanal L.N. (Eds.), North Holland, 169-191.

Arbia, G., Espa, G.(1996). Forecasting statistical models of archaeological site locations, *Archeologia e Calcolatori*, Roma.

Besag, J. (1974). Spatial interaction and the statistical analysis of lattice systems, *Journal of Royal Statistical Society*, B, **36**, 192-236.

Besag, J. (1975). Statistical analysis of non-lattice data, *The Statistician*, 24, 179-195.

Besag J. (1986). On the statistical analysis of dirty pictures, *Journal of Royal Statistical Society*, B, **48**, 259-302.

Besag, J., Green, P., Higdon, D., Megersen, K. (1995). Bayesian computation and stochastic systems, *Statistical Science*, **10**, 3-66.

Besag, J., York, J., Molliè, A. (1991). Bayesian image restoration, with two applications in spatial statistics (with discussion), *Annals of the Institute of Mathematical Statistics*, **43**, 1-59.

Cressie, N. (1991). *Statistics for Spatial Data*, Wiley.

Geman, S., Geman, D. (1984). Stochastic Relaxation, Gibbs Distributions, and the Bayesian Restoration of Images", *IEEE Transactions on Pattern Analysis and Machine Intelligence*, Vol. PAMI **6**, 721-741.

Geman, S, Graffigne, C (1987). Markov random fields and their application to computer vision, *Proceedings of the International Congress of Mathematicians*, Providence.

Gilks, W.R., Richardson, S., Spiegelhalter, D.J. (1996). Introducing Markov chain Monte Carlo, in: *Markov Chain Monte Carlo in practice,* Gilks et al. (eds.), Chapman and Hall.

Haining, R.P. (1990). *Spatial Data Analysis in the Social and Environmental Sciences*, Cambridge University Press.

Krzanowski, W.J., Marriott, F.H.C. (1995). *Multivariate Analysis. Part 2- Classification, Covariance Structures and Repeated Measurements*, Edward Arnold.

Mardia (1984). Spatial discrimination and classification maps, *Communication in Statistics. Theory and Methods* , **13**, 2181-2197.

McLachlan, G.J. (1992). *Discriminant analysis and pattern recognition*, Wiley.

Ripley, B.D. (1990). Gibbsian interaction models, in: *Spatial Statistics: Past, Present and Future,* Griffith D. (ed.), Institute of Mathematical Geography.

Strauss, D.J. (1977). Clustering of coloured lattices, *Journal of Applied Probability*, **14**, 135-143.

Strauss, D.J, Ikeda, J. (1990). Pseudolikelihood estimation for social networks, *Journal of the American Statistical Association,* 85, 204-212.

Switzer (1980). Extension of discriminant analysis for statistical classification of remotely sensed imagery, *Journal of the International Association for Mathematical Geology*, **12**, 367-376.

Tobler, W.R. (1970). A computer movie simulating urban growth in the Detroit region, *Economic Geography,* supplement n.46, 234-240.

Some Aspects of Multivariate Geostatistics

Sandra De Iaco
Facoltà di Economia, Università degli Studi di Chieti
Via dei Vestini, 66013, Chieti, Italy

Donato Posa
Facoltà di Economia, Università degli Studi di Lecce
Via per Monteroni, Ecotekne, 73100 Lecce, Italy

Abstract: Cokriging allows the use of data on correlated variables in order to enhance the estimation of a primary variable or more generally to enhance the estimation of all variables. However, in order to apply the estimation procedure, some problems must be solved: the variogram matrix must be conditionally negative definite and it is necessary to model the cross-variogram. The aim of this paper is to present a flexible solution in order to solve the mentioned problems. A case study has been presented.

Keywords: Cokriging, Cross-correlation, Multivariate Geostatistics.

1. Introduction

Classical multivariate techniques do not explicity incorporate the locations of the samples nor the spatial correlation; univariate geostatistics does have these properties but it does not incorporate intervariable correlation. Multivariate geostatistics provides tools which take into account the important features of both approaches. Cokriging is the extension of kriging to the situation when auxiliary variables can be used to improve the estimation of a primary variable or more generally to enhance the estimation of all variables (Myers, 1983).

Most of the problems encountered in the practice of cokriging are related to the estimation and modeling of cross-covariances or cross-variograms (Wackernagel, 1995):

- in order to estimate the cross-correlation, it is necessary to have a sufficient number of sample locations where both variables are sampled (for each pair of variables): this condition is frequently not satisfied in practice and in particular in the undersampled problem;
- in the case of variograms or covariances, it is easy to ensure that the positive definiteness condition is satisfied by using positive linear combinations of valid models (Christakos, 1984); moreover, the parameters of the models are at least partially identified from the graph of the sample variogram or covariance. However, positive definiteness is not a property of a cross-variogram, which

has a more complex structure and one cannot easily identify the components or the parameters from the graph of the sample cross-variogram.

The aim of this paper is to present a flexible solution to solve the second mentioned problem: the linear model of coregionalization provides a method for modeling the variograms and cross-variograms of two or more variables so that the variance of any possible linear combination of these variables is always positive. Moreover, a case study has been given.

2. Cokriging

Let $Z_1, \ldots, Z_m$ be m spatial random fields defined in the domain $D \subseteq \Re^d$, ($d \leq 3$); hence:

- the expected value, for each spatial random field, exists and does not depend on the spatial location:

$$E(Z_i(\mathbf{u})) = m_i, \qquad i = 1, \ldots, m, \forall \mathbf{u} \in D;$$

- the cross-covariance function depends only on the separation vector $\mathbf{h}$:

$$C_{ij}(\mathbf{h}) = E((Z_i(\mathbf{u}+\mathbf{h}) - m_i)(Z_j(\mathbf{u}) - m_j)), \quad i, j = 1, \ldots, m, \forall \mathbf{u}, \mathbf{u}+\mathbf{h} \in D;$$

- the cross-variogram function depends only on the separation vector $\mathbf{h}$:

$$2\gamma_{ij}(\mathbf{h}) = Cov[Z_i(\mathbf{u}+\mathbf{h}) - Z_i(\mathbf{u}), Z_j(\mathbf{u}+\mathbf{h}) - Z_j(\mathbf{u})] =$$
$$= E((Z_i(\mathbf{u}+\mathbf{h}) - Z_i(\mathbf{u}))(Z_j(\mathbf{u}+\mathbf{h}) - Z_j(\mathbf{u}))), \quad i, j = 1, \ldots, m, \forall \mathbf{u}, \mathbf{u}+\mathbf{h} \in D.$$

The cross-covariance function is not a priori an even or an odd function with respect to a change in the order of the variables or a change in the sign of the separation vector **h**, particularly:

$$C_{ij}(\mathbf{h}) \neq C_{ji}(\mathbf{h});$$
$$C_{ij}(\mathbf{h}) \neq C_{ij}(-\mathbf{h}).$$

The cross-variogram is an even function with respect to a change in the order of the variables or a change in the sign of the separation vector **h**:

$$\gamma_{ij}(\mathbf{h}) = \gamma_{ji}(\mathbf{h});$$
$$\gamma_{ij}(\mathbf{h}) = \gamma_{ij}(-\mathbf{h}).$$

In practice, asymmetry of the cross-covariance function is most often ignored for two main reasons:

1. the usual descriptive tools are the direct and cross-variograms which are symmetric;
2. modeling asymmetric cross-covariances, although possible, is difficult (see Journel & Huijbregts, 1981).

Moreover, in order to describe spatial processes for which the covariance function does not exist, it is assumed that each random field $Z_i, i = 1,\ldots,m$ satisfies the intrinsic hypothesis. Particularly, define:

$$\overline{\mathbf{Z}}(\mathbf{u}) = [Z_1(\mathbf{u}),\ldots,Z_m(\mathbf{u})]$$

then $\overline{\mathbf{Z}}$ is assumed to satisfy the intrinsic hypothesis, if:

$$E\left[\overline{\mathbf{Z}}(\mathbf{u}+\mathbf{h}) - \overline{\mathbf{Z}}(\mathbf{u})\right] = [0,0,\ldots,0] \qquad \forall \mathbf{u}, \mathbf{u}+\mathbf{h} \in \mathrm{D} \tag{1}$$

and the cross-variogram:

$$2\gamma_{ij}(\mathbf{h}) = Cov\left[\mathbf{Z}_i(\mathbf{u}+\mathbf{h}) - \mathbf{Z}_i(\mathbf{u}), \mathbf{Z}_j(\mathbf{u}+\mathbf{h}) - \mathbf{Z}_j(\mathbf{u})\right] \qquad \forall \mathbf{u}, \mathbf{u}+\mathbf{h} \in \mathrm{D} \tag{2}$$

exists and depends only on **h** for *i,j=1, . . .,m*. Two spatial random fields Z_i and Z_j which satisfy the intrinsic hypothesis are said to be *uncorrelated* if:

$$\gamma_{ij}(\mathbf{h}) = 0, \qquad \forall \mathbf{h}$$

Let $\mathbf{u}_i \in D, i = 1,\ldots,n$ be the sample spatial locations with data

$$(z_j(\mathbf{u}_i); i = 1,\ldots,n, j = 1,\ldots,m)$$

If (1) is satisfied, then (2) may be written as:

$$\overline{\gamma}(\mathbf{h}) = \left[\gamma_{ij}(\mathbf{h})\right] = 0.5E\left[\overline{\mathbf{Z}}(\mathbf{u}+\mathbf{h}) - \overline{\mathbf{Z}}(\mathbf{u})\right]^T \left[\overline{\mathbf{Z}}(\mathbf{u}+\mathbf{h}) - \overline{\mathbf{Z}}(\mathbf{u})\right] \tag{3}$$

The objective is to estimate the vector $\overline{\mathbf{Z}}(\mathbf{u})$ by a linear combination of the data vectors $\overline{\mathbf{Z}}(\mathbf{u}_1)$, . . . , $\overline{\mathbf{Z}}(\mathbf{u}_n)$, such that the estimator is unbiased and the estimation variance is minimized. The linear estimator can be written in the form:

$$\overline{\mathbf{Z}}^*(\mathbf{u}) = \sum_{k=1}^{n} \overline{\mathbf{Z}}(\mathbf{u}_k)\Lambda_k \tag{4}$$

where each Λ_k is an m x m matrix; the element λ_{ij}^k of the matrix Λ_k represents the contribution of the *ith* variable at location $\mathbf{u}_k$ to the estimate of the *jth* variable; in nonmatrix form this would be:

$$\mathbf{Z}_j^*(\mathbf{u}) = \sum_{i=1}^{m}\sum_{k=1}^{n} \mathbf{Z}_i(\mathbf{u}_k)\lambda_{ij}^k$$

Then, by requiring that the estimator is unbiased and the estimation variance is minimized, the following cokriging system (Myers, 1982) is obtained:

$$\sum_{j=1}^{n} \bar{\gamma}(\mathbf{u}_i - \mathbf{u}_j)\Lambda_j + \bar{\mu} = \bar{\gamma}(\mathbf{u}_i - \mathbf{u}), \quad \sum_{j=1}^{n} \Lambda_j = \mathbf{I} \qquad i = 1,\ldots,n$$

where $\bar{\mu}$ is an m x m matrix whose elements are the Lagrange multipliers and $\mathbf{I}$ is the identity matrix.

3. The linear model of coregionalization

One of the difficulties occurring in the use of cokriging is modeling the cross variograms. It is well-known that in order for a function to be a variogram it must satisfy a conditional positive definite condition (Journel & Huijbregts, 1981), which is easily ensured by using nested models, i. e., positive linear combinations of valid models, such as the spherical, exponential and gaussian (Posa, 1995). Unfortunately, the cross variogram has a more complex structure and it is not easy to identify the components or the parameters from the graph of the sample cross variogram.
If $\bar{\gamma}(h)$ is the m x m variogram matrix defined in (3) with $\bar{\gamma}(0) = 0$, then $-\bar{\gamma}$ is conditionally positive definite if for all vectors $\mathbf{a}^T = [a_1, \ldots, a_m]$, $-\mathbf{a}^T\bar{\gamma}\mathbf{a}$ is conditionally positive definite. In particular this means that:

$$-\sum_{r=1}^{m}\sum_{s=1}^{m}\sum_{i=1}^{n}\sum_{j=1}^{n} a_r a_s \lambda_i \lambda_j \gamma_{rs}(\mathbf{u}_i - \mathbf{u}_j) > 0$$

for any choice of $\mathbf{a}$, any choice of $\mathbf{u}_1, \ldots, \mathbf{u}_n$ and any choice of the $\lambda_1, \ldots, \lambda_m$, with

$$\sum_{i=1}^{n} \lambda_i = 0$$

Then one might attempt to characterize valid variogram matrix by requiring that the entries be valid models: for the diagonal entries this is appropriate and each

corresponds to the variogram of a single component. For the off diagonal entries the problem is not so simple: in a number of istances, authors have used variogram models for cross variograms, but in general this will lead to an invalid model in view of the definition of positive definite given above.
The linear model of coregionalization provides a method for modeling the auto and cross variograms of two or more variables so that the variance of any possible linear combination of these variables is always positive.
Let (Z_k , $k=1, \ldots, K$) be K spatial random fields defined in the domain $D \subseteq \Re^d$, representing the variables of interest and satisfying the intrinsic hypothesis. Suppose that (Y_i, $i=1, \ldots, N$) be N orthogonal random fields satisfying the intrinsic hypothesis: let $\gamma_i(\mathbf{h})$ be the variogram of Y_i and $\gamma_{ij}(\mathbf{h})$ the cross variogram between Y_i and Y_j, with $\gamma_{ij}(\mathbf{h})=0$, $\forall \mathbf{h}$, $i \neq j$. Suppose that Z_k could be written as:

$$Z_k(\mathbf{u}) = \sum_{i=1}^{N} a_{ki} Y_i(\mathbf{u}) \qquad k = 1, \ldots, K$$

It is easy to show that the cross variogram between Z_k and $Z_{k'}$ is:

$$\psi_{kk'}(\mathbf{h}) = \sum_{i=1}^{N}\sum_{j=1}^{N} a_{ki} a_{k'j} \gamma_{ij}(\mathbf{h}) = \sum_{i=1}^{N} a_{ki} a_{k'i} \gamma_i(\mathbf{h}) = \sum_{i=1}^{N} b_{kk'}^{i} \gamma_i(\mathbf{h})$$

where:

$$b_{kk'}^{i} = a_{ki} a_{k'i}, \qquad \forall i = 1, \ldots, N$$

The matrix of the auto and cross variograms can be written in the following way:

$$\begin{bmatrix} \psi_{11} & \cdots & \psi_{1K} \\ \psi_{21} & \cdots & \psi_{2K} \\ \vdots & \vdots & \vdots \\ \psi_{K1} & \cdots & \psi_{KK} \end{bmatrix} = \begin{bmatrix} b_{11}^{1} & \cdots & b_{1K}^{1} \\ b_{21}^{1} & \cdots & b_{2K}^{1} \\ \vdots & \vdots & \vdots \\ b_{K1}^{1} & \cdots & b_{KK}^{1} \end{bmatrix} \begin{bmatrix} \gamma_1 & \cdots & 0 \\ 0 & \cdots & 0 \\ \vdots & \vdots & \vdots \\ 0 & \cdots & \gamma_1 \end{bmatrix} + \cdots$$
$$+ \begin{bmatrix} b_{11}^{N} & \cdots & b_{1K}^{N} \\ b_{21}^{N} & \cdots & b_{2K}^{N} \\ \vdots & \vdots & \vdots \\ b_{K1}^{N} & \cdots & b_{KK}^{N} \end{bmatrix} \begin{bmatrix} \gamma_N & \cdots & 0 \\ 0 & \cdots & 0 \\ \vdots & \vdots & \vdots \\ 0 & \cdots & \gamma_N \end{bmatrix}$$

and it is positive definite if all the matrices $[b^{i}_{kk'}]$ are positive definite: remember that this last condition is satisfied if the eigenvalues of these matrices are positive. Then the linear model of coregionalization is very convenient because the conditions of permissibility are readily satisfied. Moreover, each auto and cross variogram model must be constructed using the same basic variogram models.

5. A case study

The multivariate approach has been applied to estimate the dissolved oxygen (DO) values using DO and temperature measurements taken in the first bay of the Mar Piccolo of Taranto. A comparison between cokriging and ordinary kriging has been provided. The data set consists of DO and temperature values measured the 20th of November 1986 in 49 survey stations, located at $1m$ under the sea level (Fig. 1).

Figure 1: *Postplots of* (a) *DO,* (b) *temperature measured the 20th of November 1986 in 49 survey stations of Mar Piccolo of Taranto,* (c) *DO measured in 7 survey stations chosen for the structural analysis.*

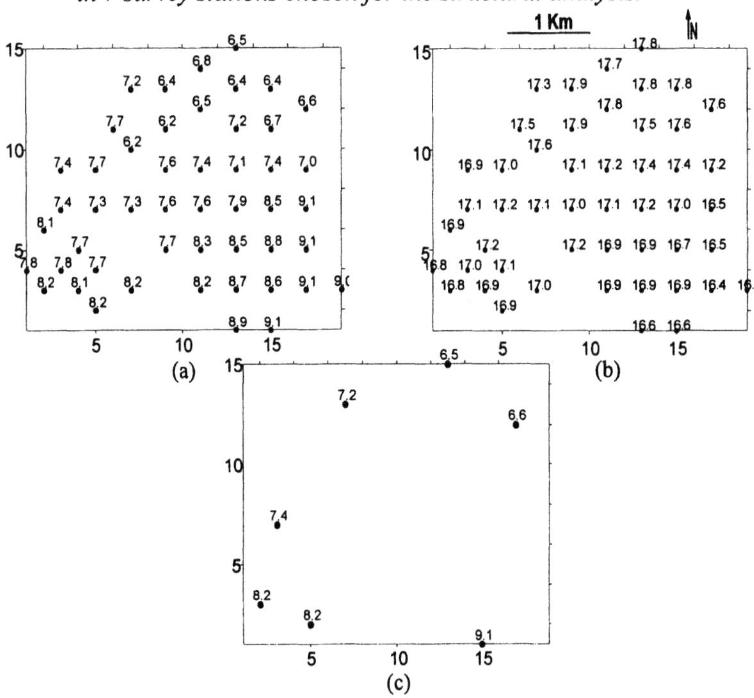

As Fig.1(a) shows, the lowest values of DO are localized in the Northern part of the bay. This is because of some submarine springs. The negative correlation between DO and temperature is confirmed by the posting map of the temperature values of Fig. 1(b), where the highest values have been measured in the same Northern part of the bay.

In order to compare cokriging and kriging DO estimates, it has been convenient:

- to choose only 7 DO sampled stations, in such a way that the estimates can be obtained in interpolation mode (Fig. 1(c));
- to take all of the temperature sampled stations (Fig. 1(b)).

The aim of this application is to illustrate how cokriging can improve an ordinary kriging estimate by using information about other variables, spatially cross-correlated with the one under study.
Fig. 2 shows the histograms, respectively, of DO and temperature daily values observed in 49 sampled stations. Fig. 3 shows the sample variograms and cross-variogram of the two variables and their models. The parameters of these models have been obtained by the weighted-least squares method (Cressie, 1993).

Figure 2: *Histograms of 49* (a) *DO and* (b) *temperature values.*

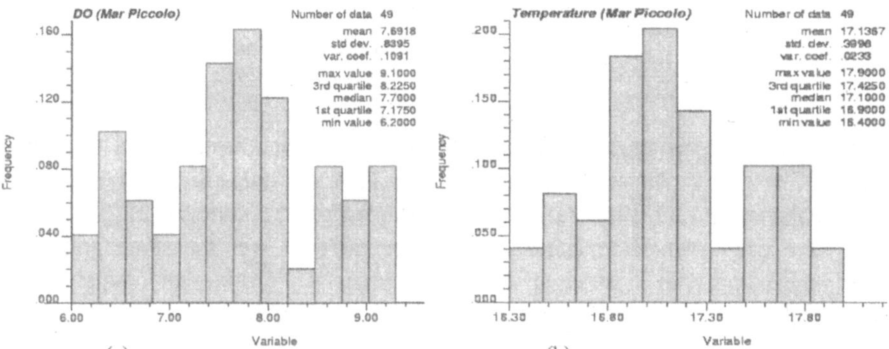

(a) (b)

Figure 3: *Sample variograms and their models of* (a) *DO and* (b) *temperature;* (c) *sample cross-variogram and its model of DO and temperature.*

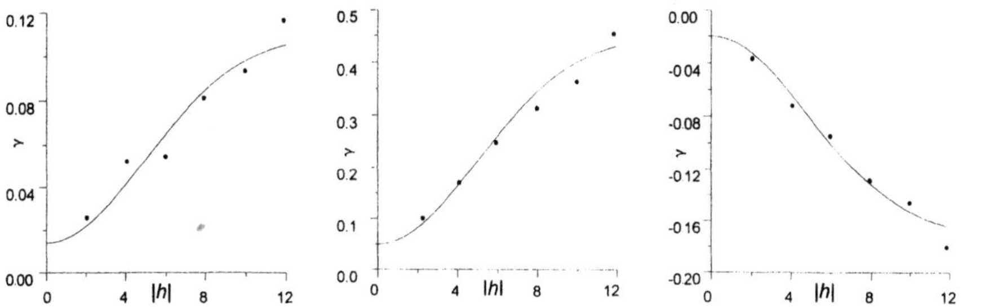

(a) (b) (c)

The linear model of coregionalization has a nugget effect and a gaussian structure with range equals to 12:

$$\gamma_Z(|\mathbf{h}|)=0.05+0.4(1-exp(-|\mathbf{h}|^2/144)) \quad |\mathbf{h}|>0 \tag{10}$$

$$\gamma_Y(|\mathbf{h}|)=0.014+0.09(1-exp(-|\mathbf{h}|^2/144)) \quad |\mathbf{h}|>0 \tag{11}$$

$$\gamma_{ZY}(|\mathbf{h}|)=-0.02-0.152(1-exp(-|\mathbf{h}|^2/144)) \quad |\mathbf{h}|>0 \tag{12}$$

where $\gamma_Z(|\mathbf{h}|)$ and $\gamma_Y(|\mathbf{h}|)$ are, respectively, the sample variograms of DO and temperature, and $\gamma_{ZY}(|\mathbf{h}|)$ is the cross-variogram of the two variables. Moreover, the admissibility of the model is ensured because the determinants related to the nugget effect coefficients and the sills values are positive:

$$\begin{vmatrix} 0.050 & -0.020 \\ -0.020 & 0.014 \end{vmatrix} = 7 \times 10^{-3} - 4 \times 10^{-3} > 0,$$

$$\begin{vmatrix} 0.400 & -0.152 \\ -0.152 & 0.096 \end{vmatrix} = 0.0384 - 0.0231 > 0.$$

Firstly, the variogram model (10) has been used to estimate the variable DO, over a grid 37 x 29, by ordinary kriging, using both 7 values only (Fig. 1(c)) and the entire data set of DO (Fig. 1(a)). Secondly, the variogram models (10), (11) and (12) have been utilized to estimate the same variable over the same grid by cokriging, using both 7 values of DO (Fig. 1(c)) and 49 values of temperature (Fig. 1(b)).

Figure 4: *Contour maps of* (a) *kriging estimates (using 7 DO values),* (b) *cokriging estimates (using 7 DO values and 49 temperature values),* (c) *kriging estimates (using 49 DO values).*

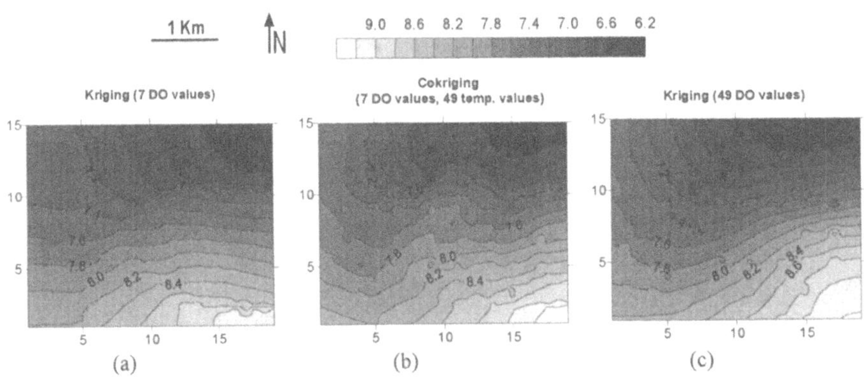

The contour maps of Fig. 4 highlights that using temperature, as a secondary variable, improves the estimates of DO, considered as a primary variable, since the two variables have a strong correlation and the secondary variable is available at many more locations than the primary variable.

Figure 5: *Postplots of the standard error* (a) *by cokriging;* (b) *by kriging.*

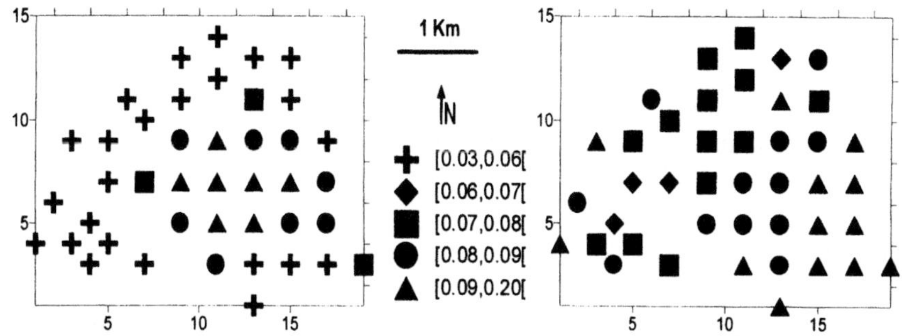

(a) (b)

Since 42 data points have not been used for structural analysis and for the estimation process of DO, a comparison between the true values, available at these 42 locations, and the estimated values, at the same locations, has been performed in order to check the goodness of cokriging. Fig. 5 shows the postplots of the standard error (square root of the error variance). Fig. 6 shows the scatterplots of the true values towards the estimated ones.

Figure 6: *Scatterplots of the true values, measured in the 42 stations not used for the estimation, towards* (a) *the cokriging estimates;* (b) *the kriging estimates (using 7 DO values).*

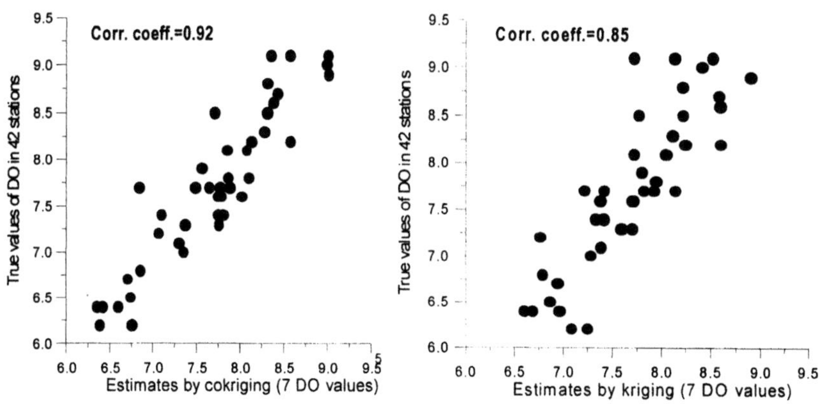

(a) (b)

Hence, it is clear from the last three figures that:

- cokriging is more efficient than kriging;
- the correlation coefficient between true values and estimated values is higher when the estimates are obtained by cokriging.

6. Conclusions

In this paper some partial solutions to some problems frequently met in the practice of cokriging, have been given. These problems are related to the estimation and modeling of cross-covariances or cross-variograms. The linear model of coregionalization provides a method for modeling variograms and cross-variograms of two or more variables. An application has been discussed.

References

Christakos, G. (1984). *On the problem of permissible covariances and variograms models*, Water Resources Research, Vol. 20, No 2, p. 251-265

Cressie, N. (1993). *Statistics for Spatial Data*, Wiley \& Sons, Inc., New York

Journel, A. G. & Huijbregts, C.J. (1981). *Mining Geostatistics*, Academic Press, New York

Myers, D.E. (1982). *Matrix Formulation of Cokriging,* Mathematical Geology, 14, 3, 249-257.

Myers, D.E. (1983). *Estimation of linear combinations and cokriging*, Mathematical Geology, 15, 633-637.

Posa, D. (1995). *Introduzione alla Geostatistica*, Adriatica Ed. Sal., Lecce.

Wackernagel, H. (1995). *Multivariate Geostatistics*, Springer, New York.

Exploring Multivariate Spatial Data: an Application to Election Data

Giovanna Jona-Lasinio, Fabio Mancuso
Dipartimento di Statistica, Probabilità e Statistiche Applicate
University of Rome "La Sapienza"
e-mail: jona@pow2.sta.uniroma1.it

Abstract: In this paper we report a brief account of an application to spatial data, observed on an irregular grid, of an exploratory technique based on the diagonalization of cross-variogram matrices. Our aim is to describe the behavior of a multivariate set of spatial data in a dimensionally reduced space in such a way that the information on the spatial variation is preserved. We adapt an exploratory technique built for the analysis of quantitative data to frequency data. We give special attention to the choice of a distance measure that well describes the type of "connection" between sites we consider for the analysis of this specific situation. The application aims to the characterization of districts in the city of Rome according to the electoral behavior of their inhabitants, special attention is given to the increasing phenomenon of abstentions.

Keywords: Cross-variogram, Principal component analysis, singular value decomposition, spatial data, undirected graphs, GIS.

1. Introduction and methods

The technique we apply has been developed by Capobianchi and Jona Lasinio (1998) for multivariate spatial data observed on a transect. Here we deal with a more complex arrangement of the spatial units. Our data are the results from several electoral competitions held in Rome (Italy) during the years from 1994 to 1996 and several socio-economical variables observed during the 1991 Italian census. Then the spatial scheme can be represented as an irregular grid whose elements are the districts in which the town is divided. The analysis of this dataset could be (in principle) developed through a three-way matrix technique, however because of the large number of spatial observations these techniques are too complex from a computational point of view, further more the procedure here adopted gives to the analyst a better inside on the spatial variation of the multivariate dataset.

In order to include information on the spatial arrangement of the observations we apply Principal Component Analysis (PCA) to the *cross-variograms* matrices (see also Tailiang Xie and Myers (1995a, 1995b)). More precisely, let $\mathbf{Z}(s_j)$ be a k dimensional vector of observations taken on the spatial unit s_j (j=1,...,n), under stationarity assumptions (at least the hypotheses of an intrinsic stationary processes generating the data should be made), we define the *cross-variogram* matrix at *lag h* as :

$$\Gamma(h) = \frac{1}{2}\mathrm{E}((\mathbf{Z}(s_j) - \mathbf{Z}(s_i)(\mathbf{Z}(s_j) - \mathbf{Z}(s_i))^T$$

where h is the value of the distance between unit s_j and unit s_i (direction is not taken into account). Our main interest is the joint variation in space of the multivariate dataset for all values of the distance between sites, then we need to synthesize the information contained in the m cross-variogram matrices ($h=h_0,\ldots,h_m$). The most natural choice in this direction is to build a *synthesis* matrix and use it to perform PCA. We build the following synthesis matrix:

$$\Gamma = \sum_{h=1}^{h_m} \Gamma(h) \tag{1}$$

In the univariate setting it is of great interest to evaluate the range (h^*) of the variogram. In other words, we can consider observations taken at locations, say, s_i, s_j, such that their distance ($d(s_i, s_j)$) is larger then h^* as almost uncorrelated. In the multivariate setting no equivalent tool has been defined. However in Capobianchi and Jona Lasinio (1998) it is proposed the use of Γ in order to identify a multivariate equivalent of the variogram range. More precisely we write (1) in terms of its eigenvalues and eigenvectors, *i.e.* $\Gamma=\mathbf{Q}\Lambda\mathbf{Q}^T$ where $\mathbf{Q}$ is the k x k matrix of its normalized eigenvectors and Λ is the diagonal matrix of its eigenvalues. Then we can decompose each eigenvalue of Γ in terms of *contributions* given to its value by each chosen spatial lag h. More precisely, if second order stationary assumptions hold, we have:

$$\lambda_i=\mathbf{q}_i^T\Gamma\,\mathbf{q}_i=\mathbf{q}_i^T\sum_{h=1}^{h_m} \Gamma(h)\,\mathbf{q}_i= \sum_{h=1}^{h_m} \mathbf{q}_i^T(\mathbf{C}(0) - \mathbf{C}(h))\mathbf{q}_i \tag{2}$$

where $\mathbf{C}(\cdot)$ is the covariance matrix computed at lag h. Now we can define:

$$h_i=\arg\max_h \mathbf{q}_i^T(\mathbf{C}(0) - \mathbf{C}(h))\mathbf{q}_i \tag{3}$$

Expression (3) can be written as $h_i=\arg\min_h \mathbf{q}_i^T\mathbf{C}(h)\mathbf{q}_i$, and h_i is the spatial lag corresponding to a minimum of the spatial variation projected on the i-th axes obtained from the singular value decomposition of Γ. As we just pointed out this quantity can be computed for each eigenvalue giving information on different patterns in the spatial variation. However as the first eigenvalue accounts for the largest amount of variability, in what follows we choose $h^*=h_1^*$ as the "global range" for the multivariate spatial phenomenon.
Notice that the univariate variogram computed on the first principal component obtained from Γ's decomposition is $\gamma^1(h)= \mathbf{q}_1^T\Gamma(h)\,\mathbf{q}_1$, then h_1^* (in our words the

"range of the cross-variogram matrix") is the range of the univariate variogram of the first principal component.
Among the many possible use of this quantity (see for instance Jona Lasinio, 1999) an interesting one is the definition of an *influence region* for each considered spatial unit, *i.e.* the influence region (*IR*) of unit s_j is the set of all units located within a distance $h<h^*$ from s_j itself:

$$IR(s_j) = \{s_i \in D : d(s_i, s_j) < h^*, i \neq j\}$$

For instance by using this definition we can (empirically) verify assumptions on the neighborhood system between sites, we can map the spatial relationships between observed units and we can try to model the dependence structure of spatial units.

2. The application

Our aim is to verify from a quantitative point of view what claimed in sociological studies of the abstention behavior in the city of Rome (see Mancuso, 1998 and references therein) and, to adapt to frequency data observed on a complex spatial scheme the exploratory technique proposed in Capobianchi and Jona Lasinio (1998). The dataset is composed of 110 spatial units on which we observe 50 demographical, social and economical variables from the 1991 census and 11 variables related to the electoral competitions, more precisely: scores of political groups in each election, level of "true" abstention (the citizen do not attend the electoral competition, ASTNV), level of "non valid vote" or "protest abstention" (the citizen invalidate its vote attending the competition, ASTVNV). As the electoral competitions taken into account are of very different nature (local, national and European elections) we analyze each year separately, building the analysis' protocol on the political elections held in Rome in 1996. The detailed analysis of these data can be found in Mancuso (1998). Almost all the variables in the dataset are expressed in terms of frequency of occurrence which can be seen as measures of the intensity of each phenomenon. The census variables standardized with respect to the population of the district (older then 18 years) and the election data are standardized with respect to the total number of votes (including abstentions and non-valid votes). We applied a logarithmic transformation to all variables in order to smooth them and better compare their behavior. Furthermore is necessary to take into account the some how arbitrary scale in which our variables are measured, then we divided the transformed variables by their standard deviation so to represent them on the same scale.
The second step has been to deal with the very high dimension of the problem. We chose to apply a standard PCA to the 50 census variables and use this artificial variables in the exploration of the spatial variation. The first two components account for 85.6% of the total variation and are easily characterized as follows:

1st component: highly educated individuals (university level), prevalence of one member families, high income, older residents (CENSUS1).

2nd component: medium to low education, prevalence of large families (3 or more members), low income, younger residents (CENSUS2).

Our analysis is based on these two "new" variables instead of the 50 census variables, this choice being supported by a later analysis, in which all the 50 census variables were used, that gave the same results.

The first step in including spatial information in the analysis is the definition of neighborhood relationships between districts. We consider units to be neighbors if they have a common boundary and at least one road to connect them. We denote by **S** the corresponding contiguity matrix. This definition can be completed using data on the availability and efficiency of public transportation. In fact units can be seen as neighbors when easily connected by a "good" public service.

In the analysis of spatial data based on cross-variogram matrices most of the spatial information is taken into account by the chosen distance measure and then the choice of such a measure becomes of relevance. In this study we considered relevant, as the sociological point of view prevails, "how easily" one unit is connected to another and less important the unit's dimension, two districts are "neighbors" if they are connected with a high-speed road or a good service of buses etc.. In fact previous studies showed how the abstention phenomenon has a higher intensity among people of relatively high social status and people connected (in a very broad sense) to them (this behavior started changing in recent years). Then, taking into account the type of information available, only how many "steps" are necessary to go from one location to another should be included in the evaluation of the distance. This last consideration led us to the choice of the *accessibility distance* (instead of the Euclidean one) which is given by $d(s_j, s_i)=\min\{b: s_{ji}^{(b)}>0\}$ where $s_{ji}^{(b)}$ is the i,j-th entry of the b-th power of matrix **S**. With this choice the distance between two location s_j, and s_i , if we look at them as knots on a undirected graph, is the measure of the length of the minimal path connecting two units, measured in terms of the number of knots one has to touch before reaching s_j, from s_i. As an example in Figure1 we show the *influence regions* for the Quartiere Nomentano (Q5). The lighter grey part contains units belonging only to the region built using the accessibility distance, the darker grey part contains units belonging only to the one obtained using Euclidean distance, the remaining colored units are common to both regions. The region resulting from Euclidean distance is smaller in terms of surface and includes mainly units closer to the center of the city, while the region resulting from the accessibility distance develops more towards the suburbs regardless to the units surface, well representing the daily movement of commuters from the suburbs to the ancient part of the city where many working places are concentrated. The central part of Rome is, roughly speaking, sociologically characterized by the presence of small family, high income and high cultural level. However we have to remember that a socio-economical characterization of Rome's districts (in particular in the center) can never be very precise, as several type of social realities are present simultaneously.

Figure 1 : *Influence regions of Quartiere Nomentano (Q5) built using the euclidean and accessibility distance (1996 elections).*

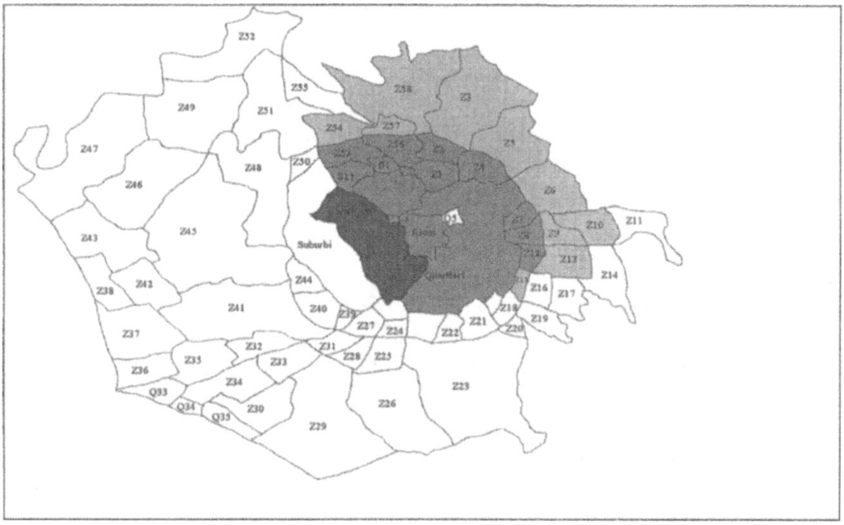

Once the distance matrix is computed we bin locations in classes of distance in order to guarantee a consistent estimate of the cross-variogram matrices at each chosen lag (we considered 17 different values of h) and we compute the empirical version of $\Gamma(h)$. The data matrices, defined on each electoral competition, we use in building the matrix Γ have 110 rows (districts) and 13 columns (variables). In table 1 results from the analysis of 1996 elections are reported. Three components account for more then 70% of the total variation. In order to make the table more readable consider that AN, FI, CCD and DINI can be considered center-right political groups, PPI, PDS, PRC and VERDI center-left political groups, while OTHERS collect a miscellanea of small groups of very different political nature. The value of the range (constant for all the 3 components and in each electoral competition) points out how all the units are spatially highly interconnected from both a social and a political point of view. The spatial pattern seems constant as it does not change from one component to the other, while the characterization of axes slightly changes in the political characterization from one type of election to the other. From Table 1.b we can deduce two patterns each connected to one of the census variable. On the first P.C. we have the association between right center groups, abstention (ASTNV) and the first census variable ("the upper class") in opposition to left-wing groups and the protest abstention. On the second P.C. the second census variable with the two abstention variable has opposite behavior with small political groups and the right wing.

Table 1 *Analysis of the 1996 election data and the two census components*

a) *Eigenvalues, range and accounted variation*

Eigenvalue	h^*	% of variation	% of cumulated variation
1	5	34.60	34.60
2	5	23.49	58.09
3	5	15.17	73.26

b) *Characterization of the first three Principal Components*

Variables	Scores 1° P.C.	Variables	Scores 2° P.C.	Variables	Scores 3° P.C.
CENSUS1	-0.4309	ASTNV	-0.5192	FI	-0.4981
DINI	-0.3642	CENSUS2	-0.4011	ASTVNV	-0.2454
PPI	-0.3245	CENSUS1	-0.0475	CENSUS2	-0.1948
OTHERS	-0.2882	ASTVNV	-0.0346	ASTNV	-0.1875
ASTNV	-0.1933	PRC	0.0582	AN	-0.1598
CCD	-0.1904	PDS	0.0618	ALTRI	-0.0673
FI	-0.0493	DINI	0.0959	CENSUS1	0.1569
CENSUS2	0.0033	CCD	0.1202	CCD	0.1875
VERDI	0.0778	PPI	0.1970	PPI	0.2485
AN	0.0838	VERDI	0.2218	PRC	0.2521
ASTVNV	0.3172	OTHERS	0.2604	DINI	0.2522
PDS	0.3755	FI	0.3584	PDS	0.3086
PRC	0.4031	AN	0.5009	VERDI	0.4946

We can build a map of the city of Rome according to the analyzed variables by binning into class values assumed by each district on each principal component. In figure 2 we report the mapping of the first and second principal component obtained from the analysis of 1996 election data together with the census variables (see table 1). The coloring proceed as follows: on the 1st PC light grey - highly educated individuals (university level). prevalence of one member families. high income. older residents. high level of abstention (ASTNV) and vote for moderate (center) political groups (CCD. PPI. DINI); as the coloring darkens the characterization moves towards a smaller level of abstentions. lower income and educational level, protest abstention and more "extreme" political choices; on the 2nd PC light grey corresponds to high level of abstention (ASTNV) in relation to the second census component and as the coloring darkens a lower level of abstention connected to right wing political choices is enhanced.

Figure 2: *Map of Rome's districts (inner loop) based on the analysis of 1996 election data and census variables*

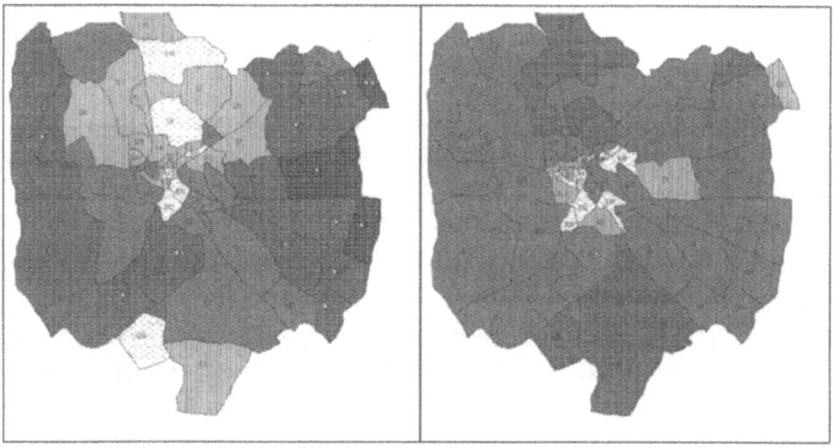

a) *1st PC* **b)** *2nd PC*

From Figure 2 we see that this technique can be successfully used to build a GIS for multivariate phenomena.

Further studies have been carried out (Mancuso. 1998). We compared the above results with the separate analyses of the census variables and the election data. In both cases the influence region is the same. (h^*=5) as well as the characterization of Rome's districts in terms of socio-economical and political variables.

3. Concluding Remarks

We consider very satisfactory the description of Rome districts here obtained. Results are highly consistent with the sociological analysis and with previous studies using multiway techniques. Furthermore our method proved to be computationally efficient. This particular application showed how the use of the proposed technique is not confined to the study of strictly quantitative phenomena. However in the study of frequency data. special care should be used in the choice of the distance measure used in the computation of cross-variogram matrices. as our example clearly pointed out. When the main interest of the study involves the surface dimension of the spatial units. the Euclidean distance is certainly appropriate. while when the "connection" between locations prevails. the accessibility distance proved to be a better choice. The joint analysis of the census variables and the election data allowed a more synthetic mapping of results. Then if the aim of the study is the building of a district's map this approach is preferable to the separate analysis one. Finally our choice of PCA as the main tool for dimensional reduction of the problem is enforced by the fact that this ordination procedure is the one that mostly preserves information based on second moments of the dataset.

References

Capobianchi. A., Jona Lasinio. G. (1998). Exploring Multivariate Spatial Data: Line Transect Data. in *Classification and Data Analysis: Theory and Application.* Eds. M. Vichi. O. Optiz. Series: Classification. Data Analysis and Knowledge Organization. Springer and Verlag.

Jona Lasinio. G. (1999). Modeling and Exploring Spatial Variation. Technical Report 5-99. Department of Statistics. Rice University.

Mancuso. F. (1998). *Analisi Esplorativa di dati spaziali multivariati: I dati elettorali a Roma dal 1994 al 1996.* Master thesis Dipartimento di Statistica. Probabilità. e Statistiche Applicate. University of Rome "La Sapienza" (in italian)

Tailiang Xie, Myers D.E.. (1995a) Fitting Matrix Valued Models by Simultaneous Diagonalization (Part I: Theory) *Mathematical Geology* **27**. N.7. 867 - 875.

Tailiang Xie, Myers D.E.. (1995b) Fitting Matrix Valued Models by Simultaneous Diagonalization (Part II: Application) *Mathematical Geology* **27**. N.7. 877-888.

Measures of Distances for Spatial Data: an Application to the Banking Deposits in the Italian Provinces

Eugenia Nissi
Università G. d'Annunzio
Dipartimento di Metodi Quantitativi e Teoria Economica
Viale Pindaro, 42 -65127 Pescara
nissi@dmqte.unich.it

Abstract: A theoretical and practical approach is presented for extending both the Mahalanobis and the Euclidean distances to spatially correlated data.
Departing from the consideration that in some cases the choice of a particular distance is suggested by the nature of the data, we propose new measures of distance for spatial observations based on the distances between the interpretative models of the data.
An application to the banking deposits of the Italian provinces in 1995 is given.

Keywords: simultaneous autoregressive models, spatial regression models, Mahalanobis distance, Euclidean distance, spatially correlated data, banking deposits.

1. Introduction

Many techniques of multivariate statistical analysis require the use of a distance function. This is the case for example of cluster, discriminant and canonical analysis.
Sometimes the distance is naturally suggested by the statement of the problem. This is the case of environmental, geostatistical data as well as spatial data in general.
In these cases a flexible and general approach uses a measure of distance based on the interpretative model of the data rather that the classical distances on the statistical units.
In this study the aim of this paper is to propose and discuss an extension of the Mahalanobis and Euclidean distances that take into account the spatial autocorrelation of the data.
In particular we develop an extension of the above mentioned distances when the data can be represented by a simultaneous autoregressive spatial model or by a regression spatial model.
An application to the banking deposits of Italian provinces in 1995 is given.

2. Introduction to spatial models

Suppose that the observations are available for i =1,2,...,n regions or sites in a study area over t=1,2,....,T time periods of time. Let the value of a single dependent variable Y observed in the j-th region for the i-th time denoted as y_{ji}. If T=1 (that is, the data are available for one period of time) we have solely spatial or cross sectional data).

Following the notation developed by Box and Jenkins (1970), we may incorporate a spatial autoregressive, moving average component and regressive component into a SARMAR model:

$$\mathbf{y} = \rho\mathbf{W}_1\mathbf{y} + \mathbf{X}\beta + \varepsilon \tag{1}$$

with $\varepsilon = \lambda\mathbf{W}_2\varepsilon + \mu$ where $\mu = MNV(0,\sigma^2(I))$.

In this general specification, **X** is a matrix of known non stochastic regressors, β is a vector of unknown parameters associated with the exogenous variables X, ρ is an autoregressive coefficient for the lagged spatial variables, λ is a coefficient of the spatial autoregressive structure of the residual component and ε is a disturbance term with zero mean.

$\boldsymbol{W}_1$ and $\boldsymbol{W}_2$ are known matrices of standardised weights (contiguity matrix) associated with the spatial autoregressive components of the equation (1) and with disturbance term.

For simplicity, we suppose that matrices $\boldsymbol{W}_1 = \boldsymbol{W}_2 = \mathbf{W}$ so we can rewrite the model (1) as follows

$$\mathbf{y} = \rho\mathbf{W}\mathbf{y} + \mathbf{X}\beta + \varepsilon \tag{2}$$

with $\varepsilon = \lambda\mathbf{W}\varepsilon + \mu$.

From model (1) we can obtain a wide class of models; e. g. if ρ=0 and λ=0 we obtain the classical regression model

$$\mathbf{y} = \mathbf{X}\beta + \mu \tag{3}$$

Further constraints can be imposed on the models (2); in particular we can consider the simplification

$$\mathbf{y} = \lambda W\mathbf{y} + \mu \tag{4}$$

when β=0 and λ=0, the model (3) is known in the literature as a Spatial Autoregressive Model. This model is purely spatial autoregressive, that is we are postulating that the value of **y** in a region is dependent upon the values of **y** in some subset of other regions of the study area.

The last element in the basic model (1) is the moving average term. That is if $\beta = 0$ and $\rho = 0$, the models in (1) are reduced to a spatial moving average scheme (SMA – Spatial moving average)

$$y = \lambda \mathbf{W} \varepsilon + \mu \tag{5}$$

The estimation of parameters of the model of the form (1) raises several problems which are discussed in Cliff and Ord (1973). In particular, the estimation procedures assume that the processes concerned are weakly spatially stationary (second order stationary). That is, we require that the quantities,

$$Cov(Y_i, Y_j) = \sigma(i, j) \tag{6}$$

to depend only upon the relative position of locations i and j. Verbally Granger (1969) has summarized spatial stationarity as "an assumption" that the relationship between values of the processes is the same for every pair of point whose relative position are the same.

3. Measures of distances for spatial models

For the above mentioned class of models (1), as for the ARIMA models (Piccolo,1983), it is possible, to define a set ℓ, whose elements are represented by certain models.
The set whose elements M_1, M_2, ...M_i,....M_m are called models allows us to define a metric space as ℓ where we can assign a real number $d(M_i, M_j)$ called distance between models.
On the basis of this consideration, departing from the interpretative model of data, a measure of Euclidean distance can be expressed as follows

$$d_2(M_i, M_j) = \sqrt{\sum_{j=1}^{m} \left(m_{aj} - m_{bj}\right)^2} \tag{7}$$

and the Mahalanobis

$$M_2(M_i, M_j) = \sqrt{\sum_{j=1}^{m} \sum_{h=1}^{m} \mathbf{S}_{hj} \left(m_{aj} - m_{bj}\right)\left(m_{ah} - m_{bh}\right)} \tag{8}$$

where $\mathbf{S}_{hj}$ is a generic element of the inverse covariance matrix and m_{aj}, m_{bj}, m_{ah} and m_{bh} are the computed value from fitted model.

In particular for the spatial regression model (2), the Euclidean measure of distance between models called SRM (Spatial Regression Metric) has the following form:

$$d_2 = (M_i, M_i) = \sqrt{\sum_{j=1}^{m} \left(\mathbf{y}_{aj} - \mathbf{y}_{bj}\right)^2} \tag{9}$$

where:

$$\mathbf{y}_{aj} = \rho \mathbf{w}^{t}_{aj} \mathbf{y}_{aj} + \sum_{p=1}^{kj} \mathbf{x}_{ajp} \beta_{jp} + \lambda_j \mathbf{w}^{t}_{a} \varepsilon_j + \mu_{aj} \tag{10}$$

$$\mathbf{y}_{bj} = \rho \mathbf{w}^{t}_{bj} \mathbf{y}_{bj} + \sum_{p=1}^{kj} \mathbf{x}_{bjp} \beta_{jp} + \lambda_j \mathbf{w}^{t}_{b} \varepsilon_j + \mu_{bj} \tag{11}$$

while the Mahalanobis distance assumes the following form

$$M_2 = (M_i, M_j) = \sqrt{\sum_{j=1}^{m} \sum_{h=1}^{m} \mathbf{s}_{hj} \left(\mathbf{y}_{ah} - \mathbf{y}_{bh}\right)\left(\mathbf{y}_{aj} - \mathbf{y}_{bj}\right)} \tag{12}$$

where

$$y_{ah} = \rho_h \mathbf{w}^{t}_{a} \mathbf{y}_{ah} + \sum_{p=1}^{kj} \mathbf{x}_{ahp} \beta_{hp} + \lambda_h \mathbf{w}^{t}_{a} \varepsilon_h + \mu_{ah} \tag{13}$$

$$y_{bh} = \rho_h \mathbf{w}^{t}_{b} \mathbf{y}_{bh} + \sum_{p=1}^{kj} \mathbf{x}_{bjp} \beta_{hp} + \lambda_h \mathbf{w}^{t}_{b} \varepsilon_j + \mu_{bh} \tag{14}$$

$$y_{aj} = \rho_j \mathbf{w}^{t}_{a} \mathbf{y}_{aj} + \sum_{p=1}^{kj} \mathbf{x}_{ajp} \beta_{jp} + \lambda_j \mathbf{w}^{t}_{a} \varepsilon_j + \mu_{aj} \tag{15}$$

$$y_{bj} = \rho_j \mathbf{w}^{t}_{b} \mathbf{y}_{bj} + \sum_{p=1}^{kj} \mathbf{x}_{bjp} \beta_{jp} + \lambda_j \mathbf{w}^{t}_{b} \varepsilon_j + \mu_{bj} \tag{16}$$

Similar measures of distance can be calculated for spatial autoregressive model SAR. In particular, for this model the Euclidean distance between models called SARM (Spatial Autoregressive metric) model take the following form

$$d_2 = (M_i, M_j) = \sqrt{\sum_{j=1}^{m} \left(\mathbf{y}_{aj} - \mathbf{y}_{bj}\right)^2} \tag{17}$$

with:

$$y_{aj} = \rho \mathbf{w}^{t}_{a}\mathbf{y}_{aj} + \mu_{aj} \tag{18}$$

$$y_{bj} = \rho \mathbf{w}^{t}_{b}\mathbf{y}_{bj} + \mu_{bj} \tag{19}$$

and the Mahalanobis distance is

$$M_2 = (M_i, M_j) = \sqrt{\sum_{j=1}^{m} \sum_{h=1}^{m} \mathbf{s}_{hj} (\mathbf{y}_{ah} - \mathbf{y}_{bh})(\mathbf{y}_{aj} - \mathbf{y}_{bj})} \tag{20}$$

$$y_{ah} = \rho \mathbf{w}^{t}_{a}\mathbf{y}_{ah} + \mu_{ah} \tag{21}$$

$$y_{bh} = \rho \mathbf{w}^{t}_{b}\mathbf{y}_{ah} + \mu_{bh} \tag{22}$$

$$y_{aj} = \rho \mathbf{w}^{t}_{a}\mathbf{y}_{aj} + \mu_{aj} \tag{23}$$

$$y_{bj} = \rho \mathbf{w}^{t}_{b}\mathbf{y}_{aj} + \mu_{bj} \tag{24}$$

It should be pointed out that the above introduced measures of distances represent empirical distance estimates of the true distance between models, obviously for a different data set one might have different values of distance function.

4. An application to the banking deposits of the Italian provinces in 1995

In order to verify the performance of the measures of distance proposed in the application of multivariate techniques, an application is given.
In particular, the spatial series of the banking deposits in the 95 Italian provinces in 1995 has been considered.
The spatial series are referred to the monetary value of deposits, for each institution in the following categories: the banking deposits in major banks, the banking deposits in large banks and the banking deposits in medium banks following the official classification of the Central Bank of Italy (Bollettino Statistico Italiano – Banca d'Italia 1995).
The banking institutions are observed in the 95 Italian provinces whose shape and area of the provinces are not the same.
The contiguity matrix has been built considering as neighbourhood of a zone those areal units that share common boundary. That is a generic element of a contiguity matrix W is:

$$w_{ij} = \begin{cases} 0 & \text{if the the zones i and j are not contiguos} \\ 1 & \text{otherwise} \end{cases}$$

The three spatial series of banking deposits can be easily fitted with a first order spatial autoregressive models.
The results of the likelihood estimation for the parameter ρ of these models are shown in the Table 1.

Table 1: *Likelihood estimation of simultaneous autoregressive models for three spatial series of banking deposits.*

	Parameter estimated	Standard Error	T-value	P-value
Spatial series of banking deposits in major banks	0,7485	0,0625	11,983	0,0000
Spatial series of banking deposits in large banks	0,5002	0,0960	5.212	0,0000
Spatial seriesof banking deposits in medium banks	0,5834	0,0865	6.7436	0,0000

According with the nature of the data object of study the measures of the Euclidean and Mahalanobis distances between the Spatial Autoregressive Model have been calculated.
After we applied the classical Multidimensional Scaling on the dis(similarity) matrix obtained departing from the data matrix of banking deposits. In particular the Multidimensional Scaling has been applied considering the following measures of distance: the classical Euclidean distances, the Euclidean distances between models and the Mahalanobis distance between models.
The results are shown in Fig. 1, 2 and 3.
All the three solutions give good results, that are confirmed by the values of S-STRESS shown in Table 2.

Table 2: *Valued of S-Stress.*

S-STRESS of MDS based on Euclidean distance	0.30581
S-STRESS of MDS based on Euclidean distance between models	0.07202
S-STRESS of MDS based on Mahalanobis distance between models	0.10408

As you can see from the comparison of Fig. 1,2 and 3 the results of MDS are strongly influenced by the kind of distances.
In particular the use of a measures of distances that take into account of the spatial autocorrelation of the data produces modification of the representation of points in the bidimensional space, for example the provinces of Gorizia, Cosenza,

Trapani, Teramo and Rieti result very close in Fig. 1. These provinces probably are very similar with reference to the structure of the banking deposits.
The same conclusions can be made for the provinces of Como, Rieti, Torino and Milano Fig. 3.

Figure 1: *Graphical representation of results obtained with MDS using Mahalanobis distances between models.*

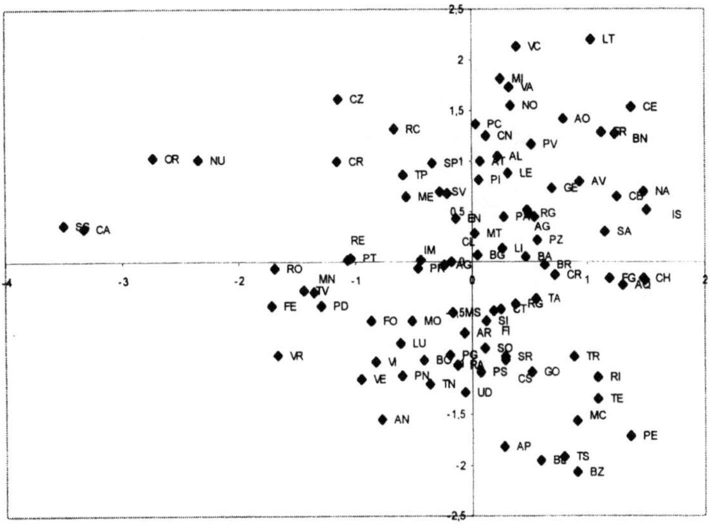

Figure 2: *Graphical representation of results obtained with MDS using classical Euclidean distance.*

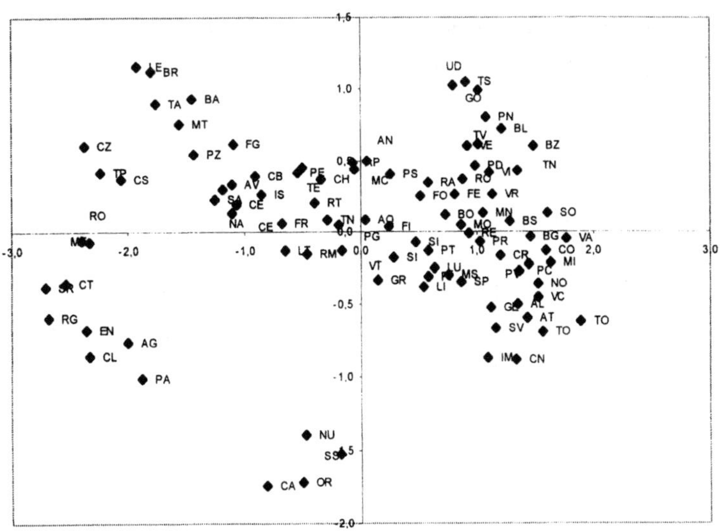

Figure 3: *Graphical representation of results obtained with MDS using Euclidean distances between models.*

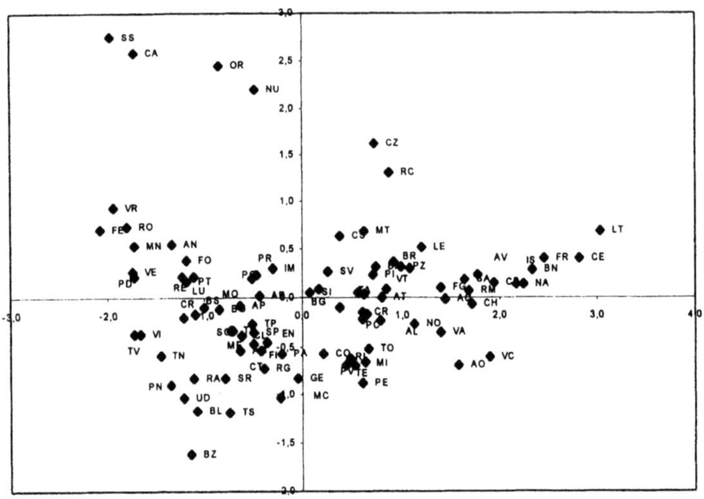

References

Anselin L. (1988), *Spatial Econometrics: model and applications*, Kluwer Dodrect.

Box G.E.P, Jenkins G.M. (1970), *Time series analysis, forecasting and control*, Holden Day, San Francisco.

Cliff A. D., Ord J.K. (1973), *Spatial Autocorrelation*, Pion Limited London.

Corduas M. (1992), Indici di distanza per processi gaussiani, *Atti XXXVI Riunione Scientifica Società Italiana di Statistica*, Pescara 21-24 aprile 1992, ppg. 111-118.

Granger C.W. J. (1969), Spatial and time series analysis, in *Studies in Regional Science*, pp. 1-24.

Leti G. (1979), *Distanze ed indici statistici*, Ed. La Goliardica Roma.

Piccolo D. (1984), Una topologia per la classe ARIMA, Statistica, XLIV, pp.47-59.

Vichi M, Di Battista T., Nissi E., (1998), Analisi di matrici a tre indici per dati di prossimità con struttura spaziale e/o temporale, *Atti XXXIX Riunione Scientifica Società Italiana di Statistica,* pp. 253-263.

Part IV

Applications and Case Studies

Statistical Analysis of Papal Encyclicals

Bruno Bisceglia* Alfredo Rizzi**
* Pontificia Università Gregoriana - Via del Seminario 120 - 00186 Roma
e-mail: bisceglia@unigre.urbe.it
** Dipartimento di Statistica, Probabilità e Statistiche Applicate
Università degli Studi di Roma "La Sapienza", piazza A. Moro 5, 00185 Roma
e-mail: rizzi@pow2.sta.uniroma1.it

Abstract: The purpose of this work is to demonstrate some possibilities offered by a statistical investigation of theological matters. The object of this paper is a textual analysis of the corpus obtained from the encyclical letters of the last 5 popes. It is possible by statistical analysis of the documents to arrive at conclusions or questions about "theology" in the encyclical letters.

Keywords: text analysis, encyclical letters, correspondence analysis, language, christological titles

1.1 The encyclical letters

In this work we show some of the results of research on the words used by the last 5 popes in their encyclicals: Pius XII (1939 - 1958), John XXIII (1958 - 1963), Paul VI (1963 - 1978), John Paul II (1978 —); John Paul I, given the shortness of his pontificate (33 days), did not produce encyclicals.

Our analysis studied either the words used in the various encyclicals of a pope, or made a comparison between popes.

What is an encyclical? This is a frequently asked question and one that has had a variety of answers. Perhaps the most recent answer was given by Pope Paul VI when he announced the imminent publication of his first encyclical letter. The Latin text, he pointed out, «would begin with the words that would serve to identify it: *Ecclesiam Suam*». Although he had issued other letters, he called this his «first» since it «had the precise character of an official *Lettera Enciclica*». An encyclical letter, he went on to say, «is a document in the form of a letter sent by the Pope to the bishops of the entire world»: *encyclical* means circular.

Although the encyclical form had been used earlier, it was not commonly adopted by the popes before the eighteenth century. Benedict XIV (1740 - 1758) was the first «to revive the ancient custom» of using the encyclical form as a means of insuring unity within the Church. «The Roman Pontiff, as the successor of Peter, is the perpetual and visible principle and foundation of unity of both the bishops and of the faithful».[1] «Nor must it be thought that what is expounded in Encyclical Letters does not of itself demand consent, since in writing such Letters the Popes do not exercise the supreme power of their Teaching Authority. For

[1] Second Vatican Council, Dogmatic Constitution *Lumen Gentium*, 23, 1964.

these matters are taught with the ordinary teaching authority, of which it is true to say: "He who hears you, hears me" (*Lk* 10, 16); and generally what is expounded and inculcated in Encyclical Letters already for other reasons appertains to Catholic doctrine».[2]

Since the latter part of the nineteenth century the encyclical letter and the encyclical epistle have come into favor as instruments of the ordinary teaching of the pope, and have been used extensively to treat not only dogmatic questions but also problems that are primarily social and ethical in scope. Pius XI, however, was the first to make a clear distinction in the use of these terms; this distinction has been maintained by his successors. Today the *encyclical letter* is used almost exclusively to indicate a circular letter addressed to all bishops on matter affecting the Church at large, while the *encyclical epistle* is reserved for letters in the nature of exhortations concerning a particular need or for those letters addressed primarily to bishops of a particular country or to a specific group of the hierarchy.

The last encyclical letter *Fides et Ratio*, was promulgated by John Paul II on 14 September 1998 on the relationship between faith and reason: «My Venerable Brother Bishops, health and the Apostolic Blessing! Faith and reason are like two wings on which the human spirit rises to the contemplation of truth; and God has placed in the human heart a desire to know the truth; in a word, to know himself, so that, by knowing and loving God, men and women may also come to the fullness of truth about themselves».[3]

1.2 The language

The official language is Latin with some exceptions: Pius XI in 1939 wrote the encyclical *Non abbiamo bisogno*, in Italian, on Catholic Action in Italy and in 1937 the encyclical *Mit brennender sorge*, in German, to the bishops of Germany on the Church and German Reich; Pius XII in 1957 wrote the encyclical *Le pelerinage de Lourdes*, in French, warning against materialism on the centenary of the apparitions at Lourdes. The documents available on magnetic support are in Italian, so the results refer to this version; the equivalent English term will be found in brackets.

1.3 Methods

The methods of data analysis used are:

- cluster analysis
- correspondence analysis
- specificity rate
- dissimilarity index

In this paper we report only some results about the analysis between popes.

---

[2] Pius XII, *Humani generis*, 22, 1950.

[3] John Paul II, *Fides et ratio*, Introduction, 1998.

2.1 Data analysis

In the following table we show the number of encyclicals and the number of occurrences, forms and important words for each of the popes examined.

Table 1

	number of encyclicals	number of occurrences	number of forms	number of imp. words
Pius XII	41	213534	18208	999
John XXIII	8	71851	9260	490
Paul VI	7	54139	7834	494
John Paul II	13	324932	17345	721

In table 3 we report some words, in order of decreasing frequency, used in the encyclicals of the last five popes. In the next to last column we have the absolute frequency in all the analyzed encyclicals. The word *Dio* [God] occurred 3119 times; this is 2.90% of all the important words used in all the encyclicals of the 4 popes. The serial number is 1: this word is the most frequently used in the encyclicals of these popes. We made the same analysis for all the important words of these encyclicals. For each Pontiff and for each word we report its series number in the whole corpus of encyclicals, the absolute frequency and percentage frequency.

2.2 Some remarks

The analysis of the above table and the other data processing that are not reported here for reasons of space lead to the following conclusions about the words used by the Pontiffs in these documents:

- In all 4 corpora analyzed, the most frequently used words are: *Dio* [God], *chiesa* [church], *Cristo* [Christ], *vita* [life], *uomo* [man].
- Paul VI and John Paul II do not frequently use the word *venerabili* [venerable], with reference to the hierarchy.
- The use of terms *Vergine* [Virgin] and *Maria* [Mary] is very common in Pius XII and John Paul II (respectively about 1% in Pius XII, 0.26% in John XXIII, 0.39% in Paul VI, 0.84 in John Paul II).
- The word *uomo* is frequently used in John Paul II, with a percentage frequency 3.23, 2nd place (after the word *Dio*) in the decreasing series list of words that occur in these documents.
- Our investigation shows also that *Figlio* [Son] is used more in the encyclicals of the present pope than in those of his predecessors.
- The word *clero* [clergy] seldom occurs in John Paul II.
- The word *mistero* [mystery] is very frequent in John Paul II (0.746%) and in Paul VI, while it is not much used by Pius XII and John XXIII.

- Also the word *predecessore* [predecessor] is not used very much by the current Pontiff. John XXIII was accustomed to refer in this way to Pius XII, his predecessor.
- John XXIII made explicit reference to *penitenza* [penance]: this term occurs with a frequency much higher (0.56%) than that of the other Pontiffs (less than 0.1%).
- The term *concilio* [council] is present in all the encyclicals, but has the maximum of relative frequency in Paul VI and the minimum in Pius XII; this matter is interesting if we take into account the historic epoch in which this important event in the life of the church happened.
- The word *celibato* [celibacy] seldom occurs in John Paul II and John XXIII, while it is much present in Paul VI.
- The word *sviluppo* [development] is much used by Paul VI (a little less than 1%), (0.54%) in John Paul II.
- The current Pontiff, in his encyclicals, often refers to problems about *lavoro* [work] (1.12%); the other popes used this word less frequently. However, the more frequent use of the word did increase over time. In fact we go from the 0.13% of Pius XII, to the 0.31% of John XXIII, to the 0.27% of Paul VI.
- The word *pace* [peace] is very often used by John XXIII (0.73%) and by Paul VI (0.71%).

3. Results

In table 2 we report some christological titles (or names of God) that occur in the important-word category.

Table 2

	Christological titles [freq %]			
	Pius XII	John XXIII	Paul VI	John Paul II
Dio [God]	2.16	2.22	2.60	3.42
Cristo [Christ]	1.55	1.61	2.95	2.27
Gesù [Jesus]	0.79	0.76	0.49	0.96
Signore [Lord]	0.48	0.68	0.70	0.76
Santo [Holy]	0.34	0.52	0.31	0.86
Padre [Father]	0.36	0.40	0.31	0.77
Figlio [Son]	0.28	0.10	0.08	0.80
totals	5.96	6.29	7.44	9.84

The results show that the christological titles are repeated strings that have a significant semantic role in emphasizing one aspect or another of the identity and function of Jesus Christ. The results also indicate the value of a method which conjoins the search for a more accurate understanding of theological concepts on the one hand, and the criteria of statistical analysis of textual data on the other. Many titles of Christ are proved to be syntheses of ancient Christian conceptions. Such syntheses arise from the blending of historical data and faith experiences. Jesus the Christ is viewed as the fullness of divine Revelation, the central theme of Christian theological speculation. An exhaustive study of christological titles comprises an integration of "what Jesus says about himself" and "what people say about him".

In the following table we have a kind of synthesis:

N — serial number
F — absolute frequency
F% — relative frequency
"-" — $F \leq 3$

Table 3

	Pius XII		John XXIII		Paul VI		John Paul II		Totals		
Words	N	F %	N	F %	N	F %	N	F %	N	F	F %
DIO	**1**	2.16	**2**	2.22	**3**	2.60	**1**	3.42	**1**	3119	2.90
CHIESA	**2**	1.86	**1**	2.55	**1**	4.10	**4**	2.63	**2**	2705	2.52
VITA	**4**	1.07	**4**	1.37	**4**	1.99	**3**	2.82	**3**	2296	2.14
CRISTO	**3**	1.55	**3**	1.61	**2**	2.95	**5**	2.27	**4**	2227	2.07
UOMO	**51**	0.35	**90**	0.28	**6**	1.28	**2**	3.23	**5**	2180	2.03
SPIRITO	**39**	0.41	**16**	0.80	**11**	0.90	**6**	1.83	**6**	1358	1.26
VERITÀ	**10**	0.66	**5**	1.09	**25**	0.59	**7**	1.63	**7**	1309	1.22
FEDE	**6**	0.94	**7**	1.01	**9**	0.95	**8**	1.22	**8**	1180	1.10
AMORE	**5**	1.00	**91**	0.26	**8**	1.07	**12**	1.06	**9**	1077	1.00
MONDO	**22**	0.54	**17**	0.78	**5**	1.57	**9**	1.17	**10**	1067	0.99
GESÙ	**9**	0.79	**19**	0.76	**35**	0.49	**10**	1.14	**11**	1028	0.96
UOMINI	**17**	0.55	**29**	0.57	**7**	1.24	**13**	0.96	**12**	900	0.84
LAVORO	**194**	0.13	**70**	0.31	**106**	0.27	**11**	1.12	**13**	758	0.71
UMANA	**53**	0.35	**123**	0.23	**12**	0.90	**14**	0.89	**14**	741	0.69
SIGNORE	**28**	0.48	**24**	0.68	**18**	0.70	**18**	0.76	**15**	715	0.67
SANTO	**54**	0.34	**36**	0.52	**83**	0.32	**15**	0.86	**16**	682	0.63
TEMPO	**31**	0.46	**66**	0.33	**29**	0.53	**19**	0.76	**17**	673	0.63
PADRE	**46**	0.36	**51**	0.40	**80**	0.32	**17**	0.78	**18**	634	0.59
CUORE	**7**	0.94	**39**	0.47	**34**	0.49	**56**	0.39	**19**	611	0.57
FIGLIO	**71**	0.28	**293**	0.10	**319**	0.08	**16**	0.80	**20**	585	0.54
MARIA	**29**	0.47	**253**	0.12	**143**	0.20	**24**	0.67	**21**	582	0.54
BENE	**50**	0.35	**28**	0.59	**587**	-	**22**	0.73	**22**	581	0.54
FRATELLI	**8**	0.80	**11**	0.89	**13**	0.85	**87**	0.28	**23**	558	0.52
MORALE	**145**	0.17	**257**	0.12	**40**	0.47	**21**	0.74	**24**	547	0.51
MADRE	**21**	0.54	**132**	0.21	**71**	0.33	**39**	0.51	**25**	526	0.49
LIBERTÀ	**97**	0.22	**189**	0.16	**66**	0.36	**23**	0.69	**26**	524	0.49
MISTERO	**421**	0.06	**1031**	-	**23**	0.62	**20**	0.75	**27**	523	0.49
STATO	**35**	0.42	**63**	0.35	**51**	0.41	**38**	0.53	**28**	514	0.48
PAROLE	**38**	0.41	**61**	0.37	**58**	0.38	**36**	0.54	**29**	513	0.48
CONCILIO	**102**	0.21	**30**	0.56	**16**	0.72	**29**	0.57	**30**	508	0.47
CRISTIANA	**15**	0.61	**13**	0.85	**37**	0.48	**70**	0.34	**31**	498	0.46

MISSIONE	167	0.15	46	0.43	61	0.37	25	0.64	32	492	0.46
UMANO	76	0.28	238	0.14	33	0.50	28	0.58	33	489	0.46
OPERA	45	0.37	60	0.37	38	0.48	41	0.49	34	479	0.45
UNITÀ	49	0.35	8	1.01	91	0.31	44	0.46	35	479	0.45
CARITÀ	12	0.65	12	0.85	14	0.83	107	0.23	36	473	0.44
DIVINA	20	0.54	58	0.37	26	0.58	63	0.36	37	465	0.43
POPOLO	25	0.50	47	0.43	32	0.50	57	0.38	38	464	0.43
POPOLI	30	0.47	49	0.42	22	0.64	61	0.37	39	456	0.42
PERSONA	152	0.17	196	0.16	180	0.16	26	0.62	40	453	0.42
LEGGE	130	0.19	435	0.07	55	0.39	27	0.58	41	448	0.42
SVILUPPO	264	0.10	479	0.07	10	0.94	35	0.54	42	446	0.42
LUCE	48	0.36	86	0.28	94	0.29	43	0.47	43	442	0.41
GRAZIA	55	0.33	41	0.47	24	0.59	52	0.41	44	434	0.40
VANGELO	104	0.21	114	0.24	140	0.21	33	0.55	45	430	0.40
FEDELI	13	0.64	10	0.90	19	0.66	134	0.17	46	427	0.40
CRISTIANI	67	0.29	35	0.52	54	0.39	46	0.44	47	425	0.40
COSCIENZA	178	0.14	688	-	30	0.52	32	0.55	48	423	0.39
SOCIALE	236	0.11	56	0.38	50	0.41	34	0.54	49	421	0.39
SAN	16	0.59	32	0.56	218	0.13	82	0.29	50	412	0.38
PACE	18	0.55	21	0.73	17	0.71	115	0.21	51	411	0.38
COMUNITÀ	191	0.13	65	0.33	117	0.24	37	0.53	52	404	0.38
STORIA	224	0.12	271	0.12	146	0.20	31	0.55	53	394	0.37
TERRA	61	0.31	50	0.42	90	0.31	55	0.40	54	391	0.36
SOCIETÀ	80	0.25	200	0.16	129	0.22	45	0.46	55	388	0.36
MORTE	96	0.22	304	0.10	213	0.13	40	0.49	56	386	0.36
DOTTRINA	41	0.40	40	0.47	28	0.54	79	0.30	57	385	0.36
CATTOLICA	23	0.53	6	1.04	77	0.32	116	0.21	58	384	0.36
PECCATO	232	0.11	1109	-	410	0.05	30	0.57	59	381	0.35
RAGIONE	126	0.19	96	0.26	172	0.17	42	0.48	60	378	0.35
FIGLI	33	0.46	20	0.73	43	0.45	109	0.23	61	368	0.34
UMANITÀ	127	0.19	237	0.14	31	0.51	50	0.42	62	368	0.34
NATURA	70	0.29	105	0.24	44	0.45	62	0.36	63	365	0.34
SALVEZZA	87	0.24	78	0.30	42	0.46	58	0.38	64	361	0.34
GIUSTIZIA	64	0.30	69	0.31	60	0.37	68	0.35	65	358	0.33
DIGNITÀ	110	0.21	184	0.16	104	0.27	48	0.43	66	357	0.33
PAROLA	107	0.21	88	0.28	41	0.46	60	0.37	67	348	0.32
DONO	239	0.11	413	0.07	86	0.31	49	0.43	68	323	0.30
FORZA	82	0.25	343	0.09	111	0.26	65	0.35	69	318	0.30
REALTÀ	394	0.07	227	0.14	47	0.41	53	0.41	70	314	0.29
CHIESE	177	0.14	128	0.21	208	0.13	54	0.41	71	313	0.29
MISERICORDIA	254	0.10	168	0.17	251	0.11	47	0.43	72	312	0.29
DIRITTI	111	0.21	93	0.26	193	0.14	66	0.35	73	303	0.28
VERGINE	19	0.55	239	0.14	147	0.20	144	0.16	74	301	0.28
VOLONTÀ	81	0.25	125	0.23	52	0.41	89	0.28	75	300	0.28
APOSTOLICA	24	0.50	34	0.52	142	0.20	158	0.14	76	299	0.28
VIRTÙ	32	0.46	33	0.56	46	0.44	184	0.13	77	299	0.28
PREGHIERA	91	0.23	43	0.47	102	0.28	84	0.28	78	298	0.28
DIRITTO	90	0.23	331	0.09	166	0.17	71	0.34	79	297	0.28
REGNO	78	0.25	48	0.43	81	0.32	97	0.26	80	290	0.27
COMUNIONE	811	0.02	210	0.14	112	0.25	51	0.41	81	285	0.27
AZIONE	162	0.16	62	0.35	93	0.29	76	0.30	82	280	0.26
NAZIONI	72	0.28	76	0.30	79	0.32	104	0.23	83	278	0.26
VIA	157	0.17	1398	-	1395	-	59	0.37	84	276	0.26
CORPO	75	0.28	53	0.38	21	0.64	136	0.17	85	272	0.25
ORDINE	86	0.24	134	0.21	98	0.28	95	0.26	86	270	0.25
PIETRO	62	0.31	22	0.71	138	0.21	127	0.18	87	267	0.25
PAOLO	301	0.09	107	0.24	170	0.17	69	0.35	88	265	0.25
DIVINO	27	0.48	75	0.30	78	0.32	232	0.10	89	264	0.25

CRISTIANO	56	0.32	57	0.37	69	0.35	133	0.17	90	262	0.24
ENCICLICA	141	0.18	161	0.17	211	0.13	77	0.30	91	261	0.24
PAESI	195	0.13	77	0.30	88	0.31	83	0.29	92	261	0.24
CROCE	260	0.10	285	0.10	346	0.07	64	0.35	93	256	0.24
APOSTOLI	118	0.20	81	0.28	148	0.19	100	0.25	94	245	0.23
ANIMO	26	0.50	25	0.61	76	0.32	483	0.03	95	243	0.23
MALE	330	0.08	299	0.10	198	0.14	72	0.33	96	242	0.23
PENSIERO	233	0.11	223	0.14	45	0.45	96	0.26	97	240	0.22
FAMIGLIA	105	0.21	216	0.14	105	0.27	108	0.23	98	238	0.22
REDENTORE	42	0.40	110	0.24	234	0.12	170	0.14	99	238	0.22
VENERABILI	14	0.64	38	0.49	1391	-	1394	-	100	236	0.22
DIALOGO	1102	-	721	-	15	0.73	86	0.28	101	235	0.22
INSEGNAMENTO	193	0.13	148	0.19	154	0.19	91	0.27	102	234	0.22
TESTIMONIANZA	304	0.09	320	0.10	183	0.16	75	0.31	103	234	0.22
SPIRITUALE	109	0.21	79	0.30	39	0.48	135	0.17	104	233	0.22
ATTIVITÀ	209	0.12	126	0.21	206	0.13	85	0.28	105	232	0.22
BENI	172	0.15	157	0.17	149	0.19	93	0.26	106	232	0.22
SERVIZIO	202	0.13	469	0.07	74	0.33	94	0.26	107	232	0.22
UNIVERSALE	258	0.10	124	0.23	132	0.22	90	0.28	108	232	0.22
AIUTO	57	0.32	80	0.28	75	0.32	174	0.14	109	231	0.22
VESCOVI	36	0.42	64	0.35	340	0.08	217	0.11	110	229	0.21
TEMPI	68	0.29	113	0.24	131	0.22	143	0.16	111	226	0.21
ESISTENZA	763	0.03	785	-	448	0.04	67	0.35	112	220	0.21
UNIONE	73	0.28	72	0.31	203	0.14	140	0.16	113	220	0.21
ESEMPIO	98	0.22	94	0.26	65	0.36	137	0.17	114	219	0.20
PERSONE	283	0.09	1126	-	137	0.21	88	0.28	115	216	0.20
SOLO	11	0.65	1292	-	1299	-	1307	-	116	214	0.20
SACRA	65	0.30	111	0.24	63	0.37	216	0.11	117	213	0.20
CULTURA	524	0.05	245	0.12	178	0.16	81	0.29	118	212	0.20
SANTA	43	0.38	122	0.23	73	0.33	282	0.08	119	212	0.20
SPERANZA	100	0.22	112	0.24	139	0.21	138	0.17	120	204	0.19
RIVELAZIONE	788	0.03	1206	-	336	0.08	74	0.31	121	203	0.19
APOSTOLO	88	0.24	100	0.24	124	0.22	150	0.15	122	200	0.19
RICERCA	784	0.03	1188	-	278	0.10	78	0.30	123	199	0.19
ROMA	93	0.23	44	0.47	281	0.10	156	0.15	124	198	0.18
FILOSOFIA	948	0.02	835	-	835	-	73	0.32	125	195	0.18
VALORE	325	0.08	1391	-	184	0.16	99	0.25	126	193	0.18

The final table shows remarkable results. The word *Dio* [God], the most used by the last five pontiffs, is not always the first word in the collected encyclical of an individual pope. The term *Dio* is the second in John XXIII, wherein the first word is *chiesa* [church]. Such a term is the second in Pius XII and the fourth in John Paul II. The term *vita* [life] is almost homogeneous (positions 4,4,4,3) in the papacies which we have considered.

The research concerns 60 years of church history, from 1939 with the election of Pius XII until today: in this period the popes produced 69 encyclicals. The table provides us with insights resulting from the evolution of some important terms in the time considered.

The behavior of the term *mistero* [mystery] is remarkable; the serial number in Pius XII is 421, while in John XXIII is 1031, 23 in Paul VI and 27 in John Paul II. John XXIII used this word two times in *Aeterna Dei Sapientia*, an encyclical commemorating the fifteenth centennial of the death of the Pope St. Leo I. «St. Leo, having made perfectly clear "what the Catholic Church universally believes and teaches concerning the mystery of the Lord's incarnation"».[4]

References

Benzécri J.P. (1980) L'Analyse des Données, in *L'Analyse des Correspondances*, vol. II, Dunod, Paris.

Bisceglia B. (1998) Statistical Analysis of Textual Data in Theological Sciences: Second Vatican Council Documents, in *JADT 1998. 4èmes Journées Internationales d'Analyse Statistique des Données Textuelles*, Mellet S. edt., Nice.

Bolasco S. (1996) Il Lessico del Discorso Programmatico di Governo, in *L'Attività dei Governi della Repubblica Italiana (1948-1994)*, Villone M. & Zuliani A. (Eds.), Il Mulino, Bologna.

Busa R. (1987) *Fondamenti di Informatica Linguistica*, Vita e Pensiero, Milano.

Enchiridion delle Encicliche (1995), Official text and Italian version, Centro editoriale dehoniano, Bologna.

Feldman A. N. (1995) Il Concetto di Sviluppo Umano secondo le Nazioni Unite: Analisi del Contenuto, in *JADT 95. Analisi Statistica dei Dati Testuali*, III Giornate Internazionali, Atti del Convegno: vol. I e II, Bolasco S., Lebart L., Salem A.(Eds.),C.N.R., Roma 11-13 dicembre 1995.

Lebart L., Salem A. (1988) *Analyse Statistique des Données Textuelles. Questions ouvertes et lexicométrie*, Dunod, Paris.

Rizzi A. (1992) Orientamenti Attuali della Statistica Linguistica, *Statistica*, 2.

Sabourin L. (1967) *The Names and Titles of Jesus*, The Mac Millan Company, New York.

[4] John XXIII, *Aeterna Dei Sapientia*:, 16, 1961.

Life Courses as Sequences: an Experiment in Clustering via Monothetic Divisive Algorithms

Francesco C. Billari
Max Planck Institute for Demographic Research,
Doberaner Str. 114, D-18057 Rostock. E-mail: billari@demogr.mpg.de

Raffaella Piccarreta
Istituto di Metodi Quantitativi, Università Bocconi,
viale Isonzo 25, I-20136 Milano. E-mail: raffaella.piccarreta@uni-bocconi.it

Abstract: We consider the problem of clustering demographic life courses, consistently with the so-called sequence representation. We use a monothetic divisive algorithm, which allows for a better readability of results with respect to the increasingly common approach based on optimal matching analysis. The algorithm eases the interpretation of splitting procedures and determinants of group membership when data represent the occurrence or non-occurrence of non-renewable events. We then apply the algorithm to the transition to adulthood in Italy, using retrospective individual-level data from a Fertility and Family Survey.

Keywords: sequence analysis, cluster analysis, monothetic divisive algorithms, event history analysis, life course, transition to adulthood, fertility and family survey.

1. The representation of life courses as sequences [1]

The question of how life courses develop themselves across time is customarily addressed by using event history methods, which prove to be useful when studying one kind of event, or the interaction between events in parallel careers. Non-parametric estimation methods, from survivor functions to more complex tools are thus used for description and basic comparison, while semi-parametric or parametric models are normally applied when trying to evaluate potential determinants. However, it is still a matter of question how to describe whole life courses, to cluster them and in case to throw light on some of their determinants. First of all, one has to make clear how to represent – in a formal but useful way – whole life courses. It is recognised that a representation in terms of event history is equivalent, when adopting a specific discrete time scale, to a representation in terms of temporally ordered state variables (see e.g. Billari and Rohwer, 1999). This is the basis of what has been recently spreading from information science and

---

(1) F.C.B. drafted §1 and §3, R.P drafted §2, §4 was jointly drafted. Disclaimer: the views expressed in this paper are attributable to the authors and do not necessarily reflect those of the institutions they belong to. We thank an anonymous referee for the useful comments.

computational biology (Sankoff and Kruskal, 1983; Waterman, 1995) to sociology (Abbott, 1995), namely *sequence analysis*. We now briefly illustrate the formal representation of life courses in such approach.

Let t refer to age, measured in some discrete unit, say months or years. Consider then a sequence of state variables such that y_{it} represents the state of individual i at time t, t=1,...,h. We assume that there is an identical state space, consisting of a finite number of states, for each individual. Let us also assume that we have a sample of n such life courses, and that there is no censoring. The life course of each individual i, i=1,...,n can then be represented by a vector $y_i=(y_{i1},y_{i2},...,y_{ih})$. It is also common to consider life courses as composed by several (say, k) parallel domains. The sequence representation for each individual becomes then a matrix with rows $y_{ij}=(y_{ij1},y_{ij2},...,y_{ijh})$, j=1,...,k. Can we find a sensible approach to clustering such a collection of life courses? In this paper we shall follow an applied perspective, and consider a typical subject of social demography, the transition to adulthood. The interest is in role transition in six separate, but connected, domains of the life course. We consider transitions towards adult roles as irreversible. The domains considered are education, the labour market, living arrangement, behaviour on sexual intercourses, union formation and parenthood.

The data set we use comes from the Italian Fertility and Family Survey INF-2 1995/96 (De Sandre *et al.*, 1997; De Sandre *et al.*, 1999). Time is measured in years, and each variable is an indicator that represents whether one has experienced the marker event for each domain at a given anniversary between the ages of 20 and 35 (so k=6, h=16 for a total of 96 variables). For sake of simplicity, we call variables EDUt, JOBt, LEAt, SEXt, UNIt, CHIt, t indicating the anniversary. As a consequence, if EDUt=1, then EDUs=1 for $s \geq t$, and this is also true for other domains (see table 1). Only people without missing values, born between 1946 and 1960 are selected. The total sample is thus composed by n=2590 people, of which 483 are men and 2107 women.

2. Current approaches for clustering life courses as sequences and the advantages of our new proposal

It is well known that several techniques have been developed to perform cluster analysis. However, only a few of such techniques are usable for time-to-event data. The most widely used algorithm in the field is derived from computational biology and known as Optimal Matching Analysis (OMA, also known as optimal alignment, see Waterman, 1995). The basic idea of OMA is to measure the dissimilarity between two sequences by properly quantifying the effort necessary to transform one sequence into the other one. In the most elementary approach, a set composed by three basic operations to transform sequences is taken into account, $O = \{\iota, \delta, \sigma\}$, where the symbols denote:

- *insertion* (ι): one state is inserted in the sequence;
- *deletion* (δ): one state is deleted in the sequence;
- *substitution* (σ): one state is substituted by another one.

Table 1: *An example of sequence representation for one individual (a matrix for six parallel sequences originated by non-renewable events).*

Domain \ *Time (years)*	20	21	22	23	24	25	26	27	28	29	30	31	32	33	34	35
EDUcation	0	1	1	1	1	1	1	1	1	1	1	1	1	1	1	1
JOB	0	0	1	1	1	1	1	1	1	1	1	1	1	1	1	1
LEAving home	0	0	0	0	0	0	0	1	1	1	1	1	1	1	1	1
SEXual intercourses	1	1	1	1	1	1	1	1	1	1	1	1	1	1	1	1
UNIon	0	0	0	0	0	0	0	1	1	1	1	1	1	1	1	1
CHIld	0	0	0	0	0	0	0	0	0	0	0	0	0	0	1	1

Suppose that k basic operations have to be performed in order to transform one sequence into another one. Then, to each elementary operation ω_i, $\omega_i \in O$, $i=1,\ldots,k$ a specific cost can be assigned, and the cost of applying a series of k elementary operations may be computed as $c(\omega_1, \omega_2,\ldots,\omega_k) = \sum_i c(\omega_i)$. The distance between two sequences a and b may thus be defined as the minimum cost of transforming one sequence into the other one. So, the computed distance takes into account the whole sequences and not only states at one point in time. Several algorithms have been developed in order to compute the minimum transformation costs (Sankoff and Kruskal, 1983; Waterman, 1995). In fact, different sets of elementary operations are conceivable, and even insertion, deletion and substitution costs may vary according to the states; e.g. it is common to give higher substitution costs when the substituting state is considered conceptually "farther" from the substituted state. Choosing a good specification of costs probably accounts as the most important theoretical problem when applying OMA. It is also possible to specify that inserting or deleting a sub-sequence of l consecutive states has a different cost with respect to inserting or deleting l times the states. The distances computed by the OMA approach can be used as inputs for subsequent analyses, for instance for hierarchical cluster analysis.

It is worthwhile at this point to introduce a new classification of hierarchical clustering techniques to highlight the motivations underlying our proposal. We distinguish between *individual*–based and *variable*–based approaches. In *individual*–based approaches attention is primarily focused on the definition of the *dissimilarity* matrix, whose entries are the dissimilarities between the individuals. Dissimilarities are nonnegative numbers measuring how much different two individuals are with respect to the characteristics taken into account. Notice that OMA is a method to build up a dissimilarity matrix. In *agglomerative* individual–based algorithms at each step the less distant clusters (eventually constituted by one individual) are joined to form a single cluster. Individual–based *divisive* algorithms proceed in the inverse order. At each step a cluster is split into two sub-clusters, which have to be as dissimilar as possible. In this setting, two clusters are considered dissimilar if they contain dissimilar individuals. We will not go into details here (for a detailed description of these algorithms see

Kauffman and Rousseeuw, 1990; Bolasco, 1999); the main point to stress is that the criterion referred to in the individuation of the sub-groups is the dissimilarity *between individuals.* A problem of individual–based approaches is that if many variables are taken into account it can be difficult to individuate which are the more relevant variables in the definition of the attained groups. Moreover, a major advantage when considering nominal and, especially, binary variables is the possibility to obtain, at least theoretically, clusters characterised by the presence (respectively the absence) of certain attributes. Individual–based algorithms do not guarantee this to happen. The criterion of joining or splitting depends in fact upon (a proper synthesis of) the dissimilarity between individuals in two clusters, which is not necessarily imputable to one variable only. In this sense, as Halpin and Chan (1998) state when reporting an OMA-based analysis "while the clusters are very easy to characterise in a general way, it is impossible to characterise them formally and exhaustively, that is, to define rules which will replicate the clusters exactly or close to exactly".

These problems can be overcome by considering a *variable*–based approach. We describe the basic idea in a divisive setting. Consider a group G of individuals which has to be split into two subgroups, G_1 and G_2; let X^* be a variable assuming values in a set $\mathbf{X}$, and let A be a subset of $\mathbf{X}$. If one case is put in group G_1 if it is characterised by $x^* \subseteq A$ and in group G_2 otherwise, the split is perfectly interpretable. Moreover, X^* appears crucial in the clustering process and we also do have information about the levels of X^* involved in the segmentation process. To apply this kind of algorithm, it is evidently necessary to measure the dissimilarity between two groups by referring to *variables* rather than to *individuals.* Moreover, it is necessary to define a criterion to select at each step of the process the cluster to be split, the variable X^* and the set A inducing the split.

We here illustrate an algorithm developed having in mind the above considerations, namely a variable–based divisive algorithm (Piccarreta, 1998). In particular, the algorithm has been developed for the case when N individuals have to be clustered on the basis of Q binary variables, $Y_1, \ldots, Y_Q$ assuming values 1 or 0 according to the presence or absence of a certain attribute. The algorithm is based on the concept of total and within-groups dispersion, measures usually referred to in the quantitative case, to evaluate the quality of a clustering and to individuate the number of clusters that have to be taken into account. It is well known that in the quantitative case the dispersion within a group is measured by "aggregating" the (sample) variances of the variables taken into account. To extend this criterion to the case when binary variables are considered, it is necessary to refer to a proper measure of the mutability of a certain variable, say Y_q. As far as concerns this point, the mutability of Y_q is measured by means of the well-known *Gini heterogeneity*, $E(Y_q)$(2). Consider now for first the group constituted by all the individuals. A suitable measure of the total dispersion in the sample is $I_1(1)=N \sum_q E(Y_q)$. Suppose now the whole sample has been divided into G groups; let

(2) Recall that if Y is a qualitative variable assuming J levels with relative frequencies $p_1, p_2, \ldots, p_J$, the Gini heterogeneity of the distribution of Y is defined as $E(Y) = 1 - \sum_j p_j^2$.

$E(Y_q|g)$ denote the heterogeneity of Y_q in the g-th group. The quantity $I_g\,(G) = N_g \sum_q E(Y_q|g)$ is a measure of the dispersion in the g-th group, and a possible measure of the within groups dispersion is $W_G = \sum_g I_g(G)$. The quality of the clustering can be evaluated by referring to the quantity $M_G = [I_1(1)-\sum_g I_g(G)]/I_1(1)$. The numerator of M_G is a measure of the dispersion between groups, and M_G can be interpreted as the "portion" of total dispersion accounted for by the G clusters. In a divisive setting, starting from G clusters, $(G+1)$ clusters have to be obtained by splitting one cluster. By construction, it will always be $M_G < M_{G+1}$. So, if one refers to the M criterion, it seems natural to select either the cluster to split and the splitting variable in such a way that $\Delta M_G = (M_{G+1} - M_G)$ is maximum. This split is thus the one assuring that M_{G+1} is maximum conditionally to the G groups. Notice that M_G is a measure of the quality of the clustering and ΔM_G can be useful in determining the number of clusters to take into account (notice, moreover, that M_G and ΔM_G are analogue to the R^2 and ΔR^2 solely referred to in the quantitative case). A sudden decrease of ΔM_G means in fact that a cluster is divided into two non-heterogeneous subgroups, so that it is not worthwhile to continue splitting (for a more detailed description of the algorithm, see Piccarreta, 1998).

We find that such algorithm is particularly useful when dealing with sequence data typically arising in demography, that is, sequence data where parallel domains are each representing states that correspond to non-renewable events. The main advantage of this algorithm is, in our opinion, the fact that it leads to easily interpretable clusters: the obtained groups are in fact perfectly characterised by the presence (or absence) of the attributes measured by the splitting variables. Moreover, it is possible both to individuate the most relevant variables in the clustering process (the splitting variables) and to rank these variables according to their importance in the clustering process. A major discriminating power can in fact be attributed to variables inducing first splits. As a minor, but still relevant property, the algorithm is characterised by a clear stopping rule and by a suitable criterion to evaluate the quality of the obtained clustering.

We applied the algorithm just described to data from the Italian Fertility and Family Survey INF-2 1995/96 (De Sandre *et al.*, 1997). In Figure 1 the splitting procedure is reproduced. We decided to consider 8 clusters, in that from the 9-th split the values of ΔM_G were very low. Table 2 reports results for the 8-cluster solution (accounting for 50.5% of the total dispersion).

3. Demographic interpretation of the results

The algorithm we presented is particularly useful for the applied researches, when one wants to analyse the results from a demographic perspective. With respect to this, we should underline least two aspects. First, it is interesting to see which variables are intervening at each stage of the split, and to learn what does this mean. Then, one may investigate the determinants of group membership. We now briefly account for the results, given the final division of the sample in 8 groups. At the first step, the splitting variable is UNI25, and the

Figure 1: *Tree representation of the procedure.*

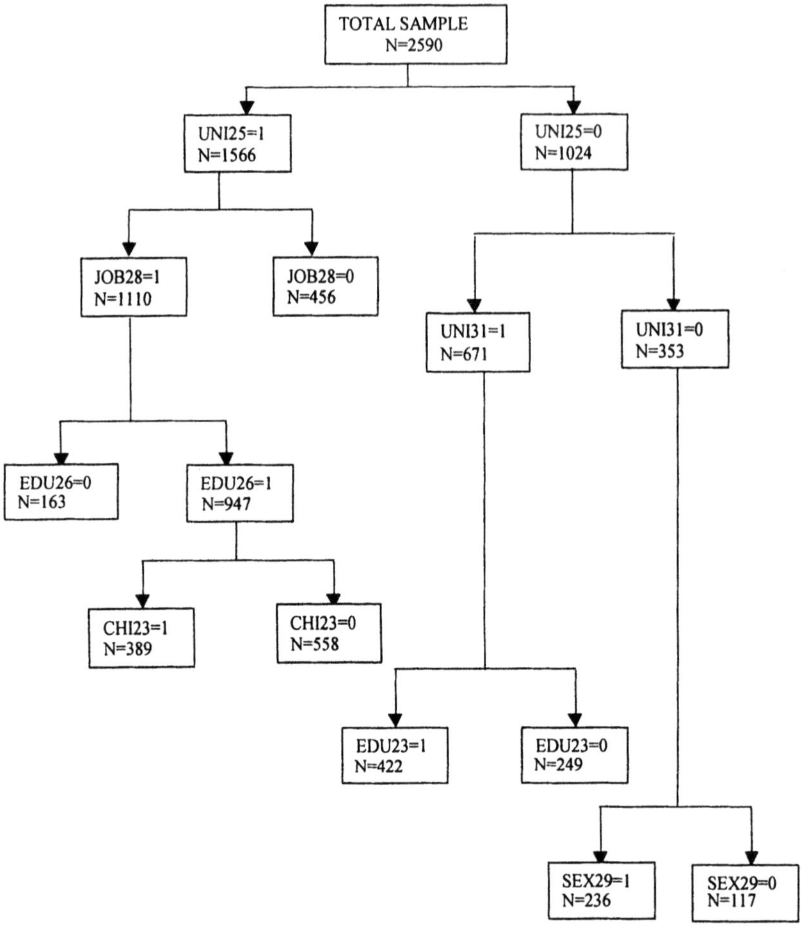

division accounts for about 20% of total dispersion. Hence, it seems that entering a union at lower or higher ages is the most important discriminating feature between Italian young adults. This is consistent with some views in the literature, where the role of unions (and unions are almost totally marriages) in Italy has been crucial. Then, the "early union" group is split according to JOB28 (11% of total dispersion); this seems to represent a traditional vs. modern pattern of female labour force participation (we shall come back to this point later when looking at the composition of this group). The third step divides people not having entered a union before 25 according to UNI31 (7% of total dispersion); again, the central role of marriage is being underlined. The fourth step splits people entering an early union and having an early job according to EDU26 (4% of total dispersion), then the group with the comparatively early end of education is divided according to CHI23 (3.3% of total dispersion). This brings to the formation of an "early-on-everything" group, and two other groups with comparatively late end of education and parenthood. At the sixth step, EDU23 is used to split the group of people

Table 2: *Results of the divisive algorithm (G=8, M_8 =0.505).*

Group	*1*	*2*	*3*	*4*	*5*	*6*	*7*	*8*
No. members (%)	456 17.6	163 6.3	558 21.5	389 15.0	249 9.6	422 16.3	117 4.5	236 9.1
Characteristics of the group	UNI25=1 JOB28=0	UNI25=1 JOB28=1 EDU26=0	UNI25=1 JOB28=1 EDU26=1 CHI23=0	UNI25=1 JOB28=1 EDU26=1 CHI23=1	UNI25=0 UNI31=1 EDU23=0	UNI25=0 UNI31=1 EDU23=1	UNI25=0 UNI31=0 SEX29=0	UNI25=0 UNI31=0 SEX29=1
% Males (*total*=18.6)	2.0	20.2	17.2	7.5	30.5	32.2	7.7	40.3
% Cohort 1946-50 (*total*=31.7)	35.3	33.1	32.8	31.4	30.1	32.2	27.4	24.2
% Cohort 1951-55 (*total*=34.6)	35.7	34.4	36.4	36.8	33.7	33.9	35.9	25.8
% Cohort 1955-60 (*total*=33.8)	28.9	32.5	30.8	31.9	36.1	33.9	36.8	50.0

entering a union between 25 and 31 (2.7% of total dispersion). The last step divides people not having entered a union at 31 according to SEX29 (2.5%); this brings to the constitution of one group of people without any sexual relation at almost their thirtieth anniversary. It is not easy to interpret every single split. However, some features appear to be constant. Union formation plays a crucial role in classifying life courses in the transition to adulthood. Leaving the parental home is not an important discriminating variable; this has certainly to be connected with the high degree of synchronisation of this event with union formation for Italian post-war cohorts. The transition to parenthood only enters marginally. Sexual behaviour matters only for people outside unions.

There is some little space here to describe the basic demographic determinants, namely gender and cohort. We will thus only refer to the most interesting clusters. Group 1, with early union and with late (or potentially absent) labour force participation, largely corresponds to a traditional female labour force pattern: there are only very few men and younger cohorts are underrepresented. Group 3 is a sort of average one, without cohort and gender particularities, and represents more than one fifth of the sample. Group 4, that we already defined "early-on-everything", exhibits the faster transition to adulthood, and it is mainly composed by women; however, the cohort pattern is not self-evident. Group 6, with unions between age 25 and 31, has comparatively more men, but no clear cohort dynamics. Group 8 has the highest share of males and of "young" members; it represents people who seem to be in line with the standard view of the transition to adulthood pattern also of more recent cohorts, having sexual relationships without union. A finer interpretation and a closer look leading to hypotheses on determinants of group membership would call for supplementary analyses.

4. Directions for further research

In this paper we sketched the opportunities of a fresh approach to the analysis of demographic life courses. It is also the purpose of the authors to pursue further studies on the subject, which is of peculiar relevance in the field. Additional research is needed both from the methodological and the applied point of view. From the methodological viewpoint, it should be interesting to compare the results with other algorithms, in case using also an OMA approach. It is however not clear which kind of index one should choose in order to evaluate the goodness of such clustering procedures, given that there is no explicit dependent variable, and that in principle one can think of different demographic outcomes as dependent variables. The robustness of the algorithm against different (e.g. monthly *vs.* yearly) specifications of the time scale and different age intervals also needs to be challenged. The consistency and robustness of the algorithm have to be confronted, also in applications. Separate analyses by gender and cohort might throw light on the different groupings between gender and cohort. Analyses of determinants and group description have also to be developed, to give further information on demographic behaviour. Such tests should give ideas on how to extend the analyses to wider parts of the life course.

References

Abbott A. (1995), Sequence analysis: new methods for old ideas, *Annual Review of Sociology*, 21, 93-113.

Billari F., Rohwer G. (1999), Some developments in methods for life course analysis and their applications", in: *Atti della XXXIX Riunione Scientifica della Società Italiana di Statistica, Comunicazioni invitate*, Napoli, 323-334

Bolasco S. (1999), *Analisi multidimensionale dei dati*, Carocci: Roma.

De Sandre P., Ongaro F., Rettaroli R., Salvini, S. (1997), *Matrimonio e figli: tra rinvio e rinuncia*, il Mulino, Bologna.

De Sandre P., Pinnelli A., Santini A (Eds., 1999), *Nuzialità e fecondità in trasformazione: percorsi e fattori del cambiamento*, il Mulino, Bologna.

Halpin B., Chan, T.W. (1998), Class careers as sequences: an optimal matching analysis of work-life histories, *European Sociological Review*, 14, 111-130..

Kaufman L., Rousseeuw, P.J. (1990), *Finding Groups in Data. An Introduction to Cluster Analysis*, Wiley, New York.

Piccarreta R. (1998), Divisive monothetic algorithms for clustering binary data: a proposal, in: *Proceedings of the NTTS '98 International Seminar*, Sorrento (NA), 405-410.

Sankoff D., Kruskal J.B. (Eds, 1983), *Time Warps, String Edits, and Macromolecules: The Theory and Practice of Sequence Comparison*, Addison-Wesley, Reading, MA.

Waterman M.S. (1995), *Introduction to Computational Biology. Maps, sequences and genomes*, Chapman & Hall.

Significance of the Classification for the Italian Service Sector Activities

Anna Rita Giorgi
ISTAT – Business Statistics Department
angiorgi@istat.it

Roberto Moro
ISTAT Staigiaire

Abstract: In current economic studies on enterprises carried out in ISTAT (the Italian Statistic Institute) the selection of the domain of interest is mainly based on the Classification of Economic Activities (ATECO'91), which is an official classification.
In this paper we analyse the coherence of the ATECO'91 in representing microdata which concern economic variables. Then we determine the statistical information, expressed by the coefficient CI, corresponding to different levels of details of the above classification and to different characters.
For this purpose we consider a set of data collected by the survey on Economic Account of the Enterprises, including both structural and economic variables.

Keywords: classification, economic activities, prevalent activity, service sector, microdata, macrodata, statistical information.

1.The official classification in ISTAT

ISTAT, the Italian Statistical Institute, currently produces a lot of information concerning enterprises. For this purpose the official classification of the economic activities ATECO '91 is used as a base for data capturing.
In this classification enterprises are grouped by their "prevalent activity", that is the business to which both economical and human resources are mainly dedicated. It is organised in a hierarchical structure, that starts with "sections", which is the most general set of activities, continues with "groups", "classes" and, finally, "categories", that is the most disaggregated level in which the prevalent activity almost coincides with the kind of product or service of each enterprise.
All the previous levels correspond to different letters (in the case of sections) or batch of digits (e.g. groups coincide with three digits, classes with four digits and so on).

ATECO '91 is a valid and objective tool, but, sometimes, it seems to be too rigid and not suitable enough to catch some relevant elements of heterogeneity among enterprises classified within the same group. Very often, precisely, this organisation of the activities is not able to reflect the structural characteristic of the enterprises, even at a certain level of aggregation. In other words, it frequently happens that enterprises belonging to the same group or class is not very similar and the only common characteristic is just the prevalent economic activity.

Generally speaking, all the aspects above illustrated become evident once we examine microdata coming from the current surveys. The information that emerges, actually, is rich and significative, but, sometimes, it does not match with that coming from the *a priori* classification. This is due to some reasons that are not only proper of each single enterprise, but also determined by specific economical or financial aspects.

A study on the breakdown of turnover for retail trade enterprises that is still ongoing in ISTAT is an example of this. From the preliminary results, actually, several misclassifications have been found, concerning the enterprises within the repair service sector. In the last years many of them have been involved in deep changes (e.g. customers' attitude, economical aspects and so on) and for this reason, very often, the greater part of their annual turnover is due to activities different from the prevalent one.

In this context it is clear that, in most cases, it is not correct to take into account the prevalent activity as the only factor that make it possible to identify an enterprise as belonging to a certain group with specific characteristics. On the contrary, it can be considered as a given attribute that can help the researcher in the selection of the units to include in a fixed domain.

2. The information given from the microdata and macrodata. Relationships with the adopted classification

In writing this paper we set out from the above consideration.[1] Our aim is mainly to examine more in detail to what extent the mentioned classification is able to correctly represent enterprises' characteristics. In other words, starting from a comparative analysis conducted both on microdata and macrodata, we attempted to verify if the classification used to select units to be studied reflects the real structural and behavioural characteristics of the enterprises.

Our current experience, actually, seems to suggest that elementary data can contain more information on the aggregation of enterprises than the macrodata, which follow the *a priori* classification. This difference could explain the greater part of lack of information registered when passing from micro to macrodata.

The enterprises considered for the analysis are those within the service sector (Sections G up to O of the ATECO'91). The reason of this choice is that these firms are very dynamic and frequently practice some activities that are as relevant as the prevalent one. In particular, from our practical experience coming from current surveys, we can say that very often these enterprises offer a mixture of services both to other enterprises and to families.

[1] This paper is the result of the common efforts of the two authors. Anyway, R.Moro was mostly involved in paragraph 2 whilst A.R. Giorgi in paragraph 3 ; both of them wrote paragraph 1 and 4.

In the table contained in the following page all the kind of activities considered and the related classification code are listed. In this table we list just the level of aggregation that corresponds to the two digits level of ATECO '91, that is represented by 18 sections.

Table 1: *Classification of the economic activities within the service sector*

ATECO '91CODE	ACTIVITY
50	Retail trade, mainteinance and repair services of motorvehicles and motorcycles
51	Wholesale trade and commission trade serv., except for m.vehicles and m.cycles
52	Retail trade exc. for mot. vehicles-mot.cycles. Repair serv. pers.-household goods
55	Hotels and restaurants
60	Inland transports
61	Sea transports
62	Air transport
63	Auxiliar transport activities. Travel agencies
64	Postal services
70	Real estate
71	Renting activities
72	Computer services and related activities
74	Other professional and management activities
80	Educational services
85	Health services and other social services
90	Waste disposal services
92	Cultural and ricreative activities and sports
93	Other service activities not elsewhere classified

As anticipated before, the elementary information used for our elaboration are those related to the single enterprises that have been interviewed for the survey on the *Economic Account of the Enterprises;* data refer to the year 1995. In particular, for each of the 46.592 units (enterprises) of the domain we heeded the following variables: id. code of the enterprise, ATECO code, number of persons employed, turnover, value added, investments, costs for personnel.
Beside this, we examined different levels of aggregations, in particular we started from Divisions (2 digits level) up to Classes (4 digits level).
The hypothesis to be verified is that if we follow a classification with a scarce level of breakdown, we risk to include in the same subset of units that are very dissimilar; on the other hand, an high level of detail can bring to a redundancy in the selection of the units to comprise in the study. Therefore, this analysis confirms the conclusion that the best classification is an intermediate one (e.g. three or four digits).

For this purpose we grouped the above enterprises according to the information coming from the mentioned current survey on the *Economic Account of Enterprises*.

The basic idea is to follow an approach that is somehow opposite to the traditional one: while usually the units (in this case the enterprises to include in the study) are chosen following the official classification, in this case we start from the elementary information and, once tested their characteristics, examine the coherence with the traditional classification. In other words the aim is to find out to what extent the groups obtained from the microdata reproduce the given partition derived from the ATECO classification.

For this aim, apart from the original characters of the data set, we consider the following variables: turnover per person employed, value added per person employed, labour costs per person employed, ratio between value added and turnover, ratio between labour costs and turnover.

We introduce these variables because they are very important in economic analysis on enterprises. The stress we put on the economic characteristics is, actually, coherent with the nature of the classification we refer to, that is based on the prevalent activity of the enterprises.

As a first step of the analysis we aggregate data at 4 digits level, that is groups, then perform a factor analysis. On the basis of the obtained results, we carry out a k-mean cluster analysis, in order to identify possible aggregations between economic sectors.

In a following stage we consider the aggregation of the sectors examined up to 2 digits. More in details, we take into account the number of economic activities or, precisely, the number of classes, included in each of the 2 digits sectors. In this phase we compare the result obtained with those coming from the application of the cluster analysis to investigate the amount of activities "well represented" by the 2 digits ATECO'91.

In the following table we indicate, for each sector, the number of activities (4 digits ATECO) that are properly classified within the correspondent sectors (2 digits ATECO).

Table 2: *Groups classified within ATECO'91 sections*

ECONOMIC ACTIVITIES (2 DIGITS)																	
50	**51**	**52**	**55**	**60**	**61**	**62**	**63**	**64**	**70**	**71**	**72**	**74**	**80**	**85**	**90**	**92**	**93**
Number of sectors represented with 4 digits																	
5	44	28	9	7	3	2	7	3	5	9	6	15	6	7	1	16	5
Number of sectors well classified																	
2	17	16	4	4	2	1	4	1	2	6	3	6	6	4	1	11	4
% of sectors well classified																	
40	39	57	44	57	67	50	57	33	40	67	50	40	100	57	100	69	80

Source: elaborations on ISTAT data

As it' clear, apart from the professional and management businesses (ATECO 74) for all the sectors there is a significative subset of activities that is spread in different groups. On this purpose we must observe that this is just a statistical analysis, even if carried out on a set of economic variables. If we want to explain the above results giving an economic interpretation, we must go into details of the classification to investigate the effective associations between economic sectors. An example of this is the case some groups of the sections 55 – "Hotels and restaurants" – that are in the same cluster of the "Travel agencies", which belong to the section 63 – "Auxiliary transport activities. Travel agencies". On the other hand, the association between "Trade activities" and "Health services" should be further investigated to find out if there are particular economic reasons which justify it.

3. Informativity of the considered classifications

Beside the analysis above illustrated, we perform a parallel study, which consists in investigating the level of informativity of a specific classification (as explained in Gismondi and Mirto, 1998). Generally speaking, if we consider *k* classifications of a quantitative character, a synthetic coefficient of "informativity" of the classifications themselves can be obtained as a mean of k coefficients CI_i, as expressed by the formula:

$$CSI \equiv \frac{\sum_{i=1}^{k} CI_i \left(\frac{2}{n_i}\right)}{\sum_{i=1}^{k} \left(\frac{2}{n_i}\right)} \equiv \frac{\sum_{i=1}^{k} \frac{var_i}{n_i (n_i - 1) m_i^2}}{\sum_{i=1}^{k} \frac{1}{n_i}} \qquad (1)$$

where

$$CI_i \equiv \frac{var_i}{(n_i - 1) m_i^2} \qquad (2)$$

is the coefficient of informativity calculated for the i-th classification, *m* is the mean of the character observed varying the modalities of the adopted classification and n is the number of the modalities of the classification itself.

The basic idea is that the 'information' given by a classification, obtained with a given criterion, can be measured by the explained variance of an ANOVA model applied on the groups of the above classification.

Once classified the domain of interest, the 'explained variance' (or variance between groups) can be represented by the variance of the observed character.

This variance, actually, depends on the mean of the classification variable. For this reason it is divided by a value which represents the maximum for the variance and depends on the product between the number of modalities and the average intensity of the classification character. In this way the coefficient *CSI* has maximum value 1 and minimum value 0.[2]

Coming back to the significance of the coefficient CI_i, we can conclude that the higher is the variance of the mentioned character when the modalities of the considered classification vary, the greater is the informative level of the classification itself. Anyway the informative function is still influenced by the number of modalities within a single partition, that is the greater is the number of the modalities, the higher is the level of informativity.

So, generally speaking, it is correct to define the classification that gives the maximum of informativity as the one that produces a certain value for the above function, given the minimum number of modalities.

In this specific case of study, we fix as characters of classification the following variables: turnover, labour costs, number of persons employed, investments persons employed, turnover per person employed, labour costs per person employed, investment per person employed.

In the table n. 3, we illustrate the values of the coefficients CI_i for the above variables.

Table 3: *Coefficient of informativity for ATECO'91 classes and different variables*

CLASSIFICATION VARIABLES	MODALITIES	COEFFICIENT OF INFORMATIVITY
turnover	178	0,23
labour costs	178	0,25
investments	178	0,37
persons employed	178	0,35
turnover per person employed	178	0,01
labour costs per person employed	178	0,05
investments per person employed	178	0,02

Source: elaborations on ISTAT data

On the other hand, we examine the same coefficient for different levels of the classification.

In the following table we indicate the coefficient we obtained.

Table 4: *Coefficient of informativity for different ATECO'91 digits*

ATECO DIGITS	MODALITIES	COEFFICIENT OF INFORMATIVITY
5 DIGITS - CATEGORIES	445	0,15
4 DIGITS - CLASSES	178	0,23
3 DIGITS - GROUPS	74	0,46
2 DIGITS - SECTIONS	18	0,61

Source: elaborations on ISTAT data

[2] The coefficient is valid only for positive variables. Negative variables should be transformed before calculating CI_i.

As it is clear from the above table when ATECO digits decrease also the number of modalities and the variability of the character decrease and CI_i increases.
So it is not adequate to use categories to classify enterprises within a certain domain. This actually confirm what anticipated in this paper and derived from our current experience: the possible gain in information coming from the huge detail is absolutely not well balanced from the costs required to investigate in such a detail. In terms of statistic analysis, considering ATECO categories means to accept too wide strata for the original domain of study, that is once more extremely reductive for the efficiency of the analysis.
On the other hand if we consider an aggregation of the units corresponding to ATECO'91 classes (4 digits), we still have a great number of modalities, even if strongly reduced compared to the 5 digits, and we can maintain a good level of detail. As for the informativity coefficient, it is not so high compared to the one for groups and sections. To sum up, looking through the results indicated in table 4 it would seem that the best levels of aggregation are groups and sections. Anyway, it is a matter of considering the aim of the single studies and above all to analyse costs and benefits both in terms of information and of resources to dedicate to the studies.
A further comparison can be done if we fix the level of aggregation to the four digits and, consequently, the number of modalities to 178. If we take into account at the same time of the results shown both in table 3 and in table 4, we can observe that the prevalent activity, that is expressed by ATECO, is not the best variable to classify enterprises. From the analysis of the values of the informativity coefficients, it emerges that characters such investments or number of persons employed seems to be more suitable to select units into subgroups.
This circumstance could be explained by the fact that, as said before, ATECO'91 does not always reflect the structural characteristics and changes of single enterprises. On the other hand, economic variables are those which reflect the continuous transformations of enterprises.
As for the last three variables – turnover per persons employed, labour costs per persons employed , investments per persons employed – it must be observed that as they are expressed by ratios, they have a lower variability and, consequentely, they give lower values for the coefficient of informativity, compared with the other variables. For this reason is more correct to explain the results concerning these ratios separately. So, in this case, we can observe that labour costs per persons employed can better put in evidence differences among enterprises.
In our opinion, the preliminary classification of the units to investigate in a study is fundamental and can influence the result of the study itself. For this reason, in the case we are describing in this paper, we think that a good approach should consider both ATECO'91 classification and other adequate economic variables. In this way it is possible to take advantage from the structure of the ATECO'91 itself and from the "sensitiveness" of further adequate variables such as the one above indicated.

4. Final considerations and perspectives

To sum up, in the light of the above explained analysis for the enterprises within the Italian service sector it is possible to put in evidence two fundamental aspects:

1. Very often the official classification, that is greatly useful at the stage of data capturing, does not reflect the real characteristics of the enterprises. This causes a lack of information when passing from micro to macrodata.
2. It is important to choose the level of partition of the activities to consider in a specific study. It has to come out from a compromise between the easiness that characterises the most aggregate levels and the redundancy of the less aggregate ones.

The analysis described in this paper are just an attempt to investigate a problem which currently affects statistics surveys or studies on enterprises.
In the light of this, it is essential to carry out further applications, with the aim to validate the obtained results and propose solutions useful to optimise the performance of economic analysis.
For this purpose it will be fundamental to consider more in detail economical factors which characterise enterprises belonging to the service sectors in the Italian frame.

References

Anitori P., Gismondi R. (1998) *Cross cutting features of the Italian Retail Trade Sector.* Paper presented at the 13th Voorburg Group Meeting –Rome – September 1998.

Giorgi A.R. (1998) *Statistics for the Breakdown of Turnover for Retail Trade Enterprises – Intermediate Report* Istat, 1998.

Gismondi R., A.P. Mirto (1998) *Classification and Heterogeneity of data in Tourism Statistics: the Italian case.* Proceedings of the Joint IASS/IAOS Conference Statistics for Economic and Social Development – September 1998.

ISTAT (1991) *Classification of Economic Activities,* «Methods and Rules», 11, Istat, Rome.

Monducci R. (1998) *Classification in the Economic Analysis on Enterprises,* paper presented at the meeting «Revision of the classification systems for products and economic activities», Istat, Rome, 19 January 1998.

A Neural Net Model to Predict High Tides in Venice

Tommaso Minerva
University of Modena, Dep. of Economics,
Viale Berengario 51, 41100 Modena, minerva@unimo.it

Irene Poli
University Ca' Foscari of Venice, Dep. of Statistics,
Campiello S. Agostin, S.Polo 2347, 30125 Venezia, irenpoli@unive.it

Abstract: In this research we design and apply a neural network model to predict the tidal levels in the Venetian lagoon. We use an evolutionary computational approach to select the net topology within the class of multilayered feedforward networks. We build a genetic algorithm, which evolves both the number of predictors and the best set of predictors for the model. The results of this approach are compared to the results we achieve with a linear model based on the same set of candidate predictor variables, whose specification is also obtained with a genetic algorithm. The predictions resulting from the genetically evolved neural net model are more accurate for both tidal levels and extreme values ("*the high waters*").

Keywords: Complex Systems, Neural Network Models, Genetic Algorithms, Model Selection.

1. Introduction

Ecosystems are complex dynamic systems, which adapt and evolve in response to their own changes and those of larger systems in which they are embedded. Modeling and predicting such systems have proved very resistant to classical mathematical methods, because of incomplete knowledge and difficulties of representation that characterize these systems, as well as the strong nonlinearities in the interactions among system elements. In the last decade, however, many new approaches have been developed to handle these problems. Important common feature of many of these approaches is that they are *adaptive* – that is, they evolve solutions rather than pre-specifying them – and they permit *parallel computation* (see for example Vapnik, 1995; Holland, 1998).

In this paper we are concerned with modelling and predicting one aspect of the dynamics of a particular environmental system: the Venetian lagoon. The Venetian lagoon, with its city and its nature, represents in fact an ecosystem of exceptional interest and importance. In the study of this system accurate predictions of the tidal levels are essential for all the activities in the historical center of the city and in the lagoon itself. For predicting the tidal levels we model the system as an artificial neural network, with a feed-forward dynamics with ordered layers. The design of the network is not assigned but evolved with a genetic algorithm (see also Anders and Korn, 1999; Rudolph, 1996) that generates a set of different network architectures, evaluates their performance, and then evolves them through a class of stochastic operators, over many subsequent generations, to reach the architecture we regard as the most appropriate. The resulting predictive system is adaptive with a high degree of parallelism.

The paper is organized as follows. In section 2, we describe the class of artificial neural networks that we propose to predict high tides in Venice, specifying the particular multi-layer topology, the activation functions and the learning algorithm for assessing the connections. Then, we present an evolutionary approach to the model selection problem and build a genetic algorithm (GENN, *Genetically Evolved Neural Networks*, as in Minerva, Paterlini and Poli, 1999) that evolves the architecture of the neural networks. In section 3, we apply the methodology to the problem of predicting the tidal levels in Venice and compare the results with the predictions obtained with more traditional tools of time series analysis – that is, ARMA and more general linear models. In this comparison, we also evolve the structure of the linear models with a genetic algorithm (GELM, Genetically Evolved Linear Models, as in Minerva and Poli, 1997).

2. The genetically evolved neural network

In prediction problems we typically consider a learning set L of data consisting of measurements on N cases, where each case consists of a response variable y_n and a prediction vector $\mathbf{x}_n=(x_{1n},\dots,x_{pn})$, with n=1,2, ...,N. We seek the function $f(\mathbf{x},L)$, which gives accurate predictions of future responses as a function of the value of the (future) prediction vector $\mathbf{x}$, based on the learning set L.

For this problem we use a neural network. That is, we model the observable variables as an information-processing system where the information in the prediction vector is transformed and distributed via a set of connections among a finite number of units (neurons) to yield the response of the system (Bose and Liang, 1996). More specifically, we adopt a neural net model based on a multilayered feed-forward architecture. It consists of three ordered layers of units: an initial input layer with p units, a single hidden or intermediate layer with q units, and a final output layer, which in the case we consider here consists of a single output unit.

The network has simple feedforward connectivity. Each input unit communicates its state to all hidden units. These units compute their states by processing the information received from the input units and then communicate the states to the output unit, which uses this information to compute the system's response. We assume a *Gaussian radial basis function* representation: that is, the network provides a map from the space $\Re^p$ to $\Re$ as follows

$$y = \frac{\sum_{j=1}^{q} w_j \exp\left(-\left\|\mathbf{x} - \boldsymbol{\mu}_j\right\|^2 / 2\sigma_j^2\right)}{\sum_{j=1}^{q} \exp\left(-\left\|\mathbf{x} - \boldsymbol{\mu}_j\right\|^2 / 2\sigma_j^2\right)} \tag{1}$$

where $\mathbf{x} \in \Re^p$ represents the input vector with elements x_i, i=1,...,p, $\boldsymbol{\mu}_j$ are vector-valued parameters called centroids that represent the *centers* of the basis functions, σ_j are positive-valued shape parameters for the Gaussian activation function, and the w_j are the scalar connection weights from the hidden units to the output unit (Poggio and Girosi, 1990; Poli and Jones, 1994).

To select the dimension of the input vector and to choose the predictors from a large set of possible predictors, we adopt the evolutionary computational approach (Fogel, 1995; Billings and Zheng, 1995). This approach developed with the idea that the evolution of natural systems could be imitated in a computer to evolve artificial systems and solve search and optimization problems. In particular, we employ *genetic algorithms* (Holland, 1992; Fogel, 1995; Mitchell, 1996; Baragona and Battaglia, 2000), procedures mainly designed to search and evolve solutions in high-dimensional space.

Therefore we adopt the GENN (Genetically Evolved Neural Networks) approach by building a genetic algorithm to select the architecture of the net, that is the set of predictors and the design of the connections. The architecture, then, is not assigned a priori, but evolved from a large set of candidates. In particular, we randomly generate an initial population of binary sequences, which identify different topologies of the network. Each topology is proposed and applied to the learning set of data, and then evaluated with respect to its fit to a validation set of data. The population is then evolved through an adaptive process, whereby the changes are produced by a set of genetic-inspired operators – *selection*, *crossover*, *mutation*. Selection uses a probabilistic sampling procedure to determine the candidates for the next generation of predictive networks. Crossover chooses two candidate topologies (at random) and exchanges (randomly chosen) components of their encoded representation. Mutation randomly selects a position in the representation of a randomly selected candidate topology, and changes the value corresponding to that position. Crossover may combine "good" features of two topologies to obtain a better one, while mutation may prevent convergence to suboptimal solutions. This cycle continues for many generations until a convergence criterion is met. The "best fit" topology of the last generation is then used to predict values for a different set of data, the test set. We then evaluate the accuracy of these test set predictions, which we compare to the values obtained by alternative methods of prediction applied to the same test set.

3. Predicting the tides of the Venetian lagoon

The tide is an essential element of the lagoon's ecosystem. It determines the variation of the water level of the lagoon. A normal tide, unperturbed by atmospheric conditions, follows the gravitational attraction of the moon and the sun and has a complicated but regular rhythm with around six hours of fall and six hours of rise. The atmospheric conditions sometimes can change considerably this regular rhythm. A special interaction of some components of the system may produce the phenomenon of "high water", where the tidal levels reach extreme values and the city of Venice is flooded, causing serious damages to its artistic patrimony and its economic and social activities.

The problem of extreme values modelling has been explored in a rich and interesting literature, including Davison and Smith (1990), Dixon and Tawn (1994) and Smith, Tawn and Coles (1997). Our concern in this research is on the prediction problem of exceedance levels for the Venetian lagoon. Our study is based on data provided by the Center for Tidal Forecasting (*Centro Previsioni e Segnalazioni Maree*) of the Venice municipal government. The Center's predictions are based upon linear models with different set of predictive variables: the *simplified model* with 50 predictors, the *complete model* with 70 predictors and the *extended model* with 70 different predictors. The description of the models, including their specification and their accuracy, is reported..in Canestrelli and Pastore (1999). Our research will consider only a small subclass of the predictors used in the Center's simplified model. More specifically, the data on which our research is based refer to the year 1984, chosen because of its relatively high number of high tide events, and consist in 8600 hourly recordings of the level of the tide at the observation station of Punta Salute, as well as the corresponding values of atmospheric pressure, recorded every three hours, at 10 Italian weather stations. We preprocessed the tidal data to eliminate the dominate lunar gravitational effect. We then partitioned the observations into a *learning set* of 4000 observations, to learn the connections of the net, a *validation set* of 2880 observations to evaluate the fitness of the net in the selection process based upon the genetic algorithm, and a *test set* of 1720 observations to evaluate the accuracy of predictions. All the comparisons between observations and predictions or between linear and neural net models have been performed on the *test set*. In the design of the predictive net, we consider a set of 182 candidate predictor variables. This set includes 132 lagged variables of the tidal level, from y_{t-1} to y_{t-132} and 5 three-hourly lagged variables, from p_{t-3} to p_{t-15}, of the atmospheric pressure considered in 10 different stations.

We then compared the performance of the neural net model to a linear model whose specification is evolved with a genetic algorithm, (*GELM: Genetically Evolved Linear Model*) as described in Minerva and Poli, (1997). As candidate predictor variables we used the same set considered for the neural net models. The comparison is expressed in terms of root-mean-square error (RMSE) evaluated on the *test set* of data, and presented in Table 1. We show the RMSE for predictions 1 hour, 3 hours, 12 hours and 24 hours ahead. While for the 1-hour prediction both models performed quite well, for the three-hour-ahead prediction we observe a markedly superior performance for the neural net model: the neural net root-mean-square-error is 3.8 cm, compared to 5.9 cm of the linear model. For the 12-hours-ahead prediction, the neural net's error increases to 6.2, and the linear model's to 7.5. For the 24-hours predictions, the neural net model continues to outperform the linear model, with errors of 8.4 cm and 9.2 cm respectively (around 12% of predictive performance difference).

Table 1: *Comparison of the prediction root mean square error (RMSE) in cm. between the Genetically Evolved Linear Model (GELM) and the Genetically Evolved Neural Network (GENN).*

Lead Time *(hours)*	**Linear Model *(GELM):* *RMSE***	**Neural Network *(GENN):* *RMSE***
1	**1.7**	**1.4**
3	**5.9**	**3.8**
12	**7.5**	**6.2**
24	**9.4**	**8.2**

The extremes values of the tides, the "high waters", are very important, but particularly difficult to forecast. In Venice, alarms are sounded when a tide greater than 110 cm. is expected. Thus, a desirable property of a predictive algorithm is that it "gets it right" when it predicts high water. Here we present the results for the high water periods in the test set, that is November 20-24, 1984, and reported in Table 2.

Table 2: *Comparison of the predictions of the tidal levels greater than 100 cm, occurred in November 20-24, 1984, between the genetically evolved linear and the neural net model.*

Day-Time	**Actual Values (cm)**	**Prediction* (cm)**							
		1 hour ahead		*3 hours ahead*		*12 hours ahead*		*24 hours ahead*	
		GELM	GENN	GELM	GENN	GELM	GENN	GELM	GENN
20 - 8:55	**115**	**112** +	**113** +	**106** –	**112** +	**98** –	**110** +	**102** –	**104** –
21 - 9:15	**119**	**117** +	**116** +	**119** +	**118** +	**117** +	**130** +	**116** +	**119** +
22 - 9:25	**111**	**111** +	**111** +	**112** +	**109** –	**106** –	**112** +	**103** –	**105** –
23 - 9:50	**103**	**103** +	**102** +	**103** +	**103** +	**90** +	**108** +	**101** +	**100** +
24 - 10:20	**100**	**100** +	**101** +	**97** +	**91** +	**91** +	**85** +	**88** +	**88** +

* *The notation + and – identifies the right and the wrong prediction with respect to the activation of the alarm system.*

On the first three of these days, an alarm should have been given, while on the last two no alarm should have sounded. Our "best" linear model (GELM) was right on 4 out of the 5 days with its 3 hour forecast, and 3 out of 5 with its 12 hour forecast. The "best" neural net model (GENN) was also correct 4 out of 5 days with its 3 hour forecast, but it hit on all 5 days with its 12 hour forecast.

References

Anders, U., Korn O., (1999), Model Selection in Neural Networks, *Neural Networks* 12. 309-323.

Baragona, R. and Battaglia, F. (2000). Some Applications of Genetic Algorithms to Statistical Inference, *XL Riunione Scientifica della Società Italiana di Statistica*, Firenze.

Billings, S. A. and Zheng, G. L. (1995). Radial Basis Function Network Configuration Using Genetic Algorithms, *Neural Networks*, 6, 877-890.

Bose, N. K., Liang, P. (1996). *Neural Networks Fundamentals with Graphs, Algorithms, and Applications*, McGraw-Hill, Inc., 1-478.

Canestrelli, P. and Pastore, F. (1999). *Modelli stocastici per la previsione del livello di Marea a Venezia,* Centro Previsioni e Segnalazioni Maree, Venezia.

Davison, A. C. and Smith, R. L. (1990). Models for exceedances over high thresholds (with discussion*). J. Roy. Statist. Soc.* Ser. B 52 393-442.

Dixon, M. J. and Tawn, J. A. (1994). Extreme sea levels: modelling interaction between tide and surge. In *Statistics for the Environment 2: Water Related Issues* (V. Barnett and K. F. Turkman, eds.) 221-232. Wiley, New York.

Smith, R. L., Tawn, J. A. and Coles, S. G. (1997). Markov chain models for threshold exceedances. *Biometrika,* 84 249-268.

Fogel, D. B. (1995). *Evolutionary computation*, IEEE Press, New York.

Holland, J. H. (1992). *Adaptation in Natural and Artificial Systems*, The MIT Press, Cambridge, Massachusetts.

Holland, J. H. (1998). *Emergence. From Chaos to Order.* Addison Wesley, Reading, Massachusetts.

Minerva, T. and Poli, I, (1997). Building an ARMA Model by using a Genetic Algorithm, *Materiali di discussione* , Università of Modena, Modena.

Minerva, T., Paterlini S. and I. Poli (1999), Algoritmi ibridi per l'analisi di serie storiche finanziarie, Scienza e Business, 1, 3-4, 56-77.

Mitchell M., (1996). *An Introduction to Genetic Algorithms.* The MIT Press, Cambridge, Massachusetts.

Poggio, T., Girosi, F. (1990) Regularization algorithms for learning that are equivalent to multilayers networks, *Science*, 247, 978-982.

Poli, I. and Jones, R.D., (1994). A Neural Net Model for Prediction, *Journal of the American Statistical Association,* 425, 117-121.

Rudolph, S. (1996), On a Genetic Algorithm for the Selectin of Optimally Generalizing Neural Network Topologies. In *Proc. of the 2nd Int. Conference of Adapive Computing in Engineering Design and Control '96,* Plymouth, United Kingdom, 76-86.

Vapnik, V. (1995). *The Nature of Statistical Learning*, Springer-Verlag. Berlin.

Acknowledgement

We thank Ing. Paolo Canestrelli of the Centro Previsioni e Segnalazioni Maree of the Venice Municipality, who provided us with the data and many helpful suggestions. We also thank the referees for the useful and interesting comments. This research has been financially supported by the University Ca' Foscari of Venice and by MURST (Italy) .

AUTHOR INDEX

KEYWORDS INDEX

GPSR Compliance

The European Union's (EU) General Product Safety Regulation (GPSR) is a set of rules that requires consumer products to be safe and our obligations to ensure this.

If you have any concerns about our products, you can contact us on ProductSafety@springernature.com

In case Publisher is established outside the EU, the EU authorized representative is:

Springer Nature Customer Service Center GmbH
Europaplatz 3
69115 Heidelberg, Germany

Batch number: 10704711

Printed by Printforce, the Netherlands